Consequences of Antisymmetry

Walter de Gruyter
1749 250 1999
Berlin · New York

Studies in Generative Grammar 46

Editors

Henk van Riemsdijk
Jan Koster
Harry van der Hulst

Consequences of Antisymmetry

Headed Relative Clauses

by

Valentina Bianchi

Mouton de Gruyter
Berlin · New York 1999

Mouton de Gruyter (formerly Mouton, The Hague)
is a Division of Walter de Gruyter GmbH & Co. KG, Berlin.

The series Studies in Generative Grammar was formerly published by
Foris Publications Holland.

♾ Printed on acid-free paper which falls within the guidelines
of the ANSI to ensure permanence and durability.

Library of Congress Cataloging-in-Publication Data

Bianchi, Valentina, 1969–
Consequences of antisymmetry : headed relative clauses / by Valentina Bianchi.
p. cm. – (Studies in generative grammar ; 46)
Based on author's doctoral dissertation.
Includes bibliographical references (p.) and index.
ISBN 3-11-016250-4
1. Generative grammar. 2. Asymmetry (Linguistics) 3. Grammar, Comparative and general – Relative clauses. I. Title. II. Series.
P158.B5 1999
415–dc21 99-34346
CIP

Die Deutsche Bibliothek – Cataloging-in-Publication Data

Consequences of antisymmetry : headed relative clauses / by Valentina Bianchi. – Berlin ; New York : Mouton de Gruyter, 1999
(Studies in generative grammar ; 46)
ISBN 3-11-016250-4

Printing: Werner Hildebrand, Berlin.
Binding: Lüderitz & Bauer GmbH, Berlin.
Printed in Germany.

Preface

In recent years, the status of X-bar theory as a set of primitive principles has been questioned. Chomsky (1995: 241-271) derives X-bar principles from basic properties of the computational operations Merge and Move. Kayne (1994) proposes a radically alternative view: X-bar principles are determined by an independent property of natural language - namely, the fact that words must be linearly ordered in a temporal sequence. Kayne proposes an invariant mapping between the hierarchical relations of nonterminal symbols and the linear ordering of terminal symbols in a tree. In virtue of this mapping, X-bar structures inherit one basic property of linear ordering, namely, antisymmetry.

With respect to standard X-bar theory, the Antisymmetry theory considerably restriots the class of possible X-bar structures. It is thus necessary to rethink the structural analysis of many syntactic constructions: in various cases, the standard analysis is incompatible with the Antisymmetry theory and it is necessary to devise a completely new approach. The theory is thus tested on empirical grounds. The minimal requirement is that it must allow for a structural analysis consistent with the data. A more ambitious goal is to show that the new analysis is even superior to previous ones in its empirical consequences.

The aim of this book is precisely to discuss the consequences of Antisymmetry in one specific empirical domain, that of postnominal headed relative clauses. Since rightward adjunction is excluded on principled grounds, the standard adjunction analysis has to be given up. Kayne (1994: 85-115) proposes an alternative "raising" analysis which combines two independent hypotheses: the DP hypothesis, according to which the determiner is a functional head that carries the referential or quantificational force of the noun phrase; and the raising hypothesis, dating back to the early Seventies, according to which the "head" of the relative clauses is generated in the gap position and raises to a pre-complementizer position.

This book critically examines the evidence in support of the raising analysis and develops it by working out various aspects that are left implicit or unexplored in Kayne's discussion. Besides these general goals, it also aims at proving that the raising analysis is empirically superior to the adjunction analysis on several grounds. First, it is argued that the raising analysis allows for a unified approach to various relativization strategies that remained unrelated in the previous approach: this has interesting consequences both on the typological and on the diachronic side. Second, the properties that distinguish appositive from restrictive relatives are reduced to different LF configurations on the basis of Chomsky's (1995: 202-210) theory of reconstruction. Third, the analysis provides a solution to some notorious problems in the morphosyntax of relative pronouns, like "Case attraction" and the PF deletion phenomena that usually go under the rubric of "doubly filled Comp" effects.

These results justify the claim that the raising analysis is empirically superior to the adjunction analysis; this in turn implies, on the theoretical side, that the restrictiveness of the Antisymmetry theory is supported over the standard X-bar theory.

This book is based on my doctoral dissertation. Various parts have been revised and extended (in particular sections III.2-4, VI.8-9, VII.1-2, and chapter VIII). However, for editorial reasons it has been necessary to leave out an entire chapter on "heavy pied piping". The interested reader may find the relevant discussion in the sixth chapter of my dissertation (Bianchi 1995).

There are a number of people whom I wish to thank for their teaching as well as for their support during my doctoral studies. Pier Marco Bertinetto taught me a lot of things, and above all, the importance of always being willing to critically rethink and even call into question one's own theoretical premises. I owe to Richard Kayne the original inspiration of this work; he supported its progression with many insightful comments and criticisms. I greatly benefitted from his rigorous way of pushing a line of reasoning to its furthest consequences. Adriana Belletti and Luigi Rizzi constantly helped me with many important suggestions since the very beginning of my studies in generative syntax: with them I learned to appreciate the value of open and collaborative exchange of ideas. Finally, I learned a lot from Rita Manzini, and in particular from her readiness in pointing out the consequences of theoretical issues on the empirical side, and vice versa.

I also wish to thank Jan Koster and Jean-Yves Pollock for carefully reading my thesis and for their stimulating comments on it.

During my doctoral studies and the subsequent period, I greatly benefitted from the interaction with students and colleagues at the Scuola Normale Superiore in Pisa, in the universities of Campinas, Florence, Florianopolis, Geneva, Milan, Padua, Siena and Venice, and in the ZAS in Berlin. Let me attempt a (surely incomplete) list: Birgit Alber, Artemis Alexiadou, Anna Cardinaletti, Carlo Cecchetto, Guglielmo

Cinque, Gloria Cocchi, Denis Delfitto, Luca Dini, Caterina Donati, Verner Egerland, Charlotte Galves, Barbara Gili, Giuliana Giusti, Alex Grosu, Maria Teresa Guasti, Liliane Haegeman, Mary Kato, Bob Learmonth, Alessandro Lenci, Giuseppe Longobardi, Michele Loporcaro, Carlos Mioto, Andrea Moro, Jairo Nunes, Yuya Perez Vazquez, Carol Peters, Cecilia Poletto, Ur Shlonsky, Mario Squartini, Vittorio di Tomaso, Giuseppina Turano, Chris Wilder, Roberto Zamparelli.

Generative syntax also gave me the occasion to meet two of my best friends: Paolo Acquaviva and Cristina Figueiredo, whom I thank heartily for their invaluable help.

A special thank to Jan Koster, who suggested that I might submit my thesis for publication in the series "Studies in Generative Grammar", to Anke Beck and Katja Huder of Mouton de Gruyter, and to Maddalena Agonigi for her crucial help in editing the manuscript.

On the personal side, acknowledgements would take the next three hundred pages. Besides all of the above mentioned people, I also wish to thank the following ones: my beloved Stevie and Isabella; my parents, Tiziano and Cioni; my sister Alessandra and Saverio; my sister Silvia and Angelo; my "acquired" family, Renata, Domenico and Alberto; Sandra Gosso; Luigi, Rosalba, Marco and Lucia Santarini; Monica, Francesco, Gilberto, Pietro, Domenico and Manuela, Giovanna, Kerstin, Grazia, Elena.

Finally, I dedicate this book to my parents, Tiziano and Cioni, with the deepest love and gratefulness.

Contents

Chapter I. The theoretical background: Kayne's (1994) "Antisymmetry of Syntax"

Chapter II. Restrictive relative clauses

Chapter IV. A reconstruction asymmetry

Chapter V. Appositive relative clauses

Chapter VI. Doubly filled Comp effects

Chapter VII. The complementizer system of headed relative clauses

Chapter I
The theoretical background: Kayne's (1994) "Antisymmetry of Syntax"

0. Introduction

Several recent studies in the Principles and Parameters framework have converged on the conclusion that the standard X-bar theory, as formulated for instance in Chomsky (1986b: 2–6),[1] is too little restrictive. Accordingly, various principles restricting the format of X-bar structures have been proposed. However, these principles are conceptually unrelated to each other and they can only be stipulated as independent axioms of X-bar theory.

Kayne's (1994) Antisymmetry theory instead attains the necessary restrictiveness on principled grounds. All of the empirically justified restrictions on the X-bar structures are derived from a single axiomatic principle, the Linear Correpondence Axiom. This principle constitutes a hypothesis on the relation between the hierarchical organization of the nonterminal symbols and the linear ordering of the terminal symbols in a tree. This is the most innovative aspect of Kayne's proposal.

However, the restrictiveness of the Antisymmetry theory has a considerable impact on the analysis of many empirical domains. One case in point is the syntax of headed relative clauses. The standard approach to them is incompatible with the antisymmetric X-bar theory; thus, an alternative must be sought for.

Kayne (1994) proposes a "raising analysis" for relative clauses. The aim of this book is to defend this proposal, elaborate it and further explore its consequences. The following chapters focus on various aspects of the syntax of headed relative clauses, showing that the raising analysis leads to a more satisfying approach to many well known problems and also reveals new interesting problems.

This introductory chapter examines the relevant literature on X-bar theory and provides a brief summary of the Antisymmetry theory, in order to lay down the theoretical background of the following discussion.

1. X-bar theory in the Principles and Parameters framework

In the Principles and Parameters framework, X-bar theory is conceived of as a component of the computational system that projects a tree structure from elements chosen from the lexicon (Chomsky—Lasnik 1991: § 3.2). The standard version of X-bar theory is given by Chomsky (1986b: 2–4) in the format of the following recursive context-free rules, where the variable X ranges over preterminal symbols, either lexical or functional:[2]

(1) X′ → X X″*
(2) X″ → X″* X′

The general format of the rules in (1)–(2) implies that all the preterminal symbols project in a uniform way. The rules express universal hierarchical relations: the two bar level projection constitutes the maximal projection of X (XP) and dominates the specifiers of X (themselves maximal projections) and the intermediate projection X′ (rule (2)). The intermediate projection X′ dominates the projecting preterminal X, called the head, and one or more maximal projections, the complements of X (rule (1)).

The rules in (1) and (2) do not exhaust the possible hierarchical relations. Besides specifiers, heads and complements, a fourth type of position has been structurally characterized by May (1985) and adopted by Chomsky (1986b: 7): the adjunct position. In the adjunction configuration, the adjoined element β is immediately dominated by a node with label α and has a sister node with the same label α. The two nodes labelled α are called the segments of the category α. The adjunct is dominated by only one segment of α, but not by the whole category α: this distinguishes it from the specifier position. Adjunction is introduced by rule (3), where α, β are both heads or both maximal projections:

(3) α → β α

The recursiveness of the rule allows multiple adjunction to a category.

1.1. Dominance versus linear order: the directionality parameters

The context free rules in (1) and (2) are meant to encode only the dominance relations between the various categories. The symbols to the right of the arrows do not constitute a linearly ordered string, but rather an unordered set;[3] the linear order of the corresponding sister categories in the tree is fixed in every language by the value of the directionality parameters. The Head-Complement parameter fixes the order of the head with respect to all of its complements: if the head precedes them, the value is head-initial; if it follows them, the value is head-final. Similarly, the Specifier-Head parameter fixes the order of the categories introduced in the right-hand side of rule (2): the specifiers may either precede or follow the sister projection X′ (and *a fortiori* the head X, which is dominated by X′).

It is the separation of the dominance relations from linear order that makes it possible to isolate the universal aspects of phrase structure from the language-specific ones. In fact, this separation is also assumed in other syntactic frameworks like GPSG (Gazdar et al. 1985: 44–50).[4, 5]

The directionality parameters are defined with respects to sister nodes. If the left value is fixed for both parameters, rules (1)–(2) generate the tree structure in (4).

(4)

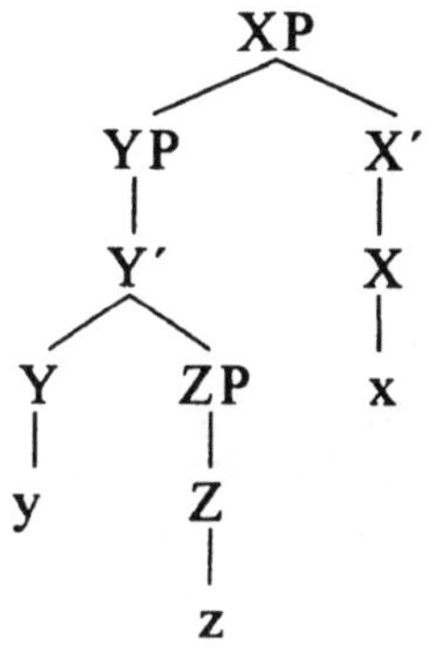

In this tree, the relative order of the sister nodes is determined by the value of the parameters: the specifier of XP, YP, is to the left of the head X, and the complement of the head Y, ZP, is to the right of Y. However, the tree does not determine the linear order of the terminal symbols *x, y, z*. The implicit assumption underlying the terms "head-initial" and "head-final" is that the string of terminal symbols is obtained by reading the leaves from left to right, rather than from right to left. This gives the string *yzx*. In graph-theoretic terms, the string is obtained by a depth-first left-to-right visit of the tree.[6] Thus, two assumptions are necessary to yield a linear order of the terminal symbols from the setting of the directionality parameters: the left-to-right orientation, and the depth-first procedure. If the tree is instead visited in a breadth-first fashion (keeping the left-to-right orientation), the output string is *xyz*, since the less embedded terminal symbol *x* is reached before *y* and *z*, and *y* is reached before *z*. These assumptions are often left implicit, which gives the misleading impression that the directionality parameters directly yield an ordered string of terminal symbols by fixing the order of sister nodes in the tree.

1.2. "Invisible" nodes

A problem with the standard X-bar theory, explicitly noted by Chomsky (1986b: 4; 1995: 242–243), is that some nodes in the tree structure are not available for syntactic manipulation. In particular, there is strong empirical motivation for restricting the application of the rule Move α to maximal projections and heads, excluding the intermediate projections;[7] but the invisibility of X′ must be stipulated.

It is also usually assumed that the lower segment(s) in an adjunction configuration cannot move (whereas the adjoined element can).[8]

Another problem with intermediate projections is their number. Jackendoff (1977) originally proposed a three level X-bar theory; rules (1)–(2) instead define the two bar level as the maximal projection. It is clear that the upper limit of projection levels must be postulated in the rules themselves;[9] it may be empirically justified, but it is not determined by any principle.

Recently, Hoekstra (1991) has proposed a one-level X-bar theory in which specifiers are structurally indistinguishable from adjuncts, but are characterized by an agreement relation with a head. Agreement relations, and licensing relations in general, are assumed to be biunique (the Uniqueness of Licensing Principle): a head can agree with only one constituent with respect to a given set of features. However, a head may be specified for more than one set of features, and thus it can enter multiple agreement relations and license multiple specifiers. It follows that the number of possible specifiers of a head is determined by the number of distinct agreement relations that the head can enter; X-bar theory does not *a priori* impose any upper bound.

1.3. Binary branching

The Kleene star in rules (1) and (2) allows an undetermined number of specifiers and complements. The sister nodes under XP and X′ are symmetric to one another with respect to the fundamental relation of c-command:

(5) α c-commands β iff neither α dominates β nor vice versa, and the first (branching)[10] node that dominates α dominates β as well.

The c-command relation enters into the definition of two significant syntactic relations, binding and government.

Kayne (1984: 129–132) proposes an original conception of the c-command requirement: its essential content is that a branching node intervening between two elements α and β blocks a syntactic relation between them. Thus in the abstract structure (6) the branching node B, which dominates A but not C, prevents A from c-commanding C; in (7), instead, no branching node intervenes and A c-commands C:

(6)

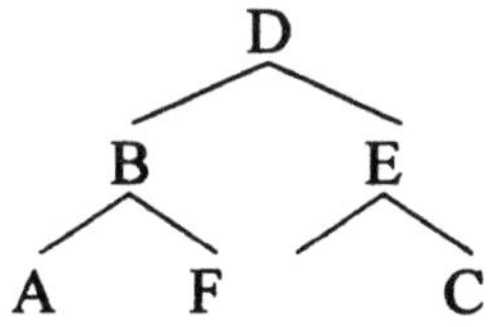

(7)

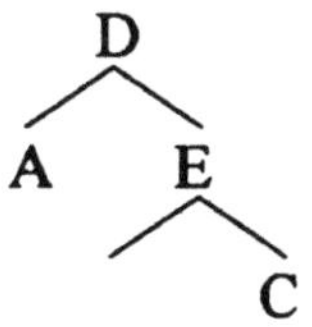

The blocking effect of branching nodes can be reformulated as a condition on the path connecting C to A (where a path is defined as a sequence of distinct and adjacent nodes in the tree). When the path reaches a branching node, two possibilities arise. If the branching node dominates exactly one node not yet included in the path, there is no choice: the path can only be extended by including the new node. This is the case in (7), where the path C, E, D reaches a branching node D, but the latter dominates only one node not yet included in the path, namely A. Kayne defines such a path from C to A unambiguous. If, on the contrary, the branching node dominates more than one node not yet included in the path, the path becomes ambiguous, for it can continue in more than one direction. This is the case in (6) when the path from C reaches the node B. B is dominated by D, which is already included in the path, but it dominates two nodes not yet included in the path, A and F: therefore, the path from C to A is ambiguous. The requirement that A c-command C can then be replaced by the requirement that there exist an unambiguous path from C to A.[11]

The unambiguous path requirement is equivalent to c-command with respect to (6) and (7), but with respect to other structures it is more restrictive. In a multiple branching structure like (8), the nodes A, B, C c-command each other, but there exists no unambiguous path linking any one of them to either of the other two.

(8)

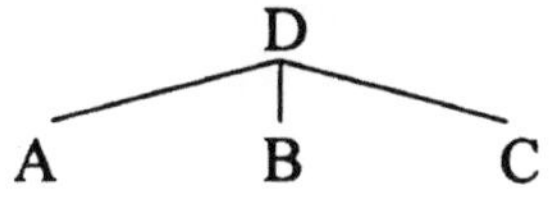

Thus, whenever a node has more than two branches, the nodes immediately dominated by it will not be reached by any unambiguous path, and consequently they will be unable to be bound or governed by any other node. This gives a strong constraint on the phrase structures of natural language. For instance, replacing the abstract symbols D, A, B, C by VP, V, DP and PP in (8), a VP structure is obtained where the verb cannot govern either one of its complements, contrary to usual assumptions (9). In order for government to hold, it is necessary to group the two complements in a category which is itself the complement of V, as in (10).

(9)

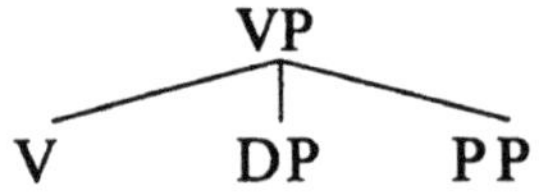

(10)

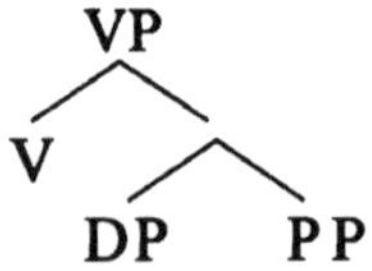

Thus the unambiguous path requirement forces binary branching.

Note that binary branching is not a general property of tree structures; it is not determined by a principle of X-bar theory, but by the need of government and binding relations, which belong to other modules of the grammar.[12] In recent studies it has become common to assume a generalized binary branching constraint: this can be easily incorporated in the standard X-bar theory by eliminating the Kleene star in (1) and (2); but this constitutes a stipulation.[13]

2. Some empirical evidence

The preceding section discussed some postulates of standard X-bar theory. The format of the context-free rules (1)–(3) imposes the separation of dominance relations from linear order and requires the assumption of independent directionality parameters; but these parameters yield a linear order of the terminal symbols of a tree only through a specific visiting procedure. Furthermore, the rule format does not determine an upper bound for the number of intermediate projections and of admissible branches, nor does it determine the visibility of the various nodes for the rule Move α. All these aspects of phrase structure must be independently postulated.

This section provides a synthesis of some empirical evidence suggesting that the standard X-bar theory is too little restrictive. Recent studies on the functional structure of sentences and noun phrases suggest the elimination of XP adjunction and the uniqueness of the specifier for any maximal projection.[14] This rules out, on the one hand, more-than-binary branching under XP and, on the other hand, recursive generation of more than one intermediate projection (§ 2.1).[15]

Furthermore, the controversy on multiple complements of a lexical head (§ 2.2) brings out two interesting problems. The empirical evidence suggests that the complements stand in asymmetric relations to each other. This can be interpreted in two opposite ways: the complements stand in asymmetric c-command relations, determined by the binary branching requirement; or else, the property which determines these asymmetries is not a structural one, but it is linear precedence. Interestingly, this controversy points out a close connection between linear precedence and asymmetric c-command.

Finally, the existence of rightward movement to an adjunct position has been recently called into question (§ 2.3). But standard X-bar theory does not exclude this possibility; then, once again, an independent postulate is required to obtain the necessary restrictiveness.

2.1. Against adjunction

Let us consider prototypical modifiers, such as adjectives and adverbs. These elements are usually taken to be adjoined to the category they modify: adverbs are adjoined to a verbal or inflectional projection, and modifying adjectives are adjoined to some nominal projection. The adjunct analysis is supported by the apparently free iteration of the modifiers.

This view of modifiers has been recently challenged by Cinque (1993; 1995), who argues that this apparently free iteration is actually an instance of asyndetic coordination; apart from coordination, the number of possible modifiers for every category is limited and subject to rigid ordering constraints.

As for nominal modifiers, Cinque (1993) notes that modifying adjectives cannot be all adjoined to one and the same projection within the noun phrase. Different types of adjective occupy distinct positions: in the Italian examples in (11), for instance, the thematic adjective *italiano* is obligatorily postnominal, whereas the adjective *mero* is obligatorily prenominal:

(11) a. *l'invasione italiana dell'Albania*
the invasion Italian of-the Albania
'the Italian invasion of Albania'
b. * *l'italiana invasione dell'Albania*
the Italian invasion of-the Albania
c. *Gianni ha fatto una mera proposta.*
Gianni has made a mere proposal
'Gianni made a mere proposal.'
d. * *Gianni ha fatto una proposta mera.*
Gianni has made a proposal mere.

The data in (11) cannot be accounted for by the assumption that the two adjectives are adjoined to different projections of the noun either. In fact, the thematic adjective in (11a–b) is assigned the external theta role by the noun, and hence it must be generated in Spec, NP. But since in Italian the specifier is to the left of the head, this should give the linear order of (11b), which is actually excluded. Moreover, in (11a) the thematic adjective appears between the noun and its PP complement, which are generated as sisters. The only possibility to derive this order while ruling out (11b) is to assume the obligatory leftward movement of the head N° past the adjective.[16] This instance of head-to-head movement forces the postulation of at least one functional head between the determiner and the noun:

(12) $[_{DP}$ D $[_{FP}$ N°+F° $[_{NP}$ thematic adjective $[_{N'}$ t_N PP]]]].

As for the contrast between (11a) and (11d), it is still possible to analyse the adjective *mero* as left-adjoined to FP, hence structurally higher than the incorporated noun. However, the existence of at least one functional head between D° and N° opens a new perspective on the syntax of adjectives: they can be taken to occupy each the specifier of a distinct functional head (Crisma 1993: 74–92). In this way, the morphological agreement between the adjective and the noun can be reduced to an instance of Spec/head agreement (assuming that the functional heads associated to the noun share its phi-features).

This analysis is supported by the observation that, abstracting from asyndetic coordination (exemplified in (13)), the number of modifying adjectives is limited and the various types of adjective must be realized in a fixed order. (14) shows the relative order of adjectives modifying an object-denoting noun:

(13) *Una lunga, piacevole festa*
'a long, agreeable party'

(14) a. evaluating > size > color
b. *a beautiful big red ball*
c. *una simpatica grande palla rossa*[17]
'a nice big red ball'

The limit on the number of non-coordinated attributive adjectives cannot be easily accounted for in the adjunct analysis, since adjunction, like coordination, allows free iteration by definition (cf. the recursiveness of rule (3)). Moreover, the adjunct analysis does not explain the fixed order of the adjectives in (14), since adjunction is usually taken to be free. As Crisma (1993: 78) observes, in order to explain the serialization of adjectives, the adjunct analysis requires two stipulations: that adjectives are always adjoined to the left, and that there is some semantic or selectional relation between an adjunct and the maximal projection it is adjoined to. The limited number and the serialization of non-coordinated adjectives instead follow straightforwardly if each adjective is taken to occur in the specifier of a nominal functional head. In fact, assuming the uniqueness of the specifier position, it follows that two adjectives of the same type cannot cooccur unless they are coordinated; moreover, functional heads occur in a fixed order determined by c-selection, and consequently their specifiers are ordered with respect to each other as well.

A similar argument holds for adverbial modifiers. In a study of the position of the past participle in Romance, Cinque (1995) shows that adverbs of various classes and floating or leftward moved quantifiers occur in a fixed order in the Romance family. Italian and French are two representative examples:

(15) a. *I bambini non hanno mica più tutti detto tutto bene alla maestra.*
the children not have at all any longer all said everything well to the teacher.
'The children haven't said everything well to the teacher any longer.'
b. *Les enfants n'ont plus tous tout bien repondu à la maîtresse.*
the children not have any longer all everything well said to the teacher.
'The children haven't said everything well to the teacher any longer.'

Note that although the order of the adverbs and quantifiers is identical in the two examples, the position of the past participle is different. Cinque proposes, following Pollock (1989) and Belletti (1990), that this is determined by the different scope of verb movement in the two languages: in Italian, the past participle can move to a higher position than in French. Since verb movement is an instance of head-to-head incorporation, it provides evidence for the existence of a head position both to the left and to the right of every one of the adverbs and quantifiers exemplified in (15). In fact, inspecting different Romance varieties, Cinque shows that the participle may intervene between any two adjacent elements of the sequence, as shown in (16):

(16) a. *I bambini non hanno detto mica più tutti tutto bene alla maestra.* (Italian)

(16) b. *I bambini non hanno mica detto più tutti tutto bene alla maestra.*
c. *I bambini non hanno mica più detto tutti tutto bene alla maestra.*
d. *I bambini non hanno mica più tutti detto tutto bene alla maestra*
the children not have (said) at all (said) any longer (said) all (said) everything (said) to the teacher.
'The children haven't said everything well to the teacher any longer.'
e. *Apo mandigadu bene.* (Logodurese Sardinian)
(I) have eaten well
'I ate well.'
f. * *Apo bene mandigadu.*
(I) have well eaten
g. *Apo mandigadu tottu.*
(I) have eaten everything
'I ate everything.'
h. *Apo tottu mandigadu.*[18]
I have everything eaten
'I ate everything.'
i. *Il a bien compris la question.* (French)
he has well understood the question
'He fully understood the question.'
l. * *Il a compris bien la question.*
he has understood well the question

The data in (16) show that the inflectional structure of the participle involves a rich array of functional heads. Since exactly one adverbial element of a specific type can occur to the left of every head position, Cinque concludes that adverbs (and quantifiers) occur in the specifier of the various functional heads.[19]

The line of the argument can be summarized as follows: each modifier is realized in a unique and specific position; since these properties are characteristic of specifiers, there is no positive evidence to the effect that modifiers are generated in positions of a different type. In particular, the properties usually attributed to the adjunct position, namely free iteration and free ordering, turn out to be empirically incorrect once the interference of asyndetic coordination is recognized. This approach to modifiers rests on the current theory of functional heads.

Two other phenomena that are often analysed as cases of adjunction are English topicalization (cf. Baltin 1982; Lasnik–Saito 1992) and scrambling (cf. Müller–Sternefeld 1993, 1994; Saito 1989).[20] In this domain too, the adjunction analysis has been recently questioned.

As for English topicalization, Authier (1992) and Rizzi (1995) argue for a CP-recursion analysis: topicalized phrases occupy the specifier of a complementizer-like functional head which occurs below the declarative C°.

Authier (1992) starts from the analysis of negative preposing, exemplified in (17). In this structure, a negative phrases is fronted and triggers auxiliary-subject inversion. Inversion is usually analysed as incorporation of I° to the immediately c-commanding head: since in (17) the preposed auxiliary follows the declarative complementizer, there must be a head position between the declarative C° and I° which hosts the auxiliary, while its Spec is the landing site of the negative phrase.

(17) *I believe* $[_{CP}$ *that* $[_{XP}$ *under no circumstances*$_i$ $[_{X'}$ *would* $[_{IP}$ *he* t_{AUX} *do that* $t_i]]]]$.

Authier analyses XP as the lower layer of a CP-recursion structure.

He then argues that English topicalization is excluded in the same contexts that exclude negative preposing, like noun complement clauses (18a–b), sentential subjects (18c–d), and factive complement clauses (18e–f). In all these contexts, the CP is an island.

(18) a. * *The fact that never has he had to borrow money makes him very proud.*
b. * *The fact that Bill, Mary likes makes John very jealous.*
c. * *That never in his life has he had to borrow money is true.*
d. * *That this book, Mary read thoroughly is true.*
e. * *John regretted that never had he seen Gone with the Wind.*
f. * *John regretted that Gone with the Wind, we went to see.*

Therefore, Authier proposes that topicalization too is an instance of CP-recursion where the topicalized phrase raises to the Spec of the lower CP level; the lower C° head, however, remains phonetically empty, since topicalization does not trigger the incorporation of the auxiliary. If topicalization were instead adjunction to IP, it would be unclear how to exclude it in (18b, d, f), since adjunction to IP is not expected to be sensitive to the properties of the CP level.

A similar approach is developed by Rizzi (1995), who actually abandons CP-recursion in favour of a "split Comp" hypothesis: the IP is introduced by several distinct functional heads, whose specifiers host specific elements. Consider for instance the Italian data in (19). The clausal particles *che* and *se* are usually taken to both occur in C°. However, they differ in relative order with respect to left-dislocated phrases: the latter obligatorily follow *che*, but they may either precede or follow *se*:

(19) a. * *Penso questo libro che lo darò a Gianni.*
(I) think this book that (I) will give it to Gianni
b. *Penso che questo libro, lo darò a Gianni.*
(I) think that this book, (I) will give it to Gianni
'I think that I will give this book to Gianni.'
c. *Non so questo libro se dovrei darlo a Gianni.*
(I) don't know this book if (I) should give it to Gianni
d. *Non so se questo libro dovrei darlo a Gianni.*
(I) don't know if this book (I) should give it to Gianni
'I don't know whether I should give this book to Gianni.'

If left-dislocated phrases are adjuncts, a contradiction arises: adjunction to CP must be impossible in (19a) but possible in (19c). Note that in both cases the CP is an argument of the matrix verb, so that Chomsky's (1986b: 6) constraint against adjunction to an argument is irrelevant. The data instead follow if *che* and *se* are analysed as two distinct functional heads, labelled C° and Focus°; the positions for left-dislocated phrases are between C° and Focus°, and between Focus° and IP:

(20) $[_{CP}$ *che* [(XP) $[_{FocusP}$ *se* [(XP) $[_{IP}$ ...]]].

This structure can also account for the different position of relative and interrogative phrases. Rizzi proposes that interrogative phrases land in the specifier of the head Focus°, which carries the feature [+wh]; this predicts that they may precede or follow left-dislocated elements. The prediction is confirmed by the data in (21). On the other hand, contrary to interrogative phrases, relative pronouns obligatorily precede any dislocated phrase: this follows if they target a higher position, namely Spec, CP, which in the standard analysis of relative clauses is adjacent to the relative pronoun's antecedent (22).

(21) a. *Mi domando, il premio Nobel, a chi lo potrebbero dare.*
(I) wonder the Nobel prize to whom (they) could award it
b. ? *Mi domando a chi, il premio Nobel, lo potrebbero dare.*
(I) wonder to whom, the Nobel prize, (they) could award it
'I wonder to whom the Nobel prize could be awarded.'

(22) a. *un uomo a cui, il premio Nobel, lo daranno senz'altro.*
a man to whom, the Nobel prize, (they) will surely award it
'a man to whom the Nobel prize will be surely awarded'

(22) b. * *un uomo, il premio Nobel, a cui lo daranno senz'altro.*
a man, the Nobel prize, to whom (they) will surely award it

In (20), left-dislocated phrases could be taken to adjoin to FocusP and IP: in fact, left dislocation exhibits the properties of iteration and free order that are usually attributed to adjunction:

(23) a. *A Gianni, questo libro, credo che nessuno glielo darebbe volentieri.*
to G., this book, (I) believe that nobody would give to-him it willingly
b. *Questo libro, a Gianni, credo che nessuno glielo darebbe volentieri.*
this book, to G., (I) believe that nobody would give to-him it willingly
'As for this book, to Gianni, I think that nobody would willingly give it.'

Nevertheless, Rizzi proposes that left-dislocated elements too are in the specifier of a recursive functional head, Topic°, which happens to be phonetically empty. Evidence for this null functional head comes from French, where a left-dislocated phrase cannot intervene between a trace in subject position and its antecedent:

(24) a. *Je ne sais pas* $[_{CP}$ *qui*$_i$ $[_{C'}$ C°$_i$ $[_{IP}$ t$_i$ *pourrait l'acheter*]]]
I not know who could it buy.
'I don't know who could buy it.'
b. *? *Je ne sais pas* $[_{CP}$ *qui*$_i$ [C°$_i$ $[_{TopP}$ *ton livre* [Top° $[_{IP}$ t$_i$ *pourrait l'acheter*]]]]].
I not know who, your book, could it buy.
'Your book, I don't know who could buy it.'

In earlier work Rizzi (1990: 51–60) proposed that the subject trace in (24) needs to be governed by a head C° endowed with agreement features in order to satisfy the Empty Category Principle. Thus, the ungrammaticality of (24b) follows from the assumption that left dislocation introduces a head position between C° and IP. This head prevents C° from governing Spec, IP, by relativized minimality.[21]

The split Comp hypothesis will be further discussed in chapters VI-VII.

Scrambling is a very complex and controversial phenomenon, but it must be mentioned that at least some authors have analysed it as involving specifier positions rather than adjunction. For instance, Sportiche (1992: 64–67) analyses Dutch scrambling as overt movement of an argument phrase to the Spec of an IP-medial functional head, the Clitic Voice.[22]

In sum, the recently developed theory of functional heads has called into question the existence of base-generated adjuncts (modifiers) and the adjunction analysis of certain instances of A' movement, such as topicalization and scrambling.[23] Many purported instances of adjunction actually involve a specifier position, which must be assumed to be unique. If empirically justified, the elimination of adjunction and of multiple specifiers constitutes a strong restriction on the phrase structure of natural language.

2.2. Against multiple sister complements

The syntax of multiple complements has been the subject of an interesting debate. Kayne (1984: 129–163) gave some empirical evidence in favour of his theoretical notion of unambiguous path, which excludes multiple sister complements (cf. § 1.3). Barss–Lasnik (1986) brought to light an asymmetry in double object constructions, where the leftmost goal argument appears to asymmetrically c-command the theme argument. Larson (1988) and Jackendoff (1990) argue that this asymmetry is shared by all double complement constructions. Since the standard formulation of X-bar theory does not provide a structural representation that may justify this asymmetry, the authors draw two opposite conclusions: Larson proposes a revision of X-bar theory, incorporating a binary branching constraint for complement positions (the Single Complement Hypothesis); Jackendoff criticizes this proposal and suggests that the relevant asymmetry concerns linear order rather than hierarchical relations. This double theoretical possibility brings to light a correlation between the two ordering principles of natural language, the hierarchical and the linear one.

Kayne (1984: 136–159) observes that nouns are unable to govern into a clausal complement. This is shown by the ungrammaticality of raising and ECM infinitival complements to a noun:

(25) a. * *Mary*$_i$ *'s appearance* [t$_i$ *to have left*]
b. *Mary*$_i$ *appears* [t$_i$ *to have left*].
c. * *Its*$_i$ *likelihood* [t$_i$ *not to be there*]
d. *It*$_i$ *is likely* [t$_i$ *not to be there*].

(26) a. *John believes* [*Mary to have left*].
b. * *John's belief of* [*Mary to have left*].[24]
c. * *Mary*$_i$ *'s belief* [t$_i$ *to have left*] *by John*

This defective behaviour of nouns also emerges in small clauses configurations, supporting Stowell's (1983) approach:

(27) a. *John believes* [*Mary a genius*].
 b. * *John's belief of* [*Mary a genius*].

This provides an empirical argument in favour of the binary branching constraint (§ 1.3). In fact, the data in (28) below show that derived nominals are unable to take more than one complement in the way that verbs do. This verb/noun asymmetry can be assimilated to those in (25)–(27) by assuming that double complements cannot be sisters to the selecting head, but they must be grouped together in a clauselike constituent. This constituent is a barrier for government by a noun head: therefore, the noun complements cannot be Case-marked by the noun itself, and moreover, an ECP violation arises if one of the complements is moved, as in (28d).

(28) a. *John robbed* [*Mary of her money*].
 b. * *John's robbery of* [*Mary of her money*]
 c. *Mary*$_i$ *was robbed* [t_i *of her money*] *by John.*
 d. * *Mary*$_i$*'s robbery* [t_i *of her money*] *by John*

An independent argument against multiple sister complements stems from Barss—Lasnik's (1986) study of double object constructions. A number of syntactic tests show that the two objects stand in an asymmetric relation: these include anaphor binding, quantifier binding, the reciprocal construction *each...the other*, and the licensing of negative polarity items. In all these phenomena, one constituent must be properly related to another one: if this relation requires c-command, it must be concluded that the leftmost goal argument c-commands the theme argument, but the reverse does not hold.

(29) a. *I showed Mary herself.*
 b. * *I showed herself Mary.*
(30) a. *I gave every worker*$_i$ *his*$_i$ *paycheck.*
 b. * *I gave its*$_i$ *owner every paycheck*$_i$..
(31) a. *I showed each man the other's socks.*
 b. * *I showed the other's friend each man.*
(32) a. *I showed no one anything.*
 b. * *I showed anyone nothing.*

This asymmetry is not predicted by a multiple complements structure like (9), where the two objects c-command each other.

Larson (1988: 338) observes the same asymmetry in the dative construction where the goal argument is realized within a dative PP:

(33) a. *I showed Mary to herself.*
b. * *I showed herself to Mary.*
(34) a. *I sent every check$_i$ to its$_i$ owner.*
b. ?? *I sent his$_i$ paycheck to every worker$_i$.*.
(35) a. *I sent each boy to the other's parents.*
b. * *I sent the other's check to each boy.*
(36) a. *I sent no presents to any of the children.*
b. * *I sent any of the packages to none of the children.*

In this case, it is possible to assume a multiple branching structure and argue that it is the PP node that prevents the second object from c-commanding the first one. This assumption, however, cannot account for the data in (37)–(40), reported by Jackendoff (1990), where two PP objects show the same left-to-right asymmetry. Clearly, here the presence of the PP node does not prevent the first complement from binding the second one:

(37) a. *I heard from John and Bill about themselves.*
b. * *I heard from themselves about John and Bill.*
(38) a. *I heard from every mother$_i$ about her$_i$ child.*
b. * *I heard from her$_i$ mother about every girl$_i$.*.
(39) a. *I talked about each boy to the other.*
b. * *I talked about the other to each boy.*
(40) a. *I talked about none of the boys to any of the girls.*
b. * *I talked about any of the boys to none of the girls.*

Jackendoff also shows that the order of the two PPs can be inverted, and the result is the same: the leftmost PP asymmetrically binds into the rightmost one.

The two authors propose two opposite solutions to the problem of double complement structures. Larson assumes a structural approach to binding and polarity items licensing, and concludes that the asymmetric relation between two complements is an argument against multiple branching. However, in the traditional analysis of the VP even a binary branching representation does not give the right structural asymmetry. In fact, the leftmost complement, which is closer to the verb, must be its sister under V'; but then it cannot be structurally higher than the rightmost complement, which must be the daughter of a higher verbal projection, as in (41):

(41)

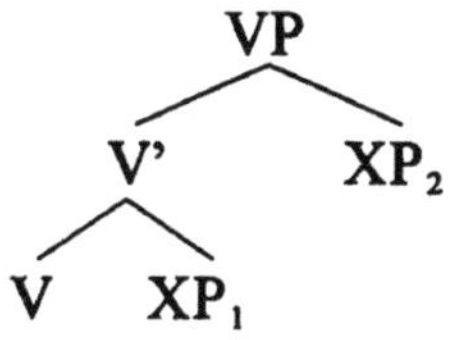

This yields an asymmetric relation which is the reverse of what we expected: here it is the rightmost argument XP_2 that asymmetrically c-commands the leftmost one.

Therefore, Larson proposes a different base structure for double complement constructions, featuring a so-called "VP shell". In the dative PP construction (e.g. (33a)), the rightmost goal argument is generated as a sister to the lexical verb, and the theme argument is instead generated in the specifier of VP. The surface order is derived by moving the verb from its base position to a higher head position, which is empty in the base: this is the head of a VP projection superposed to the VP projected by the lexical verb. In its derived position, the verb governs and Case-marks the direct object in Spec, VP. The derivation of (33a) is represented in (42).

(42)

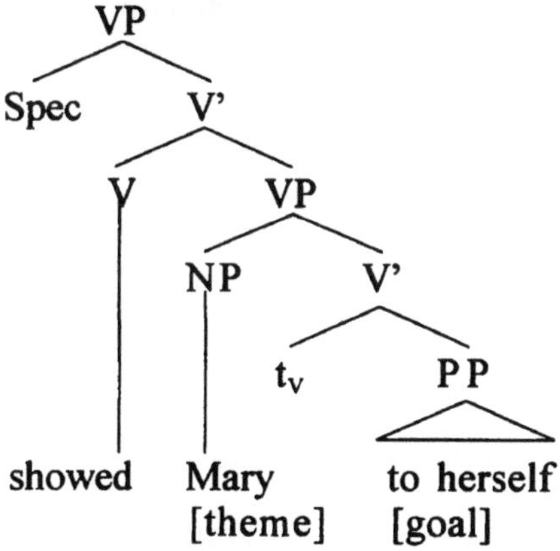

In this structure, the linear order of the complements corresponds to the desired c-command relations: the theme argument asymmetrically c-commands the PP argument. As for the double object construction, exemplified in (29a), Larson suggests that it can be derived by an operation akin to passivization. In fact, the structural relation between the two complements in the lower VP of (42) is akin to a subject/object a-symmetry. The double object construction can be derived by a process of argument demotion which assigns the theme theta role to a right-hand adjunct position; the goal argument is raised to the nonthematic

Spec of the lower VP, where it receives the structural Accusative case. (43) is the derivation of (29a):

(43)
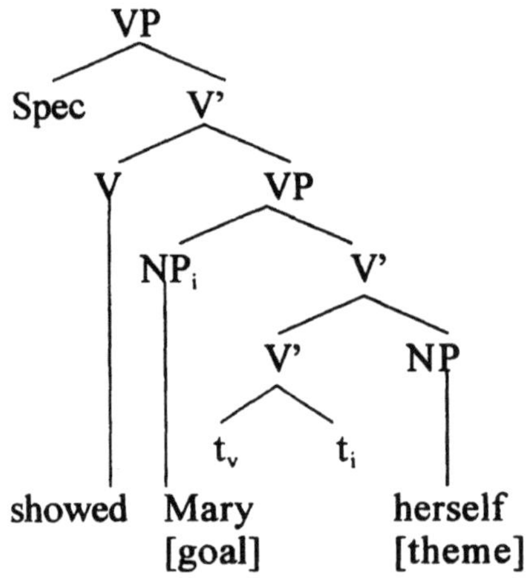

Independently of the analysis of the double complement construction, the VP shell representation of multiple complements is now widely assumed.[25] Larson (1988: 380–381) excludes the possibility of multiple branching under X' by eliminating the Kleene star in the X-bar rule expanding X' (the Single Complement Hypothesis). Thus the binary branching constraint is directly incorporated into X-bar theory.[26]

Jackendoff (1990) criticizes Larson's approach in one fundamental respect, namely the assumption that the asymmetric relation between the complements must be explained in terms of hierarchical relations. An alternative analysis, which cannot be *a priori* excluded, is that binding domains be defined in terms of linear order. In fact, linear order affects the possibility of coreference in conjoined constructions like (44), where neither of the coreferent arguments c-commands the other:

(44) a. *Fred mentioned that I saw George*$_i$ *yesterday and that he*$_i$ *looked good.*
b. * *Fred mentioned that I saw him*$_i$ *yesterday and that George*$_i$ *looked good.*

This argument is not compelling if one adopts Reinhart's (1983) proposal that the accidental coreference exemplified in (44a) is distinct from syntactic binding.

However, the generalization that in English the leftmost complement takes asymmetrically scope over the rightmost one raises an interesting theoretical problem: there is a redundancy between the two ordering relations of natural language, the linear one and the hierarchical one. Yet both of these ordering principles are necessary: on the one hand, natural languages universally have a constituent structure, and on the other hand, natural language expressions are necessarily realized as a (temporal) sequence of discrete elements. Therefore, the only way to eliminate this redundancy is to establish a principled correlation between the two types of order.

2.3. Against rightward adjunction

The possibility of rightward adjunction is exploited in the standard analysis of many "stylistic" rules, like heavy NP shift and extraposition. In a specifier-head language like English or German, rightward movement can only be an instance of adjunction.[27]

Haider (1993) gives several arguments against such an analysis of extraposition. First, he observes that extraposition does not obey the same constraints as other clear instances of movement, for instance the subject island constraint:

(45) a. $[_{DP}$ *A man* $t_i]$ *came into the room* [*with red hair*]$_i$.
b. * [*With blond hair*]$_i$, $[_{DP}$ *a man* t] *came into the room.*

Second, some constituents that can be extraposed (e.g. the PP in (46a)) cannot be extracted by leftward scrambling:

(46) a. *Ich habe gestern* $[_{DP}$ *einen Mann*] *gesehen* $[_{PP}$ *mit blauen Haaren*].
I have yesterday a man seen with blue hair
'Yesterday I saw a man with blue hair.'
b. * *Ich habe* $[_{PP}$ *mit blauen Haaren*] *gestern* $[_{DP}$ *einen Mann* t] *gesehen.*
I have with blue hair yesterday a man seen.

Thus, the class of constituents that can be extraposed does not coincide with the class of elements that can be moved. In German, for instance, DPs can move leftward but they cannot be extraposed. These asymmetries suggests that extraposition is not an instance of (rightward) A' movement.

Furthermore, there is evidence to the effect that extraposed constituents are very low in the structure. Binding phenomena show that a relative clause extraposed from the direct object is c-commanded by VP-internal complements:

(47) *I would not tell everyone*$_i$ *all the details at once* [*that he*$_i$ *might be interested in*].

This evidence is inconsistent with the view that extraposition is (even base-generated) rightward adjunction to VP[28] (or *a fortiori*, to IP).[29] This point is strenghtened by the observation that in German a constituent extraposed from the VP cannot be left behind in the left dislocation of the VP:

(48) a. [*Das Haus gezeigt, in dem ich wohne*] *das hat der Mann keinem freiwillig.*
the house shown in which I live, that has the man to-noone voluntarily
'The man hasn't voluntarily shown to anyone the house in which I live.'
b. * [*Das Haus gezeigt*] *das hat der Mann keinem freiwillig, in dem ich wohne.*
the house shown that has the man to-noone voluntarily in which I live

(49) a. [*Fremden gesagt, wo sie wohnt*] *das hat sie ja nie.*
strangers told where she lives, that has she never
'She has never told strangers where she lives.'
b. * [*Fremden gesagt*] *das hat sie ja nie, wo sie wohnt.*
strangers told, that has she never where she lives

In the light of these data, Haider argues that extraposed constituents are base-generated in a VP-internal base-generated position: the later they appear in the sentence, the deeper they are in the constituent structure.[30] Since standard X-bar theory does not exclude the possibility of rightward adjunction, Haider concludes that it is too little restrictive.

Accordingly, Haider proposes that the linear order of the symbols in the right-hand side of rules (1)–(3) must be universally fixed to the effect that the recursive node on the projection line is to the right of the non-recursive node. Thus in (2) X' must be to the right of the specifier, and in (3) the adjoined phrase β must be to the left of the lower segment of its host α. This gives a universal order specifier-head and adjunct-host.[31] This general constraint on X-bar structures is formulated as follows:

(50) Branching Constraint: the projection line is branching progressively. (Haider 1993: 9)

Haider gives an interesting cognitive motivation for this constraint in terms of optimality of parsing. In a regressively branching structure like (51) below, when the parser finds the first element V° it cannot know how deeply embedded it is, namely, how many brackets must be postulated above it, since in this structure an element "closing" a higher projection linearly follows an element closing a lower projection:

(51) $[_3[_2[_1$ V° DP] DP] PP]

Therefore, the parser cannot build the structure until the final element of the string is reached. In a progressively branching structure, instead, the first constituent of the string introduces the highest node of the projection line, the second constituent introduces an immediately lower projection, and so on: this is because a leftmost element necessarily c-commands a rightmost one, and the node that dominates the first element necessarily dominates the second element as well. Since the elements are necessarily presented in a linear, temporal order, this guarantees an optimal monotonic and incremental parsing of the string.

The essential insight of Haider's proposal is that X-bar theory must be restricted by limiting the possible linear ordering of nonterminal symbols in the tree.[32] This is also essential to Kayne's proposal, to be discussed in the following section.

3. Kayne's (1994) Antisymmetry

In the light of the empirical evidence reviewed in the preceding section, there has developed an increasing consensus on the necessity of constraining the standard X-bar theory of the Principles and Parameters framework: the works by Fukui–Speas (1986), Larson (1988), Hoekstra (1991) and Haider (1993) are representative of this tendency. However, the proposed constraints often belong to modules of the grammar independent of X-bar theory (cf. Kayne's (1984) unambiguous path requirement or Hoekstra's (1991) uniqueness of licensing principle) or, if they belong to X-bar theory, they are introduced as independent postulates (cf. Larson's (1988) single complement hypothesis and Haider's (1993) branching constraint).

An important step towards the goal of restricting the set of possible phrase structures is Kayne's (1994) Antisymmetry theory. In this theory, some of the fundamental properties of phrase structure encoded in rules (1)–(3) are derived from, while others are excluded by, one axiom and a revised definition of c-command. The following summary will be unable to do justice to the intrinsic compactness and elegance of Kayne's theory; it is only introduced for the sake of self-containment of this study.

3.1. The Linear Correspondence Axiom

As a first step, Kayne abandons the standard assumption that hierarchical structure and linear order are independent of one another. As discussed in § 1.1, in standard X-bar theory linear order is defined between sister categories introduced by the X-bar rules, and this determines the linear order of the terminal symbols that they dominate. But from a conceptual viewpoint, while linear order is an irreducible property of the string of terminal symbols, it is not quite clear why it should hold between nonterminal symbols.[33]

Kayne proposes instead to derive the linear order of terminal symbols from the hierarchical relations between the nonterminal symbols dominating them. In order to do so, it is necessary to define a hierarchical relation on the tree structure with the properties of a linear order, namely transitivity, totality and antisymmetry. The relation of asymmetric c-command has the first and the third property; however, it is not total, since it is not defined for any pair of distinct nodes in the tree. This difficulty can be obviated by defining on the tree the set A of all the ordered pairs of nonterminals $<X_i,Y_i>$ such the first element asymmetrically c-commands the second one. Assuming a relation *d* that maps every nonterminal X to the set *d*(X) of terminals that it dominates, the image under *d* of every ordered pair $<X_i,Y_i>$ can be defined as the Cartesian product of $d(X_i)$ and $d(Y_i)$, namely the set of all the ordered pairs of terminals $<x,y>$ such that $x \in d(X_i)$ and $y \in d(Y_i)$, and *d*(A) is defined as the union of the images under *d* of all the ordered pairs $<X_i, Y_i>$. Taking T to be the set of terminals of the tree, the following axiom holds:

(52) *Linear Correspondence Axiom.*
d(A) is a linear ordering of T.

Informally, the axiom implies that, although the relation of asymmetric c-command does not totally order the tree, it must hold for a number of pairs of nonterminals such that the union of the images under *d* of these pairs[34] yields a linear order of the terminals.

3.2. Consequences for phrase structure

The Linear Correspondence Axiom has two distinct sets of consequences with respect to the standard X-bar theory: some properties are derived from it, thus reducing the number of independent postulates of the theory; other properties are excluded by it, yielding a more restrictive X-bar scheme.

These consequences can be deduced by examining some elementary tree structures. For the time being, the relevant definition of c-command is the one that does not make reference to branching nodes (Kayne 1994: 7).[35]

(53)

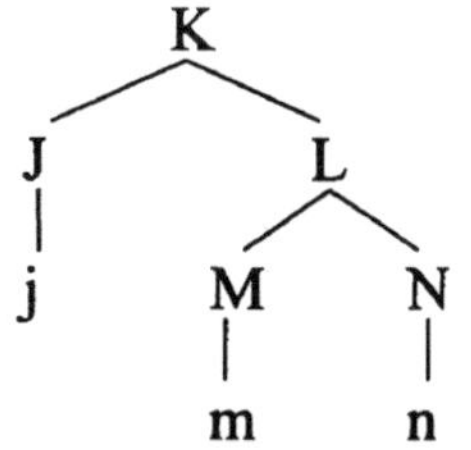

In tree (53), K dominates a preterminal J and a nonterminal L; although c-command between J and L is symmetric, J asymmetrically c-commands the daughters of L, M and N; this gives the ordered pairs <J,M> and <J,N>. On the other hand, L dominates two preterminals M and N, which c-command each other; therefore, there is no ordered pair including the nodes M and N, and the terminals dominated by them cannot be ordered with respect to each other. Therefore, the structure violates the LCA. In X-bar theoretic terms, preterminals are heads: thus the LCA derives the uniqueness of the head (which is stipulated in rule (1), by the lack of a Kleene star associated to the symbol X) and the non-head status of complements.

Another type of symmetric relation that the LCA excludes is (54):

(54)

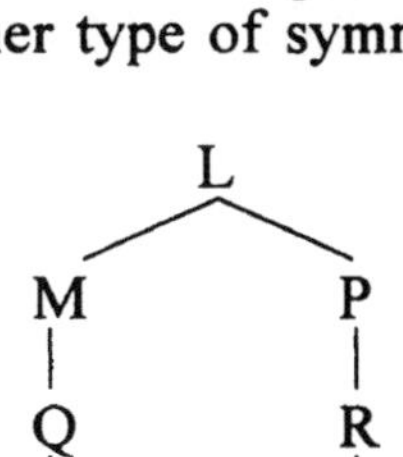

Neither of the sister nodes M and P is a preterminal. It follows that each one asymmetrically c-commands the other's daughter, giving the pairs <M,R> and <P,Q>, whose images under *d* are respectively <q,r> and <r,q>. Hence the configuration with two sister nonterminals violates the antisymmetry requirement imposed by the LCA. In X-bar theoretic terms, this has two consequences. On the one hand, there cannot be headless projections, e.g. [$_S$ NP VP], where neither NP nor VP is a preterminal.[36] On the other hand, a head cannot have more than one structural complement, since multiple complements necessarily introduce the illformed structure (54).

The impossibility of two non-preterminal sister nodes also excludes the abstract structure (55), which corresponds to the structure introduced by the X-bar rule (2):

(55)

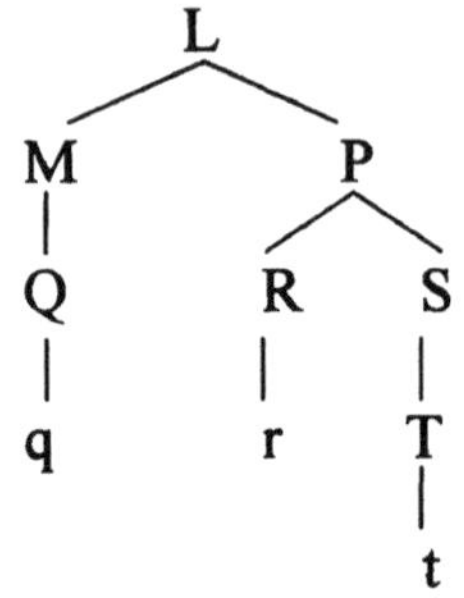

In fact, asymmetric c-command holds between M and R, but also between P and Q, which gives a violation of antisymmetry for the terminal symbols *q, r*. In order to allow a specifier position, Kayne (1994: 15–16) adopts the segment/category distinction proposed by May (1985) and gives a new definition of c-command:

(56) X c-commands Y iff X and Y are categories and X excludes[37] Y and every category that dominates X dominates Y.

Let us now consider a structure where a non-preterminal M is adjoined to another non-preterminal P:

(57)

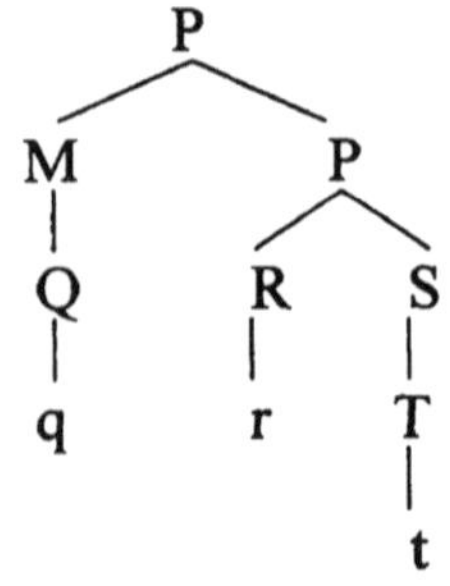

By the definition of c-command in (56), asymmetric c-command holds between M and R, but no longer between P and Q: in fact, M is a category that excludes (P and) R, but P can no longer c-command M or Q: the whole category P does not exclude them, and its lower segment is not a category, hence by definition it does not c-command anything. It follows that the adjoined phrase asymmetrically c-commands the nodes dominated by the host category, and linear order is guaranteed.[38]

Thus, the LCA rules out the phrase structure introduced by rule (2), but it allows the structure introduced by rule (3) with α, β non-preterminal nodes, under the revised definition of c-command (56). However, the possibility of adjunction is severely limited: there can be at most one adjunct per category. In fact, multiple adjuncts would be separated only by one segment of a category, and this constitutes a symmetric relation, since the internal constituents of each adjunct would asymmetrically c-command the internal constituents of the other, as represented in (58):

(58)

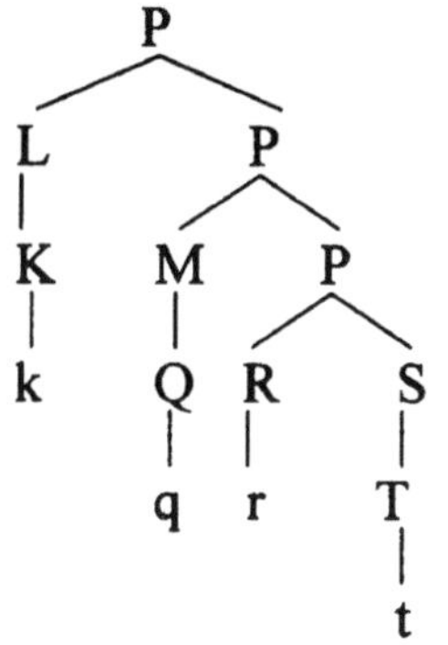

Asymmetric c-command holds for the pairs <L,Q> and <M,K>, in violation of the antisymmetry requirement. In this respect, the LCA turns out to be more restrictive than the recursive X-bar rule (3).

This proposal has several important consequences: the intermediate level X' is eliminated; there is only one projection level, which allows the adjunction of at most one constituent. This constituent has the configurational properties of an adjunct of standard X-bar theory: it is not included in the host category, and by the definition in (56), it can c-command any node which is c-commanded by the host category itself.[39] However, the adjunct position shares one important property with the traditional specifier position, namely its uniqueness.

Moreover, the elimination of X' automatically eliminates the problems related to intermediate projections (§ 1.2). The lower segment of an adjunction configuration cannot move because it is not a category, and by (56) it is unable to c-command its trace (Kayne 1994: 17).

The same reasoning holds for adjunction to a head, represented in (59):

(59)

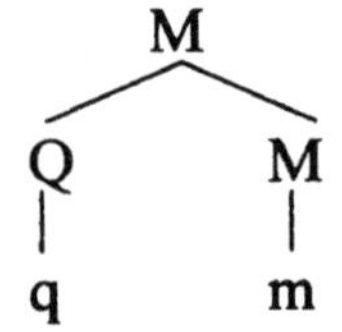

The adjoined terminal Q asymmetrically c-commands its host M, since Q excludes M, and whatever category dominates Q dominates M as well. Once again, multiple adjunction is excluded, since multiple adjoined heads would not be linearly ordered with respect to each other, and the lower segment of the host cannot move.[40,41]

To conclude, the only X-bar structure that satisfies the LCA is (60), a one-level binary branching structure (with YP and ZP optional):

(60)

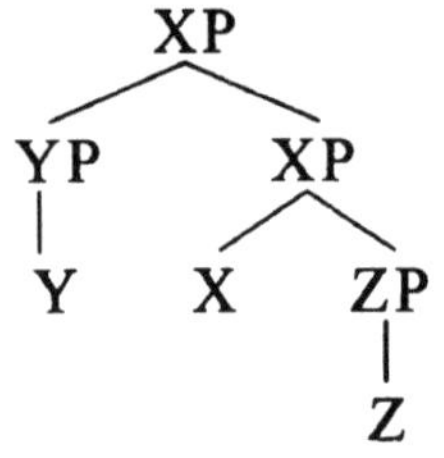

3.3. Consequences for linear order

In (60), the specifier YP asymmetrically c-commands the head X and its complement ZP, yielding the ordered pairs <YP,X>, <YP,ZP> and <YP,Z>; the head X asymmetrically c-commands the head Z of its complement,[42] yielding the pair <X,Z>. This means that in the linear order determined by phrase structure, the specifier and the complement are necessarily placed on opposite sides with respect to the head. Thus Kayne's theory severely restricts not only the possible phrase structures of natural language, but also the possible linear order of elements.

The problem is, then, how to interpret the ordered pairs of terminals <x,y> that are projected by *d* from the ordered pairs of nonterminals related by asymmetric c-command. It is possible to take the first element of every pair to precede or to follow the second element: both choices yield a linear order of the string of terminals. The first choice gives a universal order specifier-head-complement; the second choice, a complement-head-specifier order.

Kayne (1994: 36–38) argues that the first choice is likely to be correct. In fact, typological studies reveal that the specifier precedes the head in the majority of the known languages (cf. Greenberg's 1963 observation about the rarity of VOS, OSV and OVS orders); moreover, in case of incorporation the adjoined head is always to the left of its host (cf. the proposal by Kayne 1991 mentioned in fn. 5).

Kayne also proposes a motivation for the universality of linear precedence. He postulates an abstract node A that is adjoined to the root of every phrase marker,[43] so that it has the property of asymmetrically c-commanding every other node in the phrase marker: for every node X of the phrase marker there will be an ordered pair <A, X>. Supposing that A dominates an abstract terminal *a*, this element will bear the same relation to all the terminals of the phrase marker: thus, it will be an abstract beginning terminal or an abstract "closing" terminal which marks the end of the string.

Kayne proposes that *a* should be a beginning terminal. In natural language, the strings of terminal symbols are both produced and perceived in a temporal succession. Consider a sequence of successive "time slots" $t_1 \dots t_n$: intuitively, every t_i can be associated to a substring of terminals that have been produced or perceived from t_1 up to t_i: the later is a time slot, the larger is the substring associated to it. If *a* is the beginning element, *a* will be present in any substring associated with a time slot, and it will precede all the terminals in every substring. If, on the contrary, *a* is the last element, it will be present in only one string, namely the one associated with the last time slot t_n. Thus from a temporal perspective, a beginning element bears a constant relation to all the terminals of the string, but a final element does not.

If *a* is the beginning element, in the ordered pairs <x,y> x must precede rather than follow y. In fact, the abstract node A asymmetrically c-commands every node X in the phrase structure, and thus the relation *d* will yield an ordered pair $\langle a, x_i \rangle$ for every terminal symbol x_i. If the ordered pairs are interpreted with x following y, a contradiction arises, since *a* will follow all the terminals and it will not be the beginning element, but the final one. Thus if asymmetric c-command is to yield a fixed ordering relation that is preserved in a temporal dimension, the linear relation must be precedence.

With respect to X-bar theory, this implies that the linear order of constituents of a phrase marker always has the complement following the head, and the adjoined element preceding its host. This result is empirically desirable, as discussed above.

Clearly, many previous analyses based on the directionality parameters are inconsistent with this proposal, and many languages exhibit a linear order of constituents that *prima facie* contradicts Kayne's claim. For these, Kayne's theory must exploit the only other mechanism that can modify the order of the constituents, namely movement. A complex sequence of movements may yield an order of constituents that cannot be base-generated by an X-bar theory consistent with the LCA. Of course, the movement of any constituent must be independently justified.

Kayne's proposal has another interesting consequence with respect to the problem of the visiting procedure mentioned in § 1.1. Consider the phrase marker (61), which is slightly more expanded than (60). For the ease of exposition, instead of visiting (61) first with a left-to-right and then with a right-to-left orientation, its mirror image (62) has been reported, and both trees can be visited from left-to-right.

(61)

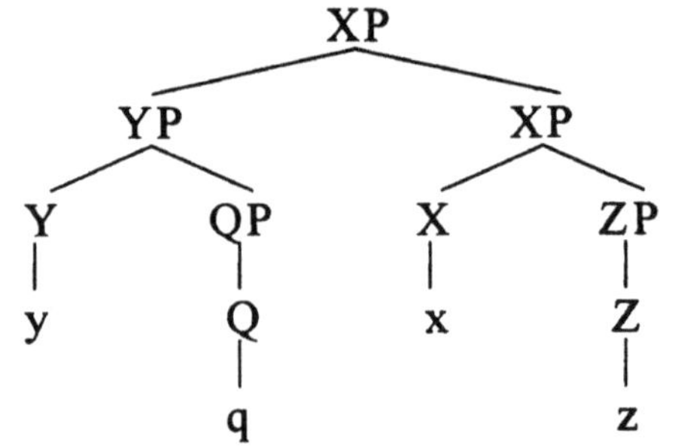

(62)

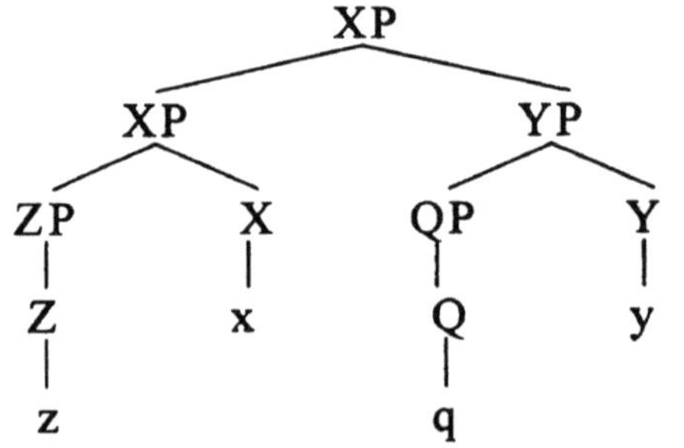

Each tree can be visited in a depht-first or in a breadth-first fashion. Starting from (61), a depth-first visit yields the string (63a), and a breadth-first visit gives the string in (63b).

(63) a. yqxz

(63) b. yxqz

For (62), a depth-first visit yields the string in (64a), and a breadth-first visit the string in (64b):

(64) a. zxqy
b. xyzq

Thus, changing the left-to-right orientation and the depht-first versus breadth-first procedure, four different strings are obtained.

From the perspective of the LCA, however, (61) and its mirror image (62) have identical asymmetric c-command relations holding between their constituent nodes: the set of ordered pairs of nonterminals is thus (65) in both cases:

(65) <Y, Q> → <y,q>
<YP, XP> → <y,x>,<y,z>,<q,x>,<q,z>
<YP, X> → <y,x>,<q,x>
<YP, ZP> → <y,z>,<q,z>
<YP, Z> → <y,z>,<q,z>
<X, Z> → <x,z>

and the resulting string is in both cases: *yqxz*.

Thus, the ordering of the terminal symbols imposed by the LCA gives one and only one string, while the visiting procedures gave four.

4. Other theoretical assumptions

The Antisymmetry theory incorporates the "traditional" conception of preterminal symbols as zero-level syntactic categories, and the assumption that every zero-level category projects a distinct maximal projection. In this respect, it is incompatible with the minimalist approach to phrase structure proposed by Chomsky (1995: 241–249).[44] Chomsky (1995: 334–340) assumes a version of the Linear Correspondence Axiom as a PF ordering principle; however, his set of assumptions yields different constraints on phrase structure than Kayne's, so that the two approaches remain distinguishable on empirical grounds. The present study aims at exploring the empirical consequences of Kayne's proposal.

The present study will also depart from the minimalist program in other respects.

(1) According to Chomsky (1995: 225–228), the features associated with the lexical items in the initial numeration undergo a limited set of operations - specifically, movement/attraction, checking, deletion and erasure; crucially, none of these operations can add to an item a feature that was not associated with it in the initial numeration. Assuming that the terminal symbol dominated by a functional head is a set of features, which is spelled out as a concrete morpheme by operations of the Morpho-phonological component of the grammar (cf. Halle–Marantz 1993), it will be proposed that (at least in this component) it is possible for an item to inherit a feature from another locally related item. This assumption will play a crucial role in the analysis of Case attraction of relative pronouns (§ III.4). Furthermore, it will be proposed that two incorporated functional heads can undergo in the Morpho-phonological component an operation of "unification" which turns them into a single set of features, to be spelled out as a single morpheme.

(2) As a more radical departure from the minimalist program, the present study remains neutral with respect to the derivational vs. representational nature of the syntactic component. Thus, we shall assume Relativized Minimality as a locality principle, instead of either of Chomsky's (1995) versions of the Shortest Movement/Minimal Link Condition (cf. Chomsky 1995: 179–186; 294–299; 355–357). Similarly, the "static" Spec/head agreement relation required by Rizzi's (1991) Wh-criterion, and related criteria, will not be reduced to the operation of feature checking.

(3) More importantly, the present study assumes head government as a significant syntactic relation, in agreement with the arguments put forth by Rizzi (1995).

Whereas with respect to the second point our analysis may ultimately turn out to be neutral (viz., it may be consistent with either a derivational or a representational version of that locality principle), the third point constitutes a radical departure from minimalist assumptions. Whenever this point becomes relevant to our analysis, alternative solutions consistent with the minimalist program will be discussed (cf. in particular §§ III.2, VI.8).

Despite these departures, the present study will borrow from Chomsky (1995) some specific assumptions:

(1) the "last resort" hypothesis, whereby movement is triggered by quasi-morphological requirements, defined in terms of specific features, to be satisfied in a local syntactic relation (construed as either checking or static agreement, in the sense discussed above);

(2) the hypothesis that parameters refer to formal/morphological features of specific items;

(3) the notion of economy of representation (cf. Chomsky 1995: 150–161), which will play a crucial role in the analysis of "PF deletion" in chapter VI;

(4) the copy theory of traces, and the related analysis of reconstruction effects (Chomsky 1995: 200–209).

It seems to us that this constitutes a coherent set of assumptions, and one that is consistent with the Antisymmetry theory.

5. Summary

This introductory chapter presented some theoretical and empirical problems with the standard X-bar theory of the Principles and Parameters framework, and synthesized the solutions that Kayne's (1994) Antisymmetry theory offers to these problems. Previously proposed solutions were partial and depended on specific X-bar theoretic postulates, or on principles independent of X-bar theory itself. The Antisymmetry theory instead provides a coherent and compact solution, based on a single axiom, the LCA, and a specific definition of c-command. The theory departs not only from the Principles and Parameters framework, but also from other syntactic frameworks, in that it does not separate the hierarchical relations from linear order by means of the ID/LP format, but relates them in an invariant way.

However, the X-bar structure (60) compatible with the LCA is much more restrictive than the currently assumed phrase structures, both with respect to the number of available positions within a maximal projection, and with respect to the possible analyses of the linear order of constituents in a given language. This leads to a revision of many largely accepted analyses of specific domains of natural language syntax. One of these domains is the syntax of relative clauses, which constitutes the topic of this book.

Chapter II
Restrictive relative clauses

0. Introduction

In the development of the Principles and Parameters framework, many syntactic analyses have been proposed which are now incompatible with the restrictiveness of the Linear Correspondence Axiom. A case in point is relative clauses. In languages like English or Italian, relative clauses linearly follow the nominal constituent that they modify: the standard analysis takes them to be right-adjoined to this nominal constituent (§ 1). But this is inconsistent with the LCA: adjoined constituents asymmetrically c-command their host, hence they invariably precede it.

Kayne (1994: chapter 8) proposes an alternative analysis consistent with the LCA. This second chapter is devoted to a discussion and an elaboration of his proposal with respect to restrictive relative clauses.

1. The adjunct analysis

In the Principles and Parameters framework, the most widely accepted analysis of relative clauses is the adjunct analysis. The literature on this topic is extensive, and it cannot be exhaustively summarized here. Therefore, this paragraph will simply present three aspects of the analysis which are relevant to the following discussion.

First, this analysis exploits the configuration of rightward adjunction, which in standard X-bar theory is introduced by rule (3) of § I.1:

(1) $[_{N^n}$ $[_{N^n}$ ... N^{n-1}...] CP]

Specific variants of this analysis differ with respect to the nature of the nominal projection N^n that the relative clause is adjoined to.[1] N^n will be called the "head" of the relative structure.[2]

A second problem is the syntactic relation linking the relative pronoun to the "head".

In the early matching analysis, the relative clause contained a NP identical to the "head", called the relative NP, which was fronted at the beginning of the relative clause and then deleted and replaced by a relative pronoun. In virtue of its identity to the NP "head", its deletion satisfied the recoverability condition (cf. Stockwell—Schachter—Partee 1973: 467–470; Chomsky 1965: 145; Ross 1967: 65).

The deletion under identity approach became subsequently obsolete. Chomsky (1977: 85–86) argued that relative clauses exhibit the diagnostic properties of *wh-* movement, in particular the sensitivity to bridge contexts and to island constraints; therefore, they involve movement of the relative pronoun to Comp, in the same way as interrogative clauses involve movement of an interrogative phrase.

In this analysis, the relative pronoun does not substitute a copy of the "head" in the course of the derivation, but it is interpreted as such by means of an interpretive relation. In itself, the pronoun lacks any intrinsic semantic content (Chomsky 1977: 86; Chomsky—Lasnik 1977: 447).[3]

As for the nature of the interpretive relation, Safir (1986) conceives it as a specific type of binding; Chomsky (1982: fn. 14) instead proposes that the relative clause is interpreted as predicated of the "head", and the relative pronoun is coindexed with the "head" by the rule of predication. Browning (1987: 52–63) further elaborates the mechanism that links the relative pronoun to the "head". The relative operator moves to Spec,CP and turns the CP into a predicate; in the configuration of predication, the NP and the predicate CP agree for phi-features;[4] thus the phi-features of the "head" are transmitted to the relative CP and percolate to its head C°; C° then agrees with the relative pronoun by specifier/head agreement. This 'agreement chain' is schematically represented in (2):

(2)

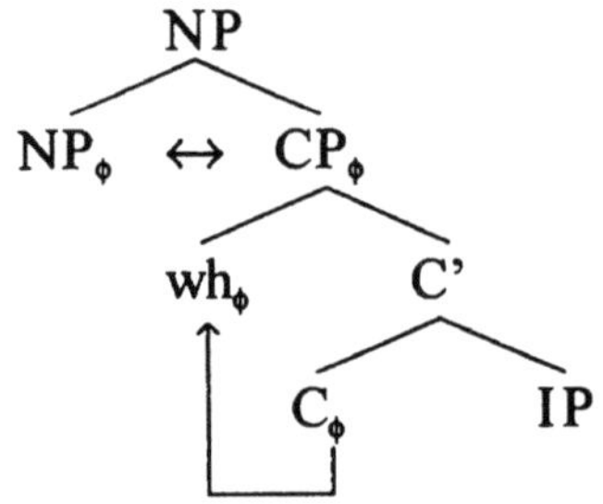

Another aspect of the syntax of relative clauses is their strong islandhood:

(3) * *Ruth liked* [*the sketch* [Op$_i$ *that the critic detested* [$_{DP}$ *the artist* [$_{CP}$ *who drew* t$_i$]]]].

This follows straightforwardly from the adjunct analysis: relative clauses occupy a non-selected position, and they can be assimilated to adjunct clauses under Huang's Condition on Extraction Domains or in Chomsky's (1986b) Barriers theory.[5, 6]

Note that the barrierhood of the relative CP does not interfere with Browning's agreement chain (2), but it would block the grammatical agreement between the relative pronoun and the "head" in a more restrictive theory of agreement, e.g. one in which grammatical agreement depended on a local relation of government (cf. Kayne 1989).

Let us now consider the consequences of the Antisymmetry theory for this analysis of relative clauses. Since an XP cannot adjoin to a head, the relative CP must be adjoined to a maximal projection. Suppose that this is NP. Then the relative CP asymmetrically c-commands NP, and by the Linear Correspondence Axiom, it should linearly precede it:

(4) $[_{NP}$ NP CP$] \rightarrow$ <CP NP>

It is not even possible to assume that the lower segment of NP in (4) moves to the left of the relative clause, since the segment is unable to c-command its trace, and hence it cannot create a wellformed chain (cf. § I.3.2):

(5) * [NP_2 ... $[_{NP1}$ CP t_{NP2}]]

Thus, the LCA excludes an adjunct configuration for postnominal relative clauses. The latter must be generated in a complement position. Since relative clauses are not theta-marked, they cannot be complements to a lexical head; the only plausible candidate is a functional head. The evidence discussed in the following subsection shows that this functional head can be identified with the Determiner.

2. Kayne's (1994) raising analysis

2.1. The DP hypothesis

Any adequate analysis of restrictive relatives must properly distinguish them from appositive ones. Some evidence shows that it is possible to characterize this distinction in configurational terms.

Consider the following examples, discussed by Stockwell—Schachter—Partee (1973: 436) and Vergnaud (1974: 28–30):

(6) a. *Mary knows few boys who enjoy knitting* ↛ *Mary knows few boys*
b. *Mary knows few boys, who enjoy knitting* → *Mary knows few boys*

The restrictive clause in (6a) falls into the restrictive term of the determiner *few*: the set denoted by the relative clause is intersected with the set denoted by the noun head *boys*, and the intersection constitutes the live-on set of the determiner (in the sense of Barwise—Cooper 1981: 178–179). In virtue of this intersection, the entailment in (6a) is not valid. In (6b), on the contrary, the appositive relative is not included in the restrictive term of the determiner: the set it denotes is not intersected with the set denoted by the lexical "head"; rather, the relative clause expresses a predicate which is true of every member of the set denoted by the "head". In this case the entailment is valid.

The question of how to derive this asymmetry has been much debated. Some authors proposed to derive it from a basic structural asymmetry between the two types of relative clause: the restrictive relative is c-commanded by the determiner, but the appositive relative is not. In other terms, appositives are adjoined to a nominal projection that dominates the determiner, whereas restrictive relatives are adjoined to a lower projection.

Among the advocates of this hypothesis are Vergnaud (1974), Jackendoff (1977) and, more recently, Fabb (1990).[7] In a two-level X-bar theory, the determiner occupies Spec,NP; then, restrictive relatives are adjoined to N', and appositive relatives are adjoined to the maximal projection NP:[8]

(7) 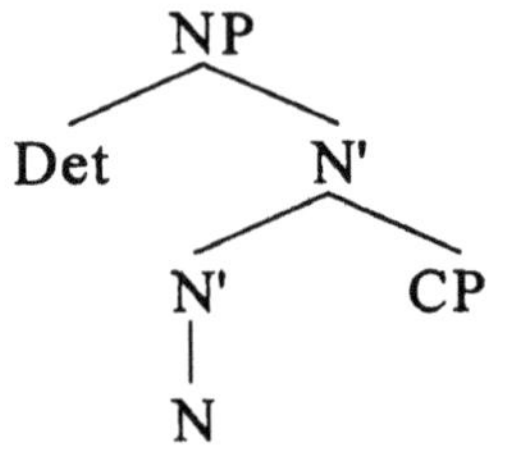

(8) 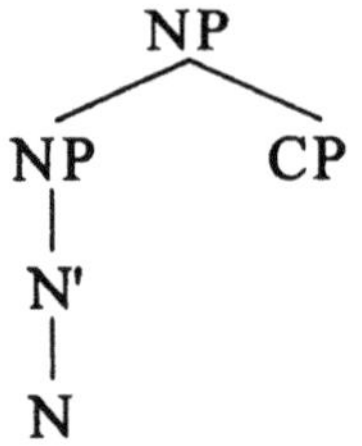

The structural definition of the restrictive term as the c-command domain of the determiner is further supported by the DP analysis, first proposed by Abney (1987), in which the determiner is not generated in Spec,NP, but it is analysed as a functional head taking NP as its complement.[9]

(9)

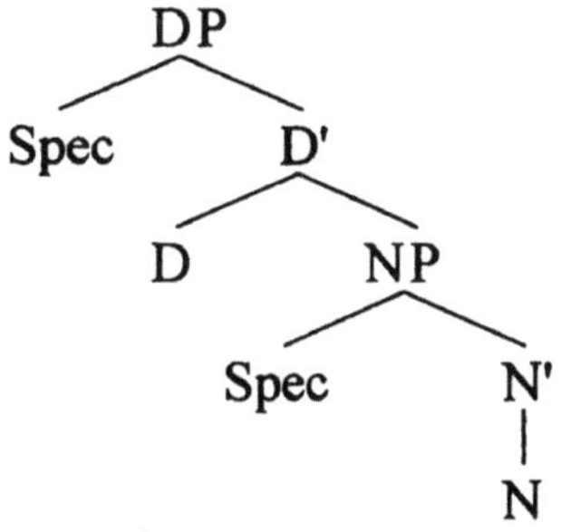

In the DP hypothesis the two maximal projections, NP and DP, have distinct semantic functions. NP is a predicative, unsaturated[10] category denoting a set; D° saturates NP and projects an argumental category, denoting a generalized quantifier. The terms 'predicative' and 'referential' are often informally used to distinguish these two functions.[11] This distinction had been already proposed by Abney (1987: 65–68; 283–284), but it has been more explicitly formulated and empirically motivated by Longobardi (1994), who proposes the following principle:

(10) DP can be an argument, NP cannot.
Reference and quantification are properties of the D position.

By this principle, argumental phrases are always introduced by a D°, although this head may be phonetically null, and the NP level is always interpreted as the restrictive term of D°. By definition, the restrictive term is c-commanded by the determiner.

Independent evidence for this structural definition of the restrictive term comes from the licensing of polarity items. Ladusaw (1980: 13) arrives at the generalization that negative polarity items are only acceptable if they are interpreted in the scope of a downward entailing expression. Note that a quantifier may be downward entailing with respect to its nuclear scope, or to its restrictive term, or both: in the first case, it is monotone decreasing, and in the second case it is anti-persistent (Barwise—Cooper 1981: 184–87; 191–193).

Two representative examples are shown in (11)–(12). The negative determiner *no* is both monotone decreasing and anti-persistent, and it can license the polarity items *anything* and *ever* both in its restrictive term and in its nuclear scope. The universal quantifier *every* instead is anti-persistent, but not monotone decreasing, and it can license the polarity items only within its restrictive term:

(11) a. [*No student who had ever read anything about phrenology*] *attended Gall's lecture.*

(11) b. [*No student who attended Gall's lecture*] *had ever read anything about phrenology.*

(12) a. [*Every student who had ever read anything about phrenology*] *attended Gall's lecture.*

b. * [*Every student who attended Gall's lecture*] *had ever read anything about phrenology.*

Thus, the distribution of polarity items is sensitive to the distinction between the restrictive term and the nuclear scope of a quantifier.

Note that the determiner is a head, but the category denoting a generalized quantifier is the maximal projection DP containing both the determiner and its restrictive term. According to May (1985), the nuclear scope of a quantified DP can be defined as its c-command domain in LF: in (11)–(12), the nuclear scope of the subject DP is the whole sentence. The restrictive term can be structurally characterized in a parallel fashion:

(13) The restrictive term of a determiner corresponds to its LF c-command domain.

This structural definition of the restrictive term is proposed by Reinhart (1987: 137–138) in her account of the phenomenon of donkey anaphora. Reinhart argues that a pronoun can be bound by a weak[12] NP that does not c-command it only if the weak NP itself is unselectively bound by a quantificational determiner. Following Heim (1982), she proposes that unselective binding is only possible if the weak NP falls into the restrictive term of the determiner. This is the case in (14a), where the weak NP *three bottles* is c-commanded by the determiner *every,* but not in (14b), where it is c-commanded by the whole DP headed by *every,* but not by the determiner itself:

(14) a. [*Every man who brought three bottles*] *put them in the refrigerator.*

b. * [*Every man who brought them*] *put three bottles in the refrigerator.*

If unselective binding simply required that the weak NP *three bottles* be in the scope of the universal quantifier, rather than in its restrictive term, (14b) should be grammatical, since in (14b) the weak NP is c-commanded by the quantified DP. The stricter condition on unselective binding is captured by the definition (13).

Note that in the simple structure (9), the restrictive term of the determiner coincides with its complement NP. In the Antisymmetry theory, the unique complement of the head[13] constitutes its c-command domain. Then hypothesis (13) can be strenghtened to (15):

(15) A determiner selects its restrictive term.

2.2. The hypothesis of the external determiner

The discussion in § 1 reached the conclusion that postnominal relative clauses must be generated as complements to some functional head. Furthermore, by definition (13) restrictive relatives must be c-commanded by the determiner of the relative "head".

In the Antisymmetry theory, there is only one structural analysis in which the relative clause is c-commanded by the determiner and linearly follows it: the relative clause is the complement of the determiner (Kayne 1994: 87).[14]

(16) [$_{DP}$ D° CP]

The D° head in (16) will be called the "external determiner" of the relative structure.

Note that the structure in (16) is compatible with the DP hypothesis. In fact, one of the problems which originally led Abney (1987) to postulate a functional head D° was exactly the occurrence of nominal specifiers, like determiners and possessives, within verbal projections, as in the gerundive construction exemplified in (17):

(17) a. *John's building a spaceship*
b. [$_{DP}$ *John's* [$_{DP}$ D° [$_{VP}$ *building a spaceship*]]]

Abney (1987: 21–25) proposes that the genitive Case is assigned by D° to its specifier; in (17b), D° selects a verbal maximal projection, rather than a nominal one.[15]

The ability of D° to select even a full CP can be observed in languages like Spanish and Modern Greek, where factive complements and subject clauses are optionally introduced by a definite determiner:

(18) a. *Me stenohori to oti efije.* (Roussou 1992)
me makes sorry the that he left
'I regret that he left.'

(18) b. *No me gusta el que tu actúes así.* (Donati 1994: 23)
not to-me pleases the that you behave like that
'I don't like your behaving like that.'

Moreover, Zaring (1992) argues that in French, the demonstrative *ce* introducing clausal complements to a preposition is a "clausal determiner" selecting CP:

(19) *Je veillerai* $[_{PP}$ *à* $[_{DP}$ *ce* $[_{CP}$ *qu'il se couche de bonne heure*]]].
I shall-see to that that he goes to bed early
'I shall make him go to bed early.'

This aspect of the analysis is (16) is thus supported by independent evidence.[16]

The structure in (16) is reminescent of an early analysis in generative grammar in which the relative clause was generated as a sister to the determiner of the NP that it modified (cf. Stockwell—Schachter—Partee 1973: 423):

(20)

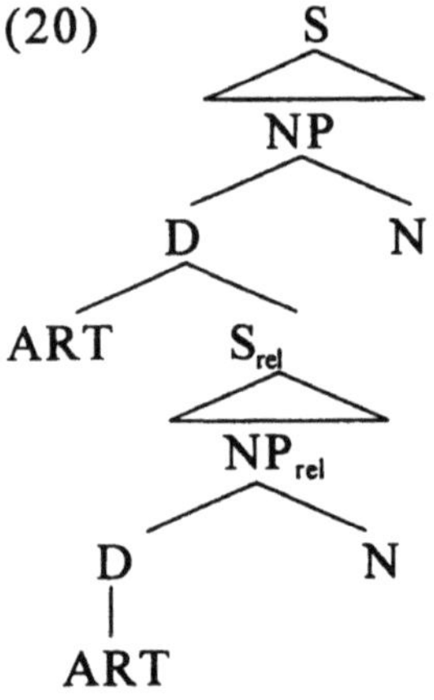

This analysis was motivated by the observation that in certain structures a determiner seems to be licensed by the presence of a restrictive relative:[17]

(21) a. *the Paris * (I love)*
b. *the three books of John's * (that I read)*
c. *the four of the boys * (that came to dinner)*

These data can be derived from Kayne's analysis: see below § 3.1. Note that the analysis in (20) requires the assumption of an obligatory process of extraposition of the relative clause in order to derive the correct linear order of terminals: this process is inconsistent with the LCA (cf. § I.2.3); however, it is not required in Kayne's analysis.

2.3. The raising of the relative "head"

In the structure (16), the determiner of the relative "head" selects the relative CP. The problem is what is the position of the residue of the "head":

(22) $[_{DP}$ *the* $[_{?}$ *book* $[_{CP}$ *that I read* t]]]

By the LCA, the residue cannot be selected by the external determiner, whose unique complement is CP. Therefore, it must originate inside the relative CP. The analysis in (16) thus necessarily incorporates the raising hypothesis originally proposed by M. Brame and developed by Schachter (1973) and Vergnaud (1974): the "residue" is generated in the position of the relative trace and moves to a position adjacent to the external determiner. From now on, the term "head" will refer to the raised category, not including the external determiner.

Note that in this analysis, the sequence *the book* in (22) is not a constituent. The "head" linearly intervenes between the external determiner and the C° head of the relative CP: hence it must be asymmetrically c-commanded by D° and asymmetrically c-command C°. The only position made available by the LCA is a specifier position; the null hypothesis is that it is in Spec,CP.[18]

(23) $[_{DP}$ *the* $[_{CP}$ *book*$_{i}$ $[_{CP}$ *that I read* t_{i}]]]

The next two sections will present evidence in support of this analysis.

3. Evidence for the external determiner

Several data show that the initial determiner of the relative structure is generated outside the relative clause; some of them are discussed in the literature, and will be only briefly summarized here. This evidence is inconsistent with the logically possible alternative that the determiner be generated within the raised "head", a possibility suggested by G. Cinque (personal communication) in relation to some data from the Venetian dialect (§ 3.6).

3.1. Licensing of a definite D°

Let us reexamine the data in (21) in the light of the analysis in (23).

(21) a. *the Paris * (I love)*
b. *the three books of John's * (that I read)*
c. *the four of the boys * (that came to dinner)*

The examples in (21) are not wellformed nominal constituents without the relative clause; this confirms the idea that in the relative structure the determiner and the following nominal expression are generated independent of each other.

In (21a), the "head" is a proper name. Let us assume, following Longobardi (1994), that proper names are generated under N°, but when they have a unique referent they raise to the D° position, which according to Longobardi is the locus of the referential properties of the noun phrase (cf. (10) above). The D° position must be empty for the N° head to incorporate to it. Put differently, a contentful D° cannot select a NP whose head is a proper name denoting a unique individual.[19] In (21a), however, the proper name lacks referential uniqueness, since it has a restrictive modifier; in this case, it does not raise to the D° position. In this sense, it is the restrictive modifier that licenses the contentful determiner. In Kayne's analysis this licensing relation is structurally implemented: in fact, the definite determiner in (21a) does not select the NP headed by the proper name, but the whole restrictive CP.

Turning to (21b), Kayne (1994: 85–87) proposes that in the genitive structure *three books of John's* the genitive preposition *of* realizes the head D°. D° selects an Inflectional head that assigns the genitive Case to the possessor in its Spec. The complement of this inflectional head is the indefinite NumberPhrase *three books*, which raises to Spec,DP. The complete structure is (24):

(24) $[_{DP}[_{NumP}$ *three books*] $[_{DP}$ *of* $[_{IP}$ *John* $[_{IP}$ *'s* $t_{NumP}]]]]$

Since the genitive preposition instantiates the head D°, this structure cannot be introduced by another determiner. In (21b), however, the definite determiner selects the relative CP and the DP *three books of John's* corresponds to the raised "head":

(25) $[_{DP}$ *the* $[_{CP}[_{DP}$ *three books of John's*$]_i$ $[_{CP}$ *that I read* $t_i]]]$[20]

A similar approach accounts for (21c), taking the partitive structure *four of the boys* to be a DP "head" raised from within the relative clause.

3.2. Indefiniteness of the relative trace

A second piece of evidence is given by Browning (1987: 129–131), who notes that the trace within a restrictive relative is interpreted as indefinite even when the modified noun phrase is introduced by a definite determiner. In fact, the relative trace can appear in a context that typically exhibits the indefiniteness effect, like the existential *there* construction:[21]

(26) a. * *There were the men in the garden.*
b. *The men that there were t in the garden were all diplomats.*

Within the adjunct analysis, Browning proposes a solution based on the level of adjunction of the relative clause. She adopts the DP hypothesis and argues that restrictive relatives are adjoined to the NP level, so that the relative operator takes the NP category as its antecedent. This category is unmarked for definiteness; the relative operator does not inherit a definiteness feature, and it is therefore compatible with the existential construction.

Note that although this hypothesis on the level of adjunction of restrictive relatives is highly plausible in the light of (13), it constitutes an independent assumption within the adjunct analysis. The analysis is (23), instead, directly derives the indefiniteness of the relative trace: the external definite determiner is not contained in the "head" raised from the postcopular position, hence it does not violate the definiteness restriction.[22]

3.3. Idiom chunks

A similar point can be made with respect to some idiomatic expressions whose direct object is obligatorily indefinite. When the idiomatic object is relativized, it can be introduced by a definite determiner (cf. Vergnaud 1974: 154; Schachter 1973: 32):

(27) a. * *Il a tiré le parti de la situation.*
he has taken the advantage of the situation
b. *Le parti qu'il a tiré de la situation nous a surpris.*
the advantage that he has taken of the situation us has surprised
'The advantage that he took of the situation surprised us.'
c. * *Ha fatto i progressi.*
he has made the headway

(27) d. *I progressi che ha fatto sono notevoli.*
the headway that (he) has made is remarkable
'The headway he made is remarkable.'
e. * *They made the fun of me.*
f. *the fun that they made of me* (Fabb 1990: 71)
g. * *John had the question for the teacher.*
h. *the question that John had for the teacher*
i. * *We made the headway on that problem.*
l. *the headway that we made on that problem*
(Browning 1987: 130)

Once again, this follows from the hypothesis that the external definite determiner is not reconstructed in the complement position of the idiomatic verb within the relative clause.

The possibility of restrictive relativization interacts with the typology of idiomatic expressions in an interesting way. In Bianchi (1993), the examination of a sample of Italian idiomatic expressions led to the distinction of two types of idiom chunks, those that have a variable determiner and those with a "frozen" determiner. A determiner is said to be frozen if it is completely fixed: for instance, in the idiomatic expression *tirare le cuoia* 'to kick the bucket' the definite article of the idiomatic object cannot be replaced by any other determiner. Idiom chunks with a frozen determiner have a completely idiosyncratic, noncompositional interpretation and cannot be modified; this suggests that the frozen determiner lacks any semantic content, and its NP complement does not have a variable position to be bound by the determiner (Bianchi 1993: 375–377).

Restrictive relativization belongs to the class of A' dependencies that only allow idiom chunks with a nonfrozen determiner: thus the idiom chunks in (28), whose determiner is variable, contrast with those in (29), whose determiner is completely frozen:

(28) a. *Il tempo che posso dedicarti oggi è ben poco.*
the time that (I) can devote to-you today is but little
'The time that I can devote to you today is little.'
b. *Gli onori che gli hanno reso erano eccessivi.*
the homages that to-him (they) paid were excessive
'The homage that they paid to him was excessive.'
c. *I debiti che ha fatto ultimamente sono ingenti*
the debts that (he) has made lately are considerable.
'The debts that he has made lately are considerable.'
d. *I progressi che sta facendo mi sorprendono.*
'The headway that he's making strikes me.'

(28) e. ? *L'attenzione che gli hanno prestato era ben poca.*
the attention that (they) to-him paid was but little
'The attention that they paid to him was little.'

f. ?? *Le grane che sta piantando mi irritano.*
the troubles that (he) is planting irritate me
'The trouble that he's causing irritates me.'

g. *Il boccone amaro cho ho dovuto ingoiare mi ha rovinato la giornata.*
The mouthful bitter that (I) have had to swallow to-me has spoiled the day
'The bitter pill that I had to swallow put me in a bad mood.'

(29) a. * *la corda che ha tagliato*
the rope that (he) has cut
the French leave that he took

b. * *il naso che ficca nei miei affari*
the nose that (he) pokes into my business
'the nose that (he) is poking into my business'

c. * *le cuoia che ha tirato ieri*
the leather stripes that (he) has pulled yesterday
'the bucket that he kicked yesterday'

d. * *il lunario che ha sbarcato*
the almanac that (he) has disembarked
'the two ends that he made meet'

e. * *la mente locale che ha fatto su quel problema*
the local mind that (he) has made on that problem
'the thoughts that he collected on that problem'

f. * *l'animo che si è messo in pace*
the mind that (he) to-himself has set in peace
'the mind that he set at rest'

This constraint can be straightforwardly recast in Kayne's analysis by disallowing a frozen determiner to select a restrictive relative CP.[23]

3.4. Scope assignment under reconstruction

The hypothesis of the external determiner is also confirmed by phenomena of scope assignment under reconstruction. Consider the following pair of sentences:

(30) a. *Ogni medico visiterà due pazienti.*
'Each doctor will examine two patients.'

(30) b. *Ogni medico visiterà i due pazienti.*
'Each doctor will examine the two patients.'

The indefinite object in (30a) is in the scope of the universally quantified subject and allows a distributive reading (namely, there may be two different patients for every doctor). In (30b), on the contrary, the definite object denotes a set with exactly two members, and the distributive reading is impossible.

Let us assume with Giusti (1991) that the numeral in (30a) is a D°, whereas the numeral in (30b) occupies a lower position.[24] Consider now the restrictive relatives corresponding to (30):

(31) a. *Ho telefonato ai due pazienti che ogni medico visiterà domani.*
'I phoned the two patients that every doctor will examine tomorrow.'
b. *Ho telefonato ai due pazienti che ogni medico visiterà domani.*
'I phoned two patients that every doctor will examine tomorrow.'

The judgements are reversed with respect to the examples in (30): this time the distributive reading is available in (31a), where the "head" is introduced by the definite determiner, but not in (31b).

This follows from the hypothesis that the determiner introducing the relative "head" is generated outside the relative clause, and hence it is not reconstructed in the trace position. The contrast between (30b) and (31a) is parallel to the contrasts in (26) and (27): once again, in (31a) the definite determiner is external to the relative clause, and the raised "head" is the indefinite phrase *due pazienti*, which can be interpreted in the scope of the universally quantified subject.

The impossibility of the distributive reading in (31b) suggests that in this example the numeral *due* necessarily realizes the external determiner, rather than Num°, and therefore it cannot be reconstructed within the scope of the universal quantifier. This may follow from the hypothesis that a numeral in Num° cannot be licensed by an empty D°.[25]

3.5. Floating quantifiers

Consider also the behaviour of the universal quantifier *tutti/e* in Italian. This quantifier introduces a definite DP, as in (32a), and allows floating quantification, as in (32b):

(32) a. *Tutti gli studenti hanno superato l'esame.*
'All the students passed the exam.'
b. *Gli studenti hanno superato tutti l'esame.*
the students have passed all the exam
'The students all passed the exam.'

According to Giusti (1991), Shlonsky (1991) and Bianchi (1992), the universal quantifier is a functional head Q° selecting a [+definite] DP complement. Adapting Sportiche's (1988) analysis, floating quantification in (32b) is derived by leftward movement of the DP complement to Spec,IP, stranding Q° in the base subject position:

(33) $[_{IP}[_{DP}$ gli studenti] $[_{I'}$ hanno superato $[_{VP}[_{QP}$ tutti $t_{DP}]$ $[_{VP}$ t_v l'esame]]]]

This Q° can also introduce the "head" of a restrictive relative, but in cannot be stranded within the relative clause:

(34) a. *Elencami tutti i libri che devi leggere per l'esame.*
tell me all the books that (you) must read for the exam
'Tell me all the books that you must read for the exam.'
b. * *Elencami i libri che devi leggere tutti per l'esame.*
tell me the books that (you) must read all for the exam
c. *Entrambe le persone che ammiravo mi hanno deluso.*
both the people that (I) admired me have disappointed
'Both the people that I admired disappointed me.'
d. * *Le persone che ammiravo entrambe mi hanno deluso.*
the people that (I) admired both me have disappointed
e. *Ho interrogato tutti gli studenti che avevano superato la prova scritta.*
(I) have examined all the students that had passed the exam written
'I examined all the students who had passed the written exam.'
f. * *Ho interrogato gli studenti che avevano passato tutti la prova scritta.*
(I) have examined the students that had passed all the exam written
g. *Tutti coloro che erano stati accusati subirono un interrogatorio.*
all those that had been accused underwent an interrogatory
'All those who had been accused were examined.'

(34) h. * *Coloro che erano stati tutti accusati subirono un interrogatorio.*
those that had been all accused underwent an interrogatory

This type of floating quantification is also excluded in French restrictive relatives according to Perlmutter (1972: fn. 2).[26] The judgement is very subtle in sentences like (34f), where an appositive interpretation is possible and makes floating quantification acceptable; it is clearcut in (34h), which only allows a restrictive interpretation.[27]

The impossibility of floating quantification in (34) follows from the fact that the distributive Q° obligatorily selects a [+definite] DP. By hypothesis, the definite article in (34b, d, f, h) instantiates the external determiner, and it is not part of the relative "head"; therefore, Q° cannot take the "head" as its complement within the relative clause, and floating quantification is excluded. Q° can only select the DP headed by the external determiner, thus appearing in the highest position of the relative structure.

Note also that Q° cannot select a relative CP in the same way that D° does. (35) shows that both in Italian and in French the universal Q° *tutto/tout* cannot directly select a relative CP, but it must be followed by a demonstrative pronoun (*quello/ce*), which is arguably of category D°.[28]

(35) a. * *Tutto che ho fatto era sbagliato.*
all that I made was wrong

b. *Tutto quello che ho fatto era sbagliato.*
all that that I made was wrong
'All I made was wrong'

c. * *tout qui était sur la table*
all that was on the table

d. *tout ce qui était sur la table* (Kayne 1975: 17)
all that that was on the table
'everything that was on the table'

3.6. An alternative analysis

In the preceding subsections, it was argued that the overt determiner introducing a relative structure realizes the external D°. However, another logical possibility is to analyse it as internal to the relative "head" and to postulate a phonetically null external D°. The DP "head" raises to the Spec of the external D° and transmits to it its definiteness value by

Spec/head agreement (G. Cinque, personal communication). Thus, (22) above would be analysed as in (36) rather than as in (23):

(36) $[_{DP}[_{DP}$ *the book*$]_i$ $[_{DP}$ D° $[_{CP}$ t_i *that* $[_{IP}$ *I read* $t_i]]]]$

This analysis can account for some data from the Venetian dialect discussed by Viotto (1994: 126–27). In Venetian, relativization of a prepositional object allows the deletion of the preposition within the relative clause if it is identical to the preposition introducing the external DP:[29]

(37) *nea situassion che semo*
in-the situation that (we) are
'in the situation in which we are'

But this deletion is impossible if there is no external preposition:

(38) * *la situassion che semo*
the situation that (we) are

Cinque suggests the following analysis: the whole PP in (37) is generated within the relative CP, and the latter is introduced by two abstract external heads, P° and D°. The PP raises to the Spec of the null P° and licenses it with an identical interpretation:[30]

(39) a. $[_{PP}[_{PP}$ *nea situassion*$]_i$ $[_{PP}$ P° $[_{DP}$ t_i [D° $[_{CP}$ t_i [*che semo* $t_i]]]]]]$

This analysis may be correct for the Venetian examples, but it is inconsistent with the Italian and English data discussed in the preceding subsections. The lack of a definiteness effect in (26b) and the relativization of idiom chunks in (27)–(28) show that the derivation in (23) is at least possible; the lack of a distributive reading in (31b) and the impossibility of floating quantification in paradigm (34) suggest that it is the only available derivation in standard Italian.

4. Evidence for raising

This section summarizes the evidence for the raising of the relative "head" presented by Schachter (1973) and Vergnaud (1974). Their arguments still hold in the current syntactic framework, although some of the underlying assumptions have changed.

4.1. Idiomatic expressions

The most famous argument in support of the raising analysis is that of idiom chunks, discussed by Vergnaud (1974). Some idiomatic expression like *prendre part* 'to take part (in something)' allow the relativization of the idiomatic object if the idiomatic verb is stranded within the relative clause, but not if it is external to the relative clause:

(40) a. *Il décrit dans son livre* [*la part qu'il a prise* t *aux travaux du 9ème congrès*].
he describes in his book the part that he has taken at the workings of the 9th conference
'He describes in his book the part that he had in the 9th conference.'

b. * *Il a pris aux travaux du 9ème congrès* [*la part qu'il décrit* t *dans son livre*].
he has taken in the workings of the 9th conference the part that he describes in his book

c. *Peu de gens ont parlé du* [*parti qu'il a tiré* t *des difficultés économiques*].
few people have spoken of-the advantage that he has taken of the difficulties economical
'Few people have spoken about the advantage that he took of the economical crisis.'

d. * *Il a tiré (des difficultés économiques)* [*un parti dont peu de gens ont parlé* t].
he has taken of-the difficulties economical an advantage of-which few people have spoken (Vergnaud 1974: 58–59; 200)

Both Schachter and Vergnaud assume that idiomatic expressions are subject to an adjacency requirement in D-structure: the idiomatic verb and object must be generated next to each other. In a matching analysis, the contrast in (40) is not expected: as the "head" and the relative NP are identical, the adjacency requirement should be satisfied both in (40a), where the idiomatic verb selects the relative NP, and in (40b), where it selects the "head". More recent variants of the adjunct analysis, in which the relative clause contains a relative operator, instead predict the opposite contrast: in (40b) the idiomatic object is generated next to the verb in the "head" position, fulfilling the adjacency requirement; in (40a), instead, the idiomatic verb selects a null relative operator in D-structure. Only the raising analysis predicts the observed pattern: if the "head" is generated in the trace position and then raised to its surface position, the adjacency requirement is satisfied in (40a), but not in (40b).

In his minimalist paper, Chomsky (1993) suggests that selectional restrictions – of which idiomatic expressions represent an extreme case – should not apply at the D-structure level. Actually, this level can be completely dispensed with, and the restrictions apply to the LF representation, which constitutes the interface with the cognitive-intentional system. Then, the idiomatic rules impose a sort of adjacency requirement in LF: if an idiom chunk has moved from the complement position of the idiomatic verb, it must be fully reconstructed in order to be interpretable. Lack of reconstruction blocks the application of the idiomatic rules, and consequently, the uninterpretable symbols violate the Full Interpretation principle.

Chomsky (1993: 34–41) proposes a restrictive theory of reconstruction, based on the assumption that traces are copies of the moved category that fail to be spelled out in the phonetic representation, but are present in the LF derivation.[31] Reconstruction is the consequence of an LF process of partial deletion which splits an A' chain into an operator and a variable. An A' moved category, e.g. *which pictures of John,* contains two parts: the *wh*-word is the operator, and the rest is its restrictive term (cf. (13)). Thus, two possibilities arise: if the whole A' moved category is retained in the highest position and the lower traces are completely deleted, the resulting configuration is one of restricted quantification, as in (41b). However, it is also possible to selectively delete the restrictive term in the highest position of the chain, and the operator in the trace positions: the resulting LF structure (41c) is interpreted as an unrestricted quantification:

(41) a. [*Which pictures of John*] *do you like* [$_t$ *which pictures of John*]?
b. [*Which pictures of John*] *do you like* [$_t$ e]?
c. [*Which* e] *do you like* [$_t$ e *pictures of John*]?

Chomsky shows that the obligatory reconstruction of idiom chunks has detectable effects on binding. The expression *take pictures* allows both an idiomatic interpretation and a nonidiomatic one. In (42), the direct object contains an anaphor and undergoes *wh*-movement: in this configuration, two binding possibilities arise. If the *wh*-phrase is partially reconstructed in the intermediate Spec,CP, the anaphor is bound by the matrix subject (42b); if it is fully reconstructed to the base position, the anaphor is bound by the embedded subject (42c):

(42) a. [*Which pictures of himself*]$_i$ *does John think* [*that Bill took* t$_i$]?
b. [*Which* e] *does John*$_j$ *think* [$_{CP}$ [e *pictures of himself*$_j$] [$_{CP}$ *that Bill took* t]]?
c. [*Which* e] *does John think* [$_{CP}$ *that Bill*$_k$ *took* [e *pictures of himself*$_k$]]?

The idiomatic interpretation of the phrase *take pictures* is only compatible with the second binding possibility (42c): this confirms that it requires reconstruction in LF.[32]

Note that this interaction of the idiomatic interpretation with binding is not predicted by the D-structure adjacency requirement. In every other respect, Chomsky's theory of reconstruction subsumes the D-structure adjacency requirement in a derivational framework lacking this level of representation. In fact, reconstruction presupposes the presence of a trace: it is the presence of a trace of the idiom chunk following the idiomatic verb that allows reconstruction in (40a), but not in (40b). This implies that only in (40a), but not in (40b), the idiomatic object was inserted in the tree in the complement position adjacent to the idiomatic verb. Conversely, the ungrammaticality of (40b) implies that in LF the relative "head" obligatorily undergoes reconstruction in the base position, so that in LF it is not adjacent to the external idiomatic verb (cf. chapter IV).

In sum, the idiom argument can be recast in the current syntactic framework lacking the D-structure level; Chomsky's restrictive theory of reconstruction singles out the raising analysis of relative clauses as the only one that explains the contrasts in (40).

4.2. Predicate nominals

Another argument put forth by Vergnaud (1974: 63–68) concerns the relativization of predicative DPs. This is impossible if the features of the matrix and the embedded subject do not match:

(43) * *Ce ne sont pas les comédiens que leur père était.*
it not (they) are not the comedians that their father was

This immediately follows from the raising analysis (44a), since the predicative "head" cannot carry two opposite feature specifications:

(44) a. ... [$_{DP}$ *les* [$_{CP}$ *comédiens* [$_{CP}$ *que* [$_{IP}$ *leur père était* t]]]]

But the adjunct analysis too provides an explanation, under the assumption that the null relative operator in Spec,CP receives its feature specification from the "head". In (44b) below, the null operator in Spec,CP receives a plural number specification from the "head", which is incompatible with the singular number of the subject of the relative clause, with which it is coindexed by predication:

(44) b. $[_{DP}$ *les* $[[_{NP}$ *comédiens*$_{PL}]$ $[_{CP}$ Op_{PL} $[_{CP}$ *que* $[_{IP}$ *leur père*$_{SG}$ *était* t]]]]]

A more interesting variant of this argument is provided by the so called "degree relatives", to be discussed in more detail in § 6:

(45) *No puedes imaginar lo inteligentes que* pro *dicen que* pro *son.*
you cannot imagine the intelligent$_{PL}$ that (they) say that (they) are
'You cannot imagine how intelligent they say they are.'
(Rivero 1980: 437)

Here the "head" of the relative clause is an adjectival phrase obligatorily introduced by an invariable definite determiner. The point is that the adjectival "head" does not agree with any DP belonging to the matrix clause, but it agrees with the subject of the relative clause. In the adjunct analysis, the adjectival "head" must be generated in the matrix clause with a full phi-features specification; but then it is unclear how its features can be checked against those of the referential category of which it is predicated:[33] feature checking would have to be mediated by the relative null operator. This extension of the checking mechanism is not required in the raising analysis: the adjectival phrase is generated in the post-copular position, where it is coindexed with the subject of the relative clause, and then it raises to Spec,CP.

4.3. Reconstruction for binding

As discussed in § 4.1, Chomsky's (1993) analysis implies that reconstruction is a diagnostic of movement, because it depends upon the presence of a movement chain, consisting of copies of the moved category. Kayne (1994: 87) notes that this approach to reconstruction supports the raising analysis of restrictive relatives, since the latter exhibit reconstruction of the "head" (cf. Schachter 1973: 32–35; Vergnaud 1974: fn. 10 to chapter 3). This topic will be discussed in detail in chapter IV, where the behaviour of restrictive and appositive relatives with respect to reconstruction will be systematically compared.

4.4. The problem of the trigger

The evidence reviewed in the preceding subsections supports the hypothesis that the relative "head" directly raises from the trace position to Spec,CP. But an important question was left open, namely, what triggers this raising in the first place.

Note that in the raising analysis the external determiner and the "head" are not in a complementation relation, in the way they are in the adjunct analysis. From this perspective, the structural relation between the determiner and the noun phrase in the two strings *the boy* and *the boy I met* is completely different. But there are significant superficial similarities: in particular, in a language like Italian, the external determiner overtly agrees with the NP in both cases.

Since the two categories must be syntactically related, the simplest answer to the problem of the trigger is that the "head" must raise in order to establish a strictly local syntactic relation with the external determiner. Kayne (1994: 90) suggests that it must reach a position governed by the external D°.

This hypothesis is clearly inconsistent with Chomsky's (1995) minimalist assumptions, which exclude head government as a possible checking configuration. Let us assume it nevertheless, deferring the discussion of this problem to § III.2; at that point, many other relevant aspects of the raising structure will have been clarified.

5. A locality problem

In the preceding section we assumed Kayne's proposal that the raised "head" must be governed by the external determiner. Note also that the "head" appears to be transparent for the extraction of an internal constituent.[34] In fact, there is a clear contrast between the examples in (46), where a PP is extracted from the relative "head", and those in (47), where the PP is extracted from within the relative CP:

(46) a. *E' [di questo mosaico]$_i$ che abbiamo ritrovato* [$_{DP}$ *il* [*tassello* t$_i$]$_j$ [*che* t$_j$ *mancava*]].
(it) is of this mosaic that (we) have found the piece that was-missing
'It is of this mosaic that we found the piece that was missing.'

(46) b. ? *Questo è l'autore* [*di cui*]$_i$ *non conosco* [$_{DP}$ *i* [*libri* t_i]$_j$ [*che hai comprato* t_j]].
this is the author of whom (I) not know the books that (you) have bought
'This is the author the books that you bought by whom I don't know'

c. ? *Questo è il libro* [*di cui*]$_i$ *non ho ancora letto* [$_{DP}$ *la* [*recensione* t_i]$_j$ [*che hai scritto* t_j]].
this is the book of which (I) not have yet read the review that (you) have written
'This is the book of which I haven't read the review that you wrote.'

(47) a. * *E'* [*da questo mosaico*]$_i$ *che abbiamo ritrovato* [$_{DP}$ *il* [*tassello*]$_j$ [*che* t_j *mancava* t_i]].
(it) is from this mosaic that (we) have found the piece that was-missing

b. * [*A quale ragazzo*]$_i$ *non conosci* [$_{DP}$ *i* [*libri*]$_j$ [*che ho consigliato* t_j t_i]]?
to which boy not (you) know the books that (I) have recommended?

c. * *Questa è la rivista* [*per cui*]$_i$ *non ho ancora letto* [$_{DP}$ *la* [*recensione*]$_j$ [*che hai scritto* t_j t_i]].
this is the review for which (I) not have read yet the paper that (you) have written

These data follow straightforwardly from the adjunct analysis:

(48) [$_{DP}$ D° [$_{NP}$[$_{NP}$ N° PP] CP]]]

Here D° takes NP as its complement, and *a fortiori* governs it; agreement follows from whatever mechanisms allows it in simple DP structures. The strong island effect in (47) derives from the barrierhood of the relative CP, which is in a non-selected position; however, the PP complements extracted in (46) are not dominated by the CP barrier, since they are generated within the NP projection.

In the raising analysis, instead, a locality problem arises. The ungrammaticality of (47) requires the assumption that the relative CP is a barrier. But then it is necessary to exempt the Spec of the relative CP from the barrier: the relative "head" in Spec,CP must be governed by the external D°, and it must be transparent for the extraction of an internal PP complement in (46).[35]

In other terms, the pseudo-constituency of the external D° and the NP "head" can only be explained if Spec,CP is external to the CP barrier. This requires an elaboration of the locality theory for strong islands.

The following discussion explores two alternative approaches to the locality problem. Section 5.1 is devoted to the Barriers approach proposed by Chomsky (1986b) and further elaborated by Cinque (1990); section 5.2 is instead devoted to Manzini's (1994) locality theory, which dispenses with the notion of barrier and is more similar to Kayne's (1984) Connectedness Condition.

5.1. Barriers

The major problem for locality theory raised by Kayne's (1994) X-bar system is the fact that specifiers are assimilated to adjunct positions; in Chomsky's (1986b) Barriers theory, instead, the distinction between specifiers and adjuncts plays a crucial role: the specifier of an XP is dominated by the whole category XP, whereas a category adjoined to XP is only dominated by one segment of it. Given the following definitions:

(49) 1. A is a blocking category for B iff A is not L-marked[36] and A dominates B.
2. A maximal projection A is a barrier for B iff (a) or (b):
2a. A immediately dominates C, C a blocking category for B;
2b. A is a blocking category, A ≠ IP. (Chomsky 1986b: 14)
3. A excludes B iff no segment of A dominates B. (Chomsky 1986b: 9)
4. B is n-subjacent to A iff there are fewer than n+1 barriers for B that exclude A.
5. Subjacency: in a chain link (A_i, A_{i+1}), A_{i+1} must be 1-subjacent to A_i. (Chomsky 1986b: 30)

only whole categories, but not segments, qualify as barriers for Subjacency; therefore, if XP is a barrier, its specifier is included in it, but an adjunct is not.

Since in Kayne's X-bar theory the Spec and adjunct positions are not distinguished, it must be determined whether a segment may qualify as a barrier for the specifier that it dominates. If it does, then the Spec has the same status as in the Barriers system, but the possibility of evading from a barrier through intermediate adjunction is excluded. If it does not, then the Spec position is an "escape hatch" from any barrier, and this raises problems for the analysis of strong islands. Let us consider the consequences of these two possibilities for the relative structure (23), repeated here:

(23) [$_{DP}$ *the* [$_{CP}$ *book*$_i$ [$_{CP}$ *that I read* t$_i$]]]

A first possibility is to assume that a segment does qualify as a barrier, but the external determiner L-marks its complement CP (under some extended definition of L-marking). Then in (23) the relative CP is not a barrier. In this way, the external determiner governs the relative "head", since they are not separated by any barrier. The phrase moved to Spec, CP isn't a barrier either if we assume, following Chomsky (1986b: 24) that when a head L-marks a category it also L-marks the Spec of the latter. Thus, the extraction of the PP complement in (46) also follows straightforwardly: the relative CP and the NP in its Spec will be L-marked by the external D°, allowing the extraction of the N-complement PP:

(50) [$_{DP}$ D° [$_{CP}$ [N° PP] [$_{CP}$ C° [$_{IP}$ ...t...]]]]

In (47), on the contrary, the base position of the extracted PP is inside the relative clause, and it is dominated by a blocking category, IP, from which CP inherits barrierhood. The strong island effect obtains if one assumes, following Cinque (1990: chapter 1), that chain links require 0-subjacency.

Alternatively, it is possible to exploit the distinction between segments and categories: a segment does not qualify as a barrier, but only a whole category does. Then in (23), even if the relative CP is not L-marked by the external D°, the relative "head" will be governed by the latter, since it is only dominated by one CP segment.

The contrast between (46) and (47) also follows. In fact, the extraction of PP in (46) takes place in two steps: in the first step, represented in (50), the relative "head" including the PP adjoins to CP, thus crossing only one segment of it; as a second step, the PP is extracted from it, and it only crosses the upper segment of CP. Even if the relative "head" is a barrier, extraction respects 1-subjacency.[37]

In (47), on the contrary, the extracted PP has to cross the whole CP category: in fact, the LCA excludes multiple adjunction, and the relative phrase has already adjoined to CP. Thus the CP barrier is necessarily crossed, in violation of 0-subjacency; if the DP dominating CP also inherits barrierhood, there is also a violation of 1-subjacency.

Note that the assumption that a segment does not qualify as a barrier opens the possibility of evading any barrier by moving through its Spec. However, R. Kayne (personal communication) points out that this problem also arises in the Barriers system, where extraction from an adjunct island can take place through intermediate adjunction to the adjunct itself. In both systems it is necessary to exclude the possibility of voiding barrierhood through successive cyclic adjunction (a point also made by Cinque 1990: 24–30).[38] This idea can be more naturally expressed in the locality theory proposed by Manzini (1994).

5.2. Head-to-head dependencies

Manzini (1994) formulates a locality condition based on a modification of the notion of minimal domain of a head, first proposed by Chomsky (1993):

(51) The minimal domain of a head X, notated (X), includes all elements that are immediately dominated by, and do not immediately dominate, a projection of X.

The relation of domination holds of categories and not of segments: it follows that the minimal domain (H) of a head H includes its complement and, recursively, all the phrases adjoined to some member of the minimal domain: the Spec of the complement, the Spec of the Spec of the complement, and so forth. On the contrary, the Spec of HP is not included in the minimal domain of H, but in the minimal domain of the immediately higher head. Thus the Spec-head relation involves two positions belonging to two distinct but adjacent minimal domains.

Manzini's locality condition is not formulated in terms of chain links of the usual type, but rather in terms of dependencies between heads related to each other by an elementary relation: complementation or checking. The trace is included in the minimal domain of the lowest head of the dependency, and the antecedent is linked to the highest head.

The locality condition on these dependencies is formulated as follows:

(52) 1. a minimal domain (Y) is superior to a minimal domain (X) iff all categories that dominate (Y) dominate (X);
2. (Y) is immediately superior to (X) iff (Y) is superior to (X) and there is no (Z) such that (Z) is superior to (X) and (Y) is superior to (Z).

(52) 3. Let (X_i) be the minimal domain to which A_i belongs. $(A_1,...,A_n)$ is a dependency only if for every i, (X_i) is immediately superior to (X_{i+1}).[39]

Consider for instance the subject island in (53):

(53)

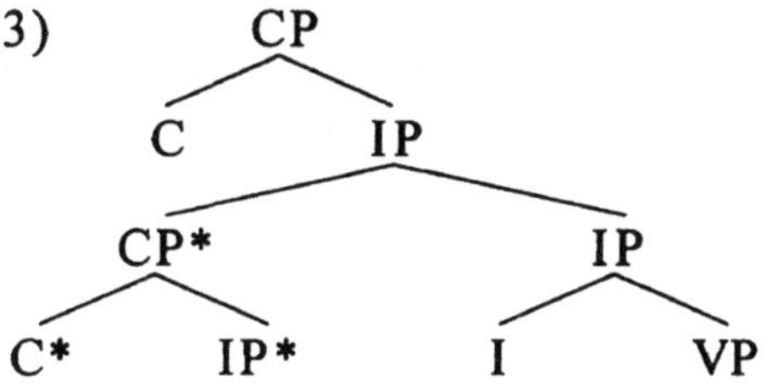

A subject CP* adjoined to IP falls into the minimal domain of the head C, but the head C* of the subject clause does not bear any relation to C (either complementation or checking); therefore, a dependency including C* and C is not licensed, and the extraction of any constituent from the subject CP* is ruled out.

Let us now consider the consequences of this locality condition for the relative structure (23), reproduced here in a tree shape:[40]

(54)

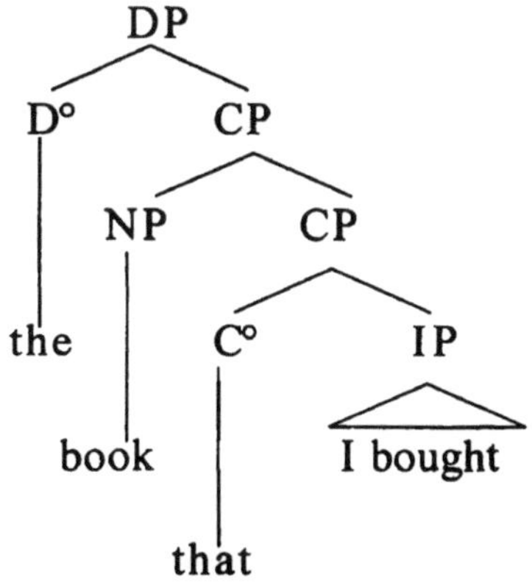

Manzini's definitions in (52) imply that the "head" in the Spec of the relative CP falls in the minimal domain of the external D°. This is a strictly local syntactic relation, corresponding to the traditional notion of head government. In this respect, the relation between the external D° and the "head" in (54) can be essentially assimilated to the relation between D° and NP in the simple DP structure (55):

(55)

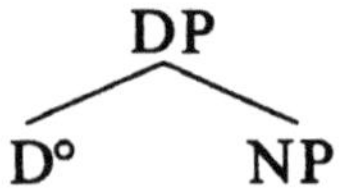

Thus, whatever the nature of the dependency between D° and N° in (55), it will also hold in (54). This point will be discussed in detail in § III.2.

Let us then us consider the extraction of the noun complement PP in (46), whose schematic representation is (50), repeated in a tree shape:

(56)

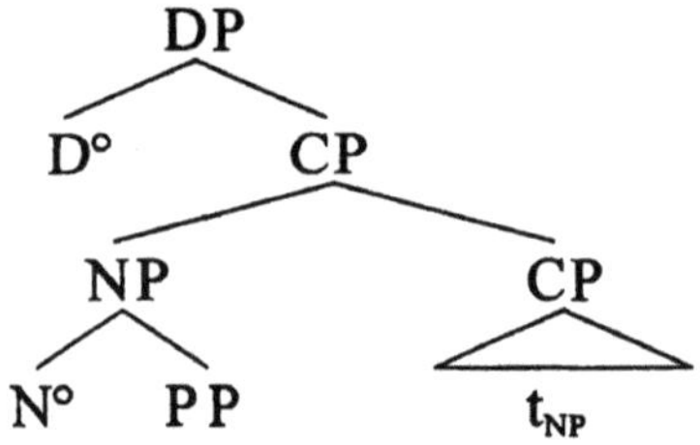

The constituent raised to Spec,CP falls in the minimal domain of the external D°; the PP to be extracted belongs to the minimal domain of N°. (D°) is immediately superior to (N°), and the two heads are linked by a wellformed dependency; thus, the extraction of PP satisfies the locality condition.

Finally, consider the strong island effect in (47), where a PP is extracted from a position embedded within the VP of the relative clause. The abstract structure is (57):

(57)

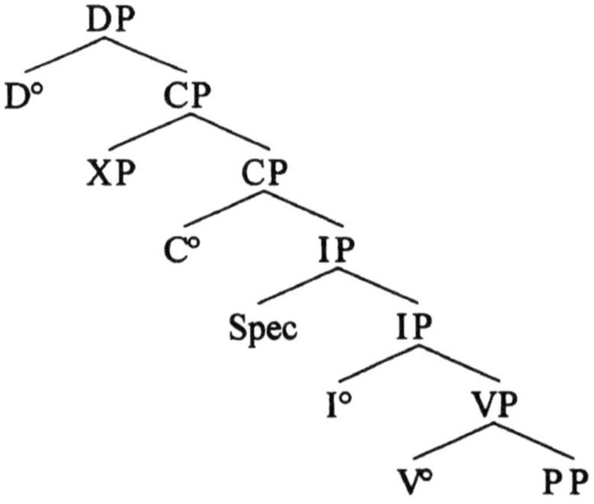

The complementation relation creates a wellformed dependency between V°, I° and C°: this corresponds to the creation of an extended projection in the sense of Grimshaw (1991).[41] The problematic link must be the one between C° and D°. Note that although there is a relation of complementation, the categorial features of the external D° do not match those of C°, so that the external DP does not qualify as an extended projection of CP. Moreover, the external D° and C° are not related by any thematic or checking relation (cf. again § III.2). Therefore, let us assume that the syntactic relation between D° and C° is not sufficient to extend farther the dependency terminating in V°: this will derive the strong islandhood of the relative CP.

5.3. Concluding remarks

The problem of formulating a locality condition for strong islands consistent with Kayne's (1994) X-bar theory exceeds the limits of this discussion. The main point of this section is that the identification of specifier and adjunct positions is problematic for a locality theory based on the notion of barrier; with respect to the analysis of the restrictive relative structure, Manzini's approach, which eliminates intermediate traces in favour of head-to-head dependencies, yields the desired results in a more natural way. R. Manzini (personal communication) observes a similarity between her conception of dependencies and Kayne's (1984: 167–76) Connectedness Condition. Note that contrary to the Connectedness Condition, Manzini's proposal does not make reference to the notion of canonical government, which is defined on the basis of the head-complement parameter, now inconsistent with the LCA (see § I.1.1).

6. Cross-linguistic evidence in favour of the raising analysis

This section discusses some relative structures which provide additional evidence in favour of the raising analysis.

6.1. Internally headed relative clauses

A relative structure which is particularly interesting in the light of the raising analysis is the so called "internally headed" relative clause, attested in various unrelated languages such as Quechua, Navajo, Lakhota, Bambara, Japanese (Cole—Harbert—Hermon 1978; Cole 1987; Cole—Hermon 1994; Culy 1990; Downing 1978: 398; Williamson 1987). The peculiarity of this structure is that the relative "head" appears in the argument position within the relative clause, i.e. it is not raised to Spec,CP, and it bears the Case assigned within the relative clause. The relative CP is usually introduced by a determiner, or else its predicate bears a nominalizing affix. As Cole—Harbert–Hermon (1978: 26) point out, this structure realizes at the surface what the raising analysis takes to be the underlying structure of headed relative clauses. Two examples are reported in (58):

(58) a. *Nuna bestya-ta rantishqan alli bestya-m.*
man horse-ACC bought good horse-EVIDENTIAL
'The horse that the man bought (is) a good horse.'
(Ancash Quechua; Cole—Hermon 1994: 248)
b. [[*Mari* [*owiza wa*] *kage*] *ki*] *he ophewathu.*
Mari quilt a make the Dem I buy
'I bought the quilt that Mari made.'
(Lakhota; Williamson 1987: 171)

The analysis of Lakhota internally headed relative clauses proposed by Williamson (1987) constitutes the null hypothesis with respect to Kayne' raising analysis. She argues that the relative clause is generated as a sister to the determiner (though for independent reasons, both the relative CP and NP surface to the left of the determiner), and the "head" remains in its base position within the relative clause in overt syntax. However, according to Williamson the "head" moves to the Comp of the relative clause in LF: in fact, it has obligatorily wide scope over the operators contained in the relative clause, e.g. sentential negation.[42] From a semantic viewpoint, Williamson (1987: 185) proposes that the relative clause denotes the intersection of the sets denoted by the "head" and by the relative IP, which in LF contains a variable in the base position of the raised "head"; this intersection constitutes the live-on set of the external determiner. The intersection interpretation is only possible if the relative DP is not intrinsically quantified: accordingly, Williamson (1987: 171; 175–77) shows that the internal "head" is obligatorily indefinite, as in (58b), and it cannot be definite as in (59):

(59) * [[*Mari* [*owiza ki*] *kage*] *ki*] *he ophewathu.*
Mari quilt the make the Dem I buy

Assuming Williamson's analysis, externally headed and 'internally headed' relative clauses turn out to differ only in one respect: in both cases the "head" originates within the relative clause; but the relative DP raises to Spec,CP overtly in the first ones and covertly in the second ones.

This type of syntactic variation is also attested by the French direct questions with *wh-* movement and *wh- in -situ*: *Qui a-t-elle rencontré?* 'who did she meet?' vs. *Elle a rencontré qui?* 'she met whom?' (Rizzi 1991: § 6).[43]

Note that the adjunct analysis of externally headed relative clauses, in which the "head" originates outside the relative CP, must treat the two types as completely unrelated. This is not very satisfying, since there are languages that exhibit both types of relative clause. For instance, Quechua has, besides the internally headed relative clause exemplified in (58a), an externally headed structure in which the "head" follows the relative CP:

(60) $[_{NP}[_{CP}$ *Nuna* t_i *rantishquan*] *bestya*$_i$] *alli bestya-m.*
man bought horse good horse- EVIDENTIAL
'The horse that the man bought is a good horse.'
(Cole & Hermon 1994: 247)

The situation of Quechua is not exceptional. Cole (1987) notes a typological correlation: internally headed relative clauses are only found in head-final languages. In order to explain this correlation, Cole assumes, contrary to Williamson, that the internally headed relative clause is not the sister to an external determiner, but rather, it is the sister of a phonetically null external "head". He assumes that in head-final languages the modifying relative clause precedes the phonetically null "head":

(61)
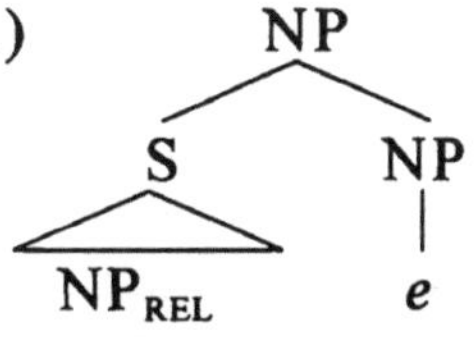

Cole proposes that the null NP is anaphoric to the internal "head". He thus derives the typological correlation from the assumption that an anaphor cannot both precede and command its antecedent (Cole 1987: 238). In the relative structure of a head-final language, like (61), the null anaphoric "head" commands its antecedent NP_{REL}, but it does not linearly precede it, and the condition is satisfied. In a head-initial language, however, the null NP "head" would linearly precede the relative clause, and NP_{REL} would be both preceded and commanded by the anaphoric null NP. This explains why internally headed relative clauses are impossible in head-initial languages.

Cole's proposal is incompatible with the Linear Correspondence Axiom, since it is based on a directionality parameter. However, Kayne (1994: 95–97) reformulates Cole's insight in a way compatible with the LCA. As for prenominal relative clauses like (60), Kayne assumes that they have the raising structure of postnominal relative clauses, and proposes the following derivation: the "head" raises to Spec,CP, as in English or Italian relative clauses (62b); the surface order is obtained by the further movement of IP of the relative clause to the Spec of the external D°, stranding a null complementizer (62c):

(62) a. $[_{DP}$ D° $[_{CP}$ C° $[_{IP}$ ... NP_{REL}...]]]
b. $[_{DP}$ D° $[_{CP}$ NP_{REL} $[_{CP}$ C° $[_{IP}$ ... t ...]]]]
c. $[_{DP}$ $[_{IP}$ ... t ...] $[_{DP}$ D° $[_{CP}$ NP_{REL} $[_{CP}$ C° t_{IP}]]]]

According to Kayne (1994: 93–94), the hypothesis that the prenominal relative clause is an IP rather than a CP explains why prenominal are never introduced by a complementizer.

Kayne also derives internally headed relative clauses from the structure in (62c). Assuming with Chomsky (1993) that traces are copies of the moved category deleted in the PF representation, he proposes that internally headed relative clauses correspond to a structure like (62c) where the trace of the raised "head" is spelled out, and the occurrence of the "head" in Spec,CP is instead deleted:

(63) $[_{DP}[_{IP}$ ... $[_{t}$ NP$]_{i}$ ...] $[_{DP}$ D° $[_{CP}$ $[_{NP}$ $e]_{i}$ $[_{CP}$ C° t_{IP}]]]]

Cole's generalization is then reformulated as follows: it is possible to delete the highest element of the chain in Spec,CP and to overtly realize the trace within the relative clause only if the relative IP containing the trace is preposed to Spec,DP, as in (63), but not if it remains in a position c-commanded by Spec,CP. This restriction derives from a principle whereby the deleted link of a chain cannot c-command the phonetically realized one. This principle excludes the deletion of the antecedent in Spec,CP in the postnominal relative clause, where the antecedent c-commands its trace, but it allows this deletion in (63), where the antecedent does not c-command its trace within the raised IP.

Note that this analysis of internally headed relative clauses involves overt raising of the NP "head", rather than LF movement: this overt movement is obscured by the fact that it is the antecedent rather than the trace that gets deleted in the phonetic representation. This implies that the raising of the "head" universally takes place in overt syntax.

6.2. Superlative and "degree" relative clauses

The preceding discussion concentrated on relative clauses headed by a nominal maximal projection. In fact, it is usually assumed that restrictive relative clauses, contrary to appositive ones, can only have noun phrases as "heads" (cf. Fabb 1990). However, R. Kayne (personal communication) points out that in English scalar adjectives and adverbs in the superlative can be "heads" of a "degree" relative:[44]

(64) a. *(This is) the fastest that he's ever run.*
b. *(This is) the most skillfully that he's ever spoken.*

A related construction is the Spanish (and Catalan) "degree relative":

(65) a. *Juan vio lo contenta que estaba María.*
Juan saw the happy that was Mary
'Juan saw how happy Mary was.'
b. *lo mal que respondió María*
the badly that answered Mary
'how badly Mary answered'
c. *lo muy cerca de la frontera que estamos*
the very close to the frontier that (we) are
'how close to the frontier we are' (Spanish; Ojeda 1982)
d. *No vulguis saber l'entremaliat que es aquell nen.*
not want to know the mischievous that is that child
'You cannot know how mischievous that child is.'
e. *Pensa en lo tard que arrivaba.*
think (in) the late that (he) arrived
'(He) thinks how late he arrived.'
(Catalan; Hirschbühler & Rivero 1981: 596)

The Spanish/Catalan structure is particularly interesting in that adjectives and determiners have morphological agreement, contrary to their English equivalents. Ojeda (1982) notes that the agreement pattern in this structure is exceptional in two respects. One the one hand, the adjectival "head" shows non-local agreement with an internal constituent of the relative clause; as discussed in § 4.2, this follows straightforwardly from the raising analysis. Ojeda notes that this analysis is supported by some literary examples in which the adjectival "head" does not raise:

(66) a. *El que el oro es dúctil, el acero es resistente.*
the that gold is ductile, steel is hard
'Gold is as ductile as steel is hard.'

(66) b. ... *no dezían la meytad de lo que ella era hermosa.*[45]
(they) didn't say the half of the that she was handsome
'They didn't describe half of her beauty.'

On the other hand, the determiner introducing the adjectival "head" does not agree with it for phi-features, but it remains in the masculine singular form (whence the term "neuter article"). Ojeda proposes that the neuter article is actually a pronoun referring to degrees; it is clearly able to nominalize the relative structure, since the latter can undergo passivization:

(67) [*Lo contenta que estaba María*] *fue visto por todos.*
the happy that was Mary was seen by everybody
'Everybody saw how happy Mary was.'

Ojeda's analysis of the degree relative is very close to Kayne's raising analysis of restrictive relatives.[46] The possibility of a unified approach to restrictive and degree relatives constitutes an advantage of this analysis over the adjunct analysis. In order to show this, however, it is necessary to discuss in detail the parallelism in the functional structure of noun phrases and adjectival phrases. The following discussion is based on recent work by R. Zamparelli.

6.2.1.

In a comparative study of Romance and Germanic adjectival phrases, Zamparelli (1994) argues for the following functional structure:

(68) $[_{AgrP}$ Agr° $[_{DegP}$ (Measure phrase) $[_{DegP}$ Degree° $[_{AP}$ A° (XP)]]]]

The highest head Agr° contains the agreement morphology of the adjective; Degree° contains the comparative and superlative synthetic morphemes, whereas its Spec contains measure phrases (e.g. [*two meters*] *tall*).

In Romance languages, adjectives show overt agreement morphology: Agr° is strong and the lexical head A° must incorporate to it, passing through Degree°. This gives rise to a superlative form like the following:

(69) *alt-issim-i*
tall-est-M.PL.

In English, on the contrary, adjectives have no agreement morphology: consequently, they do not incorporate to Agr°, but only to Deg° when the latter contains a synthetic comparative or superlative morpheme (*tall-er, tall-est*). This accounts for the following facts:
(a) a measure phrase in Spec,DegP precedes the adjective in English but follows it in Italian:

(70) a. *The man was* $[_{AgrP}[_{DegP}[_{MP}$ *two meters*$]$ $[_{AP}$ *tall*$]]]$.
b. *L'uomo era* $[_{AgrP}$ *alto*$_i$ $[_{DegP}[_{MP}$ *due metri*$]$ $[_{DegP}$ t_i $[_{AP}$ $t_i]]]]$.
the man was tall two meters
'The man was two meters tall.'

(b) In Italian, the measure phrase intervenes between the inflected adjective and the complement it selects:

(71) *Maria era* $[_{AgrP}$ *lontana*$_i$ $[_{DegP}[_{MP}$ *due chilometri*$]$ $[_{DegP}$ t_i $[_{AP}$ t_i $[_{PP}$ *dalla casa*$]]]]]$.
Mary was distant-AGR two kilometers from the house
'Mary was two kilometers far from the house.'

(c) In Italian, adverbs have no agreement morphology, hence they do not raise to Agr° and they follow the English pattern, appearing to the right of the measure phrase:

(72) $[_{AgrP}[_{DegP}[_{MP}$ *due chilometri*$]$ $[_{AdvP}$ *lontano* $[_{PP}$ *dalla città*$]]]]$ *c'era una casa.*
two kilometers far from the town there was a house
'There was a house two kilometers far from the town.'

Zamparelli further argues that scalar adjectives without overt measure phrases receive a default positive amount reading, corresponding to a phonetically empty measure phrase in Spec,DegP.

In his paper, he does not discuss analytical comparatives and superlatives forms like the following:

(73) a. *John is more intelligent than Mary.*
b. *Gianni è più intelligente di Maria.*
John is more intelligent than Mary
c. *John is the most intelligent of the students.*
d. *Gianni è il più intelligente degli studenti.*
John is the most intelligent of-the students

Note that at least in Italian the analytical comparative/superlative particle cannot be analysed as the Deg° head of (68), since it would block the raising of the adjective to Agr°. For the time being, let us assume that the Italian analytical particles realize a head Deg° above Agr° and that a superlative Deg° must be selected by a definite D°.[47]

On the basis of the structure in (68), the examples in (64) can receive an analysis parallel to that of NP-headed restrictive clauses. The relevant difference is that these relative CPs are selected by a superlative Deg° rather than by a D°. Since Deg° must bind an adjective or adverb, an adjectival or adverbial "head" raises to the Spec of the relative CP:

(74) $[_{DP}$ *the* $[_{DegP}$ *most* $[_{CP}$ *skilfully*$_i$ $[_{CP}$ *that* $[_{IP}$ *he's ever answered* $t_i]]]]]$

In (64a), the lexical head of the adverbial phrase in Spec,CP incorporates to Deg°, which contains the bound superlative morpheme, thus satisfying the morphological requirements of the latter. Recall that Spec,CP is transparent for extraction (§ 5), hence this incorporation does not violate the locality condition.[48]

The reason for generating the superlative Deg° outside the relative CP is that it takes the whole relative clause in its restrictive term and licenses polarity items within it, e.g. *ever* in (64); cf. the discussion in § 2.1.

Turning to the Spanish degree relatives in (65) and (66), Ojeda's analysis can be reformulated in terms of the structure in (68). According to him, a sentence like (65a) means "Juan saw the degree at which Maria was happy": the neuter article must then be analysed as a definite (iota) operator binding a degree variable provided by the adjectival "head". In terms of the structure (68), the neuter article is a definite Deg°, parallel to a definite D°.[49]

(75) $[_{DegP}$ *lo* $[_{CP}$ $[_{AgrP}$ *contenta*$]$ $[_{CP}$ *que* $[_{IP}$ *estaba María* t_{AgrP} $]]]]$

The degree relatives in (66) receive a parallel structural analysis, the only difference being that the adjectival "head" remains *in situ* in overt syntax.

An open problem is why the Deg° head selecting the relative CP does not agree with the adjectival "head" when the latter is inflected for phi-features. Adopting the typology of features proposed by Chomsky (1995: 276–279), it may be assumed that the agreement features of the adjectival phrase are [-interpretable] (roughly, they are irrelevant for the LF interpretation), and as such they are deleted after they have been checked against the [+interpretable] features of the subject of predication. This entails that the features of the adjective are not available to check the phi-features of an external head.[50,51]

Consider now a restrictive relative whose "head" contains a superlative adjective:

(76) a. *This is the most interesting book that I've ever read.*
b. *Questo è il più bel libro che io abbia mai letto.*
this is the most nice book that I have-SUBJ ever read

A first possibility is to extend to (76) the analysis proposed in (74): the relative CP is selected by the superlative Deg°, but in addition, in (76) the adjectival "head" pied pipes a whole noun phrase. In particular, the adjectival phrase must appear in the highest Spec of the pied piped noun phrase:

(77) [$_{DP}$ *the* [$_{DegP}$ *most* [$_{CP}$ [$_{XP}$[$_{AgrP}$ *interesting*] [$_{XP}$ *book*]][$_{CP}$ *that* ...]]]]

Alternatively, it is possible to place the superlative adjectival phrase in the Spec of the external D° selecting the relative CP, following Zamparelli (1994b):

(78) [$_{DP}$ [*the most interesting*] [$_{DP}$ D°$_{[+DEF]}$ [$_{CP}$ *book* [$_{CP}$ that ...]]]][52]

Although many details of the comparative/superlative structures remain obscure, the preceding discussion aimed at showing that the raising analysis of restrictive relatives, together with the functional structure of adjectival phrases proposed by Zamparelli (1994), offers a simple approach to the English and Spanish degree relatives, whose syntax is conceived of as essentially parallel to that of NP-headed relative clauses. This is so because Kayne's analysis can exploit the correspondence in the functional structure of nominal and adjectival phrases.

7. Summary and open questions

The standard analysis of restrictive relative clauses, which takes them to be right-adjoined to their nominal antecedent, is incompatible with the LCA. Assuming that the restrictive CP must be c-commanded by the determiner of the "head", there is only one structural analysis compatible with the LCA: the relative CP must be generated as a complement of the determiner. This structure is consistent with the DP Hypothesis.

Given the uniqueness of the complement position, the nominal "head" must be generated within the relative CP and raise to a Spec position in between the Determiner and the C° head of the relative CP. The null hypothesis is that this position is Spec,CP. This implies that the NP category modified by the restrictive relative does not form a constituent with the determiner that precedes it.

However, since the determiner selecting the relative CP agrees for phi-features with the "head" raised to Spec,CP, the two elements must be syntactically related. Thus, Spec,CP must be exempted from the islandhood of the relative CP; this raises a problem for locality theory, and in particular for the analysis of strong island effects.

The preceding discussion left open the question of whether the raising analysis can be extended to relative clauses featuring a relative pronoun:

(79) *the boy* who *I met yesterday*

The syntax of relative pronouns in the raising analysis constitutes the topic of the next chapter.

Chapter III
The syntax of relative determiners

0. Introduction

The preceding discussion left open the question of the status of relative pronouns. In the adjunct analysis, the latter are conceived as operators binding the relative trace:

(1) $[_{DP}$ *the* $[_{NP}$ $[_{NP}$ *boy*] $[_{CP}$ *who*$_i$ [*I met* t_i *yesterday*]]]]

This is obviously inconsistent with the raising analysis. If the relative pronoun is analysed as a full pronominal DP, then the relative "head" cannot have raised from within the relative clause, for at least two reasons: first, both the "head" and the pronoun would have to be generated in the direct object position; second, the two elements would both occur in the only available landing site, Spec,CP. In other terms, in the raising analysis the "head" and the relative pronoun appear to be in competition both for the base position and for the landing site.

There are two ways out of this dilemma. The first one is to maintain a raising analysis for relative clauses without a relative pronoun, and adopt instead a more traditional analysis for those with an overt pronoun. This line of analysis, which has been defended by Åfarli (1994), is critically examined in § 1.

The second solution is to abandon the assumption that *who* is an independent pronoun and to analyse it as a subconstituent of the relative "head". This line of analysis, proposed by Kayne (1994: chapter 8), is discussed in great detail in §§ 2–3. The following sections 4 and 5 show that this analysis opens a new and revealing perspective on the behaviour of relative morphemes in various relative structures, allowing a more unified approach to the phenomenon of relativization in its varied forms.

1. A mixed analysis: Åfarli (1994)

Åfarli (1994) proposes a raising analysis for Norwegian restrictive relative clauses introduced by the complementizer *som* and lacking a relative pronoun. Specifically, he analyses this relative structure as a clausal category (TP) occupying a DP position within the matrix clause. The mechanism of c-selection is modified so as to allow a lexical head to c-select not only its complement, but also the Spec of the latter. Thus in (2) the transitive verb takes TP as its structural complement, but it satisfies its selectional requirements by c-selecting Spec,TP, which hosts the raised DP "head"[1] at the end of the derivation:

(2) *Han kjøpte* $[_{TP}$ *eplet*$_i$ $[_{T'}$ *som* $[_{VP}$ *var størst* $t_i]]]$
he bought apple-the that was biggest
'He bought the biggest apple.'

Åfarli proposes instead a more traditional analysis for relative clauses with an overt relative pronoun: in this case, it is the relative pronoun that raises to Spec,TP, and it is then related to an external DP by a rule of predication (cf. Chomsky 1982: fn. 14).[2] This structural difference is justified by the observation that only relative clauses of the first type exhibit reconstruction effects with respect to anaphor binding (3) and quantifier binding (4):

(3) a. *Det av husa sine*$_i$ *som Jon*$_i$ *bor i, er ganske falleferdig.*
that of the houses his C° Jon lives in is quite miserable
b. ?? *Det av husa sine*$_i$ *der Jon*$_i$ *bor, er ganske falleferdig.*
that of the houses his where Jon lives is quite miserable
'The house of his in which Jon lives is quite miserable.'

(4) a. *Det brevet til han*$_i$ *som kvar arbeidstakar*$_i$ *trur at sjefen sender oppseiinga i, er uønska.*
the letter to him C° each employee believes that the boss sends the dismissal in, is unwanted
b. ?? *Det brevet til han*$_i$ *der kvar arbeidstakar*$_i$ *trur at siefen sender oppseiinga i, er uønska.*
the letter to him where each employee believes that the boss sends the dismissal in, is unwanted
'That letter to him that every employee believes that the boss sends the dismissal in, is unwanted.'
(Åfarli 1994: 89–90)

In the (b) examples, the presence of a relative pronoun blocks reconstruction: in fact, reconstruction depends on the direct raising of the relative "head" (cf. § II.4.3), and this does not take place in relative clauses with a visible pronoun.

The author also extends this approach to English restrictive relatives, claiming that at least some speakers find a contrast parallel to (4) between *that-* and *which*-relatives:

(5) a. *the present for him*$_i$ *that every boy*$_i$ *believed that Mary had bought* t...
b. % *the present for him*$_i$ *which every boy*$_i$ *believed that Mary had bought* t...

Furthermore, some speakers (the same ones?) perceive a weak crossover effect in *that-* relatives, but not in *wh*-relatives:

(6) a. % *the man*$_i$ *that his*$_i$ *mother loved* t$_i$ *best...*
b. *the man*$_i$ *who his*$_i$ *mother loved* t$_i$ *best...*

With respect to the contrast in (6), Åfarli adopts a proposal by Chomsky (1982: 92–93), whereby the rule of predication coindexing the antecedent and the relative pronoun in (6b) does not apply at LF, but at a later level LF', at which the principle responsible for weak crossover does not hold; thus, at LF the possessive pronoun is coindexed with the external antecedent of the relative clause, but the relative pronoun is not, and this configuration does not incur in a weak crossover violation. On the other hand, (6a) does not contain a null relative pronoun, but it is derived by raising: the raised head and its trace are coindexed, and coindexing the possessive pronoun with the "head" in LF yields a weak crossover effect.

Restricting the discussion to English, it must be pointed out that the data on which Åfarli's argument rests are not very clear. As for the weak crossover effect, Lasnik—Stowell (1991: 706–707) assert that it is only detected in restrictive relatives, but not in appositive ones, and it is stronger when the "head" is introduced by a quantificational determiner; this point is strenghtened by Postal (1993). As for binding under reconstruction, the author himself recognizes that judgements vary to a considerable extent among speakers. The contrast in reconstruction in (5) is probably flawed by various interfering factors. Consider for instance the data in (7)–(8):

(7) a. *This is the picture of himself*$_{i/*j}$ *that John*$_i$ *thinks Bill*$_j$ *likes* t.
b. *This is the picture of himself*$_{i/*j}$ *which John*$_i$ *thinks Bill*$_j$ *likes* t.

(8) a. *This is the picture of himself*$_i$ *that I think Bill*$_i$ *likes* t.
b. *This is the picture of himself*$_i$ *which I think Bill*$_i$ *likes* t.

Seven speakers out of nine accepted the coreference between the anaphor and the highest subject of the relative clause (*John*), discarding the embedded subject (*Bill*), in both the examples (7); one speaker accepted the coreference with the most embedded subject in (7a) but not in (7b); finally, only one speaker accepted the coreference with the highest subject in (7a), but not in (7b), in accordance with the pattern predicted by Åfarli's analysis. Thus, it appears that for many speaker reconstruction is possible in both types of relative clauses, and it displays a "closest antecedent" effect: coreference is only possible with the highest subject of the relative clause. The point is that this effect does not distinguish the two types of relative clause (except for one speaker). The "closest antecedent" effect is eliminated in (8), where the first person subject of the higher clause is not a potential antecedent for the third person anaphor; in this case, the two speakers whom the examples were submitted to accepted coreference of the anaphor with the most embedded subject in both relative clause types.

An independent factor that probably interferes with the speakers' judgements on (5), (7) and (8) is the marginal status of the relative pronoun *which* in restrictive relatives. In contemporary English the use of this pronoun is being gradually restricted to appositive relatives, and in restrictive ones it is perceived as literary and obsolete. Then the contrast in (5) could be related to the restrictive vs. appositive interpretation of the examples, rather than to the absence versus presence of a relative pronoun; the uncertainty of the judgement for (5b) would reflect a difficulty on the part of the speaker in the choice between a restrictive and an appositive interpretation for this example.[3] The influence of the interpretation of the relative clause on the reconstruction of the "head" will be thoroughly discussed in chapter IV.

In sum, the empirical evidence that excludes the raising of the relative "head" in the presence of a relative pronoun is not quite clear, and reconstruction examples like (8b) suggest that raising must in fact be admitted. This requires a new analysis of relative pronouns.

2. Relative pronouns as determiners

Kayne (1994: 88–91) proposes that relative pronouns are actually the original determiners of the relative "head". Thus in (9) it is the constituent [$_{DP}$ *who* [$_{NP}$ *boy*]] that raises to Spec,CP (9b); the surface order of the elements is obtained by further moving the NP [*boy*] to a position that asymmetrically c-commands the relative determiner *who* and is asymmetrically c-commanded by the external determiner. This position Kayne assumes is the Spec of the relative determiner (9c):

(9) a. $[_{DP}$ *the* $[_{CP}$ C° *I met* $[_{DP}$ *who* $[_{NP}$ *boy*]]]]
b. $[_{DP}$ *the* $[_{CP}$ $[_{DP}$ *who* $[_{NP}$ *boy*]]$_i$ $[_{CP}$ C° $[_{IP}$ *I met* t_i]]]]
c. $[_{DP}$ *the* $[_{CP}$ $[_{DP}$ *boy* $[_{DP}$ *who* t_{NP}]]$_i$ $[_{CP}$ C° $[_{IP}$ *I met* t_i]]]]

The relative pronoun will be henceforth called "relative determiner", and its maximal projection the "relative DP". The term relative "head" will be restricted to designate the NP complement of the relative determiner.

This analysis can be extended to relative clauses featuring the pied piping of a lexical element along with the relative determiner to Spec,CP. In (10), for instance, the relativized phrase is a PP properly embedding the relative DP; but strikingly, the NP "head" precedes the preposition heading the pied piped phrase:

(10) a. *la sedia su cui era seduto*
b. *the chair on which he was sitting*

Kayne (1994: 89) extends to (10) his analysis of (9), arguing that the NP "head" raises to the Spec of the pied piped constituent, stranding the relative determiner. The derivation of (10) is thus (11):

(11) a. $[_{DP}$ *the* $[_{CP}$ C° $[_{IP}$ *he was sitting* $[_{PP}$ *on* $[_{DP}$ *which* $[_{NP}$ *chair*]]]]]]
b. $[_{DP}$ *the* $[_{CP}[_{PP}$ *on* $[_{DP}$ *which* $[_{NP}$ *chair*]]] [C° [*he was sitting* t]]]]
c. $[_{DP}$ *the* $[_{CP}[_{PP}$ $[_{NP}$ *chair*] [*on* $[_{DP}$ *which* t_{NP}]]] [C° $[_{IP}$ *he was sitting* t]]]]

This new conception of relative pronouns allows the extension of the raising analysis to all types of restrictive relative. However, it reverses the traditional conception of the constituency of the modified noun phrase: the "head" does no longer form a constituent with the determiner that precedes it, but rather, with the relative determiner that follows it.

2.1. The trigger of raising

The representations in (9c) and (11c) yield the correct linear order of the elements, with the NP "head" occurring immediately after the external determiner. This is achieved in two steps: first, the relative DP or the pied piped PP move to Spec,CP (cf. (9b) and (11b) respectively); second, the NP moves out of the complement position of the relative D°, and it reaches the most prominent specifier position within the relative clause - the one that asymmetrycally c-commands everything else within the relative CP. Both these steps must be justified by identifying the relevant triggers.

2.2. Movement of DP/PP

The first step involves the movement to Spec,CP of a category properly containing the NP "head". It is possible to assume that this step is triggered by a Relative Criterion, formally parallel to Rizzi's (1991) Wh-Criterion. From this perspective, the relative DP and the C° head of the relative clause are both endowed with a feature [+rel], and they must be related to each other by Spec/head agreement. This justifies the step (9b) in the derivation (9):

(12) $[_{DP}$ *the* $[_{CP}$ $[_{DP}$ *who boy*$]_{[+REL]}$ $[$ $C_{[+REL]}$ $[_{IP}$ *I met* $t_i]]]]$

As for the pied piping structure (11), the pied piped PP can be assumed to inherit the relevant feature [+rel] from the relative D° which it dominates. This "percolation" approach to pied piping is relatively standard (cf. for instance Kayne 1984, Ortiz de Urbina 1990).

The assumption of a Relative Criterion constitutes a straightforward extension of Rizzi's (1995) "Split Comp" hypothesis, mentioned in § I.2.1: under this hypothesis, the complementizer system of a clause consists of an array of functional projections, each one providing the landing site for a specific type of A' movement. The specificity of the landing site can be expressed by a generalization of the Wh-criterion, whereby a constituent endowed with a feature [+F] targets the specifier of a Comp head endowed with the same feature, so that the two categories can establish an agreement relation. This extension will be discussed in more detail in § VI.4.[4]

Alternatively, it is possible to analyse both the movement of the whole relative DP in (9b) and the movement of PP in (11b) as instances of pied piping: the category to be raised is actually the NP "head", and the larger categories are pied piped simply because they cannot be stranded in the base position.

Note that there exist some structures in which a (weak) determiner is stranded in its base position by the leftward movement of its complement : for instance, *ne*-cliticization in Italian or "split topicalization" in English:

(13) a. *Ho letto* [$_{DP}$ *qualche* [*libro*]]
b. *Ne*$_i$ *ho letto* [*qualcuno* e$_i$] (Cardinaletti—Giusti 1991)

(14) a. *I have* [$_{DP}$ *no* [*inspiration*]].
b. [*Inspiration*] *have I* [$_{DP}$ *none* e].

However, these structures induce focalization over the stranded determiner, which is probably interpreted as a cardinality predicate. It is possible to speculate that a similar interpretation is unavailable for a relative determiner, and therefore the latter cannot be stranded in the base position, but it must be pied piped to Spec,CP.

From this perspective, the first step of the derivations (9) and (11) would be reduced to the trigger of the second step, which only involves the NP category.

2.3. The raising of the NP "head"

In § II.4.4, we adopted without discussion Kayne's (1994: 90) proposal that the NP "head" must reach a position governed by the external determiner. Note that this condition is satisfied both in (9c) and in (11c): no potential head governor intervenes between the two categories, so that relativized minimality is respected. Equivalently, by the definition of minimal domain (51) of chapter II, repeated below for convenience, the NP "head" falls in the minimal domain of the external D°, because it is dominated only by two segments: one segment of DP and one segment of CP in (9c); one segment of PP and one of CP in (11c).

(15) The minimal domain of a head X, notated (X), includes all elements that are immediately dominated by, and do not immediately dominate, a projection of X. (Manzini 1994)

Crucially, head government must be assumed to establish a checking relation between the NP "head" and the external determiner.

Following Chomsky (1993; 1995), let us assume that overt movement is always driven by the morphological requirement of checking a strong feature. The most plausible hypothesis is that the strong feature to be checked belongs to the external D°.[5] Note that the latter is not an invariable clausal determiner (§ II.2.2), but it is a nominal determiner endowed with agreement features. Furthermore, let us assume, adopting an idea of Chomsky's, that the external D° bears a strong N-feature to be checked by a [+N] category. As these nominal features cannot be checked by the CP complement, they trigger the raising of an NP to a position governed by the external D°.

This proposal raises two problems with respect to the theory of checking. According to Chomsky (1993; 1995), checking can only take place between a head and a category within its checking domain (including the specifiers and the categories/features adjoined to the head); therefore, the landing site of overt movement must belong to the checking domain of the triggering head. In the raising relative structure, however, the "head" in Spec,CP does not fall in the checking domain of the external D°.[6]

Moreover, there is a problem with respect to Chomsky's (1993: 22–23) extension condition on overt substitutions. In fact, the raising of the "head" to Spec,CP is triggered after the external D° has been merged with the relative CP, which yields a countercyclic derivation.

However, note that in Chomsky (1995: 232–235) the extension condition is no longer postulated, but it is derived from the defining property of strong features: a strong feature must be eliminated by checking as soon as it is introduced in the derivation, i.e. before the category bearing it is merged with another category. Suppose that a head can also establish a checking relation with a category that it governs (or equivalently, one that falls in its minimal domain). This entails a weaker version of the extension condition: the landing site of overt movement can be a position governed by the triggering head.

Here we shall assume, following Rizzi (1995), that head government does qualify as a proper configuration for checking (or, more neutrally, sharing of morphosyntactic features between two categories).

Nevertheless, let us also consider a slightly different analysis which is consistent with Chomsky's (1995) minimalist assumptions.[7] Suppose that the overt agreement morphology of a nominal determiner corresponds to a head Agr_D selected by D° and selecting NP (16): this would closely correspond to the clausal architecture, where a C° head selects an Agr phrase. It is Agr_D that bears strong nominal features; their checking may now take place in overt syntax, by moving the NP complement to Spec,Agr_D, as in (16b). Then Agr_D, being affixal, incorporates to D° to yield an inflected determiner:

(16) a. $[_{DP}$ D° $[_{AgrP}$ Agr_D NP]]
b. $[_{DP}$ D° $[_{AgrP}$ NP [Agr_D t_{NP}]]]
c. $[_{DP}$ Agr_D+D° $[_{AgrP}$ NP [t_{Agr} t_{NP}]]]

Similarly, it may be assumed that in the raising relative structure CP is selected by Agr$_D$. The NP "head" is then extracted from the complement position of the relative D° [8] in order to enter the checking domain of the external Agr$_D$. Instead of (9), for instance, we would adopt the derivation in (17):

(17) a. [$_{DP}$ *the* [Agr$_D$ [$_{CP}$ *I met* [$_{DP}$ *who* [$_{NP}$ *boy*]]]]]
b. [$_{DP}$ *the* [Agr$_D$ [$_{CP}$ [$_{DP}$ *who* [$_{NP}$ *boy*]]$_i$ [$_{IP}$ *I met* t$_i$]]]]
c. [$_{DP}$ *the* [$_{AgrP}$ *boy* [Agr$_D$ [$_{CP}$ [$_{DP}$ *who* t$_{NP}$]]$_i$ [$_{IP}$ *I met* t$_i$]]]]]
d. [$_{DP}$ Agr$_D$+*the* [$_{AgrP}$ *boy* [t$_{Agr}$ [$_{CP}$ [$_{DP}$ *who* t$_{NP}$]]$_i$ [$_{IP}$ *I met* t$_i$]]]]]

Note that this analysis is consistent with the standard checking theory of the minimalist program, and from a derivational perspective, it is perfectly cyclic.

However, this analysis is less "minimal" than the head government solution in that it requires the postulation of an external head Agr$_D$. Though fairly plausible, the checking process postulated in (16) and (17) requires independent empirical justification.

Much of the following discussion is compatible with either the head government solution or the alternative in (17). In some cases the latter may be preferable (e.g., with respect to the intonation breaks delimiting appositive relatives; cf. fn. 26 to chapter V; and also with respect to the coordination facts discussed in § VIII.2). Whenever the two options yield different consequences, the fact will be explicitly discussed.

2.4. Omission of the relative determiner

Let us now reconsider the status of the raised category in an example like (23) of chapter II, repeated here as (18), where the relative "head" is not introduced by an overt relative determiner.

(18) [$_{DP}$ *the* [$_{CP}$ *book*$_i$ [$_{CP}$ *that I read* t$_i$]]]

Kayne (1994: 87) states that in this case the raised category is "probably of category NP". This means that the relativized phrase has a different categorial status in relative clauses with and without an overt relative determiner: respectively, DP and NP. (Note that (18) does not seem to satisfy the Relative Criterion, since the feature [+rel] is plausibly specified on the relative D°.).[9] Alternatively, the raised constituent in (18) could be a DP headed by a silent relative D°. For the moment, the choice between the two alternatives is irrelevant; the problem will be thoroughly examined in chapter VI.

2.5. A morphosyntactic problem

The assumption that relative *who* is a determiner may look scarcely plausible, because its interrogative counterpart cannot be used as a determiner, but only as an independent pronoun. Kayne (1994: fn. 12 to chapter 8) suggests that relative *who* is a form of *which* agreeing with a [+human] NP "head" in its Spec (cf. 54c).[10, 11]

Note that it is common for a determiner to take a "pronominal" form when its complement NP moves leftwards. In the Italian *ne*-cliticization structure exemplified in (13b) above, the indefinite quantifier *qualche* assumes the form *qualcuno*, incorporating an indefinite determiner, when its complement is extracted. The same form is also used as an independent pronoun (e.g. *qualcuno è venuto* 'someone came'). Similarly, in the "split topicalization" structure exemplified in (14b) the negative determiner *no* in English takes the "pronominal" form *none* when its complement moves away. Presumably, the change in the form of the determiner is related to the necessity of licensing the empty category.

These data show that the morphological shape of an element may conceal syntactic ambiguity: even if in some syntactic contexts a given form functions as a pronoun, in other contexts the same form may have the function of a determiner.[12] Hence, nothing in principle excludes the possibility of analysing relative *who* as a determiner, rather than a pronoun.

3. The non-definiteness of the relative DP

The analysis of relative morphemes as determiners raises a problem with respect to the Full Interpretation Principle at LF. In fact, the NP "head" originates as the complement of the relative determiner, but it moves out of its c-command domain, and in LF it is interpreted as a part of the restrictive term of the external determiner. Thus, by the definition (13) of chapter II, the relative D° lacks a restrictive term in LF: therefore, it should violate the Full Interpretation principle, being interpreted as a vacuous quantifier.

The same conclusion holds if one assumes, following Higginbotham (1985), that a determiner binds an open variable position inside its complement NP: then the variable position of the NP "head" cannot be bound both by the relative determiner and by the external one, and either one of them will have no variable to bind, again violating the Full Interpretation principle.

The solution to this problem comes from a proposal by Reinhart (1987), whereby the relative DP is, from a semantic viewpoint, an indefinite in the sense of Heim (1982).[13] In the Heim-Reinhart theory, an indefinite determiner is not a function mapping NP into a generalized quantifier, but rather, it is an expression of cardinality predicated of the set denoted by the NP head. The whole indefinite DP is thus a predicative category to be bound by an external operator.

In this approach, relative *who* and *which* are analysed as indefinite determiners, uninflected for number.[14] Let us consider how the structure can be interpreted.

As already discussed in § II. 4.1, in an A' chain like (19a) below it is possible to reconstruct the lexical part of the *wh*-phrase: this yields an unrestricted quantification structure like (19b), which receives an interpretation like (19c):

(19) a. [*Which picture of John*]$_i$ *do* [$_{IP}$ *you like* [$_t$ *which picture of John*]$_i$]]?
b. [[*Which* e]$_i$ *do* [$_{IP}$ *you like* [$_t$ e *picture of John*]$_i$]]?
c. Which x is such that (x is a picture of John) & (you like x)?

A parallel LF derivation can be assumed for the raising relative structure. As it will be argued in chapter IV, in LF the relative DP is reconstructed in the trace position. The resulting configuration (20b) is parallel to (19b): the relative DP within IP has an open position and is bound by a higher operator, namely the external D°. The only difference is that here, contrary to (19b), the whole structure is c-commanded by the higher determiner, and hence constitutes its restrictive term rather then its nuclear scope (20c):

(20) a. [$_{DP}$ *the* [$_{CP}$ *boy* [$_{DP}$ *who* t_{NP}]] [$_{IP}$ *I met* [$_t$ *who boy*]]]]
b. [$_{DP}$ *the* [$_{CP}$ [$_{IP}$ *I met* [$_{DP}$ *who boy*]]]
c. the x such that (x is a boy) & (I met x)

The resulting LF interpretation is close to Reinhart's (1987) "maximal clausal set" interpretation:[15] the relative DP is bound by the external determiner in LF.[16] Thus the indefinite interpretation of the relative DP allows the relative structure to satisfy the Full Interpretation principle.[17]

3.1. Empirical evidence

We can assume that the relative D° is actually non-definite, namely, it is not positively specified [-definite]. In fact, it is possible to conceive of the definite/indefinite opposition as a privative one. In the theory of specificity and definiteness developed by Enç (1991), every noun phrase is endowed with two referential indices, one referring to the individual denoted by the whole noun phrase, and the other referring to the set which the noun's open variable ranges over. A noun phrase is definite if its first referential index is marked as [+familiar], and it is specific if its second index is marked [+familiar]. From this perspective, definiteness can be considered a feature that is added to the noun phrase's specification, and the opposite value (non-definiteness) reduces to the lack of this positive specification.

The claim that the relative DP is non-definite is controversial. In some languages, relative pronouns are morphologically definite: for instance, the Italian relative pronoun *il quale* incorporates a definite article. The origin and the morphological shape of relative pronouns in the Indo-European languages will be discussed in § 5. This section presents some syntactic phenomena in which a morphologically definite relative determiner behaves as if it were non-definite: this evidence casts doubt on the syntactic and semantic significance of the definite morphology.

A first case in point is the agreement for definiteness displayed by Hungarian transitive verbs (Ingria 1990: 196–198; Horvath 1986: 44–; Szamosi 1976). The verb takes two different forms according to the definiteness of its DP direct object:

(21) a. *Akart egy könyvet / * a könyvet.*
he wanted$_{indef}$ a book / * the book
b. *Akarta a könyvet / * egy könyvet.*
he wanted$_{def}$ the book/ * a book

Restrictive relative clauses are obligatorily introduced by a relative determiner, which consists of an interrogative determiner preceded by the definite article *a.* The nonspecific interrogative *mi* 'who' yields the form *ami*: interestingly, this form triggers the indefinite conjugation on the verb of the relative clause:

(22) a. *egy könyv amit akart t*
a book which he wanted$_{indef}$
b. *a könyv amit akart t*
the book which he wanted$_{indef}$

(22) c. * *egy könyv amit akarta t*
a book which he wanted$_{def}$
d. * *a könyv amit akarta t*
the book which he wanted$_{def}$[18]

This suggests that the definite article incorporated in the relative determiner does not carry any definiteness feature, but it is a sort of expletive (see § 5).[19]

The phenomenon of verbal agreement is actually more complex.[20] Hungarian has another relative determiner, *amely*, derived from the interrogative *mely* 'which'. According to Szamosi and Horvath, this form triggers the definite conjugation, like its interrogative counterpart:[21]

(23) a. *Ez az a könyv amelyiket akarta.*
this that the book which he wanted$_{def}$
b. * *Ez az a könyv amelyiket akart.*
this that the book which he wanted$_{indef}$
(24) *Melyik könyv akarta/ * akart?*
which book did he want$_{def / *indef}$ (Szamosi 1976: 410–411)

The definiteness of the relative determiner in (23) is inconsistent with the Reinhart analysis of the relative DP.

An alternative solution is provided by the theory of specificity proposed by Enç (1991), mentioned above. Roughly, in this theory definite DPs introduce a "familiar", presupposed referent, whereas specific DPs introduce a variable whose range is presupposed. Suppose that the Hungarian verbal agreement is sensitive to specificity, rather than to definiteness:[22] then the relative DP headed by *amelyik* introduces a variable to be bound by the external determiner, whose range is presupposed. Note that even if the verbal agreement is determined by specificity rather than definiteness, the point concerning (22) still holds: in fact, in Enç's theory definiteness implies specificity, and therefore, if the definite article incorporated in the relative pronoun *amit* were not expletive, it should trigger the "specific" conjugation on the verb within the relative clause.

Next, let us consider the Albanian relative pronoun. According to Morgan (1972: 64–65), this consists of an interrogative pronoun (itself composed of a morpheme *cil-* combined with the enclitic definite determiner) preceded by an "adjectival article". This is an independent morpheme which introduces adjectival or genitive modifiers and agrees in definiteness with the main determiner. Thus, there are two series of adjectival articles, a definite and an indefinite one (Morgan 1972: 64–

66):[23] interestingly, the adjectival article introducing the relative form belongs to the indefinite series (G. Turano, personal communication): [24]

(25) *Qyteti në të cilin banonte tër jetën...* (Morgan 1972: 63)
city-the in ADJ.ART. which (she) lived all life-the
'the city where she lived all her life'

The fact that the adjectival article belongs to the indefinite series suggests that the whole relative determiner is not definite, despite the fact that the word *cil-* apparently incorporates a definite article.

Other evidence suggests that even in relative clauses lacking a relative determiner the raised "head" is non-definite (cf. § II.3.2). This obviously follows from the hypothesis provisionally adopted in § 2.1, whereby in these structures the "head" is an NP category. However, this hypothesis will be revised in chapter VI, where it will be argued that the "head" is actually introduced by a phonetically null relative determiner: the latter too must be analysed as non-definite.

Consider first the phenomenon of double definiteness in Swedish, discussed by Santelmann (1993). According to this author, the enclitic definite article of Swedish is not generated in D°, but in a lower functional head called Art°. In DPs without a restrictive modifier, N° incorporates to Art°, and the resulting form raises to D°:

(26)

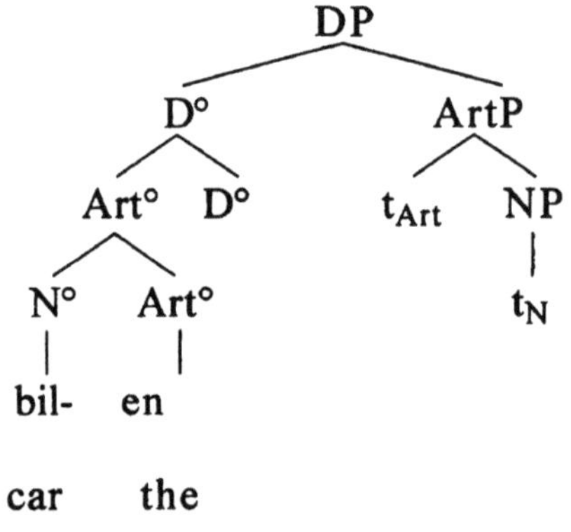

(Santelmann 1993: 165)

However, when a prenominal modifier intervenes between Art° and D°, as in (27a) below, N°+Art° cannot raise to D°[25] and the latter is filled by a non-clitic definite determiner *den*, as in (27b). The non-clitic determiner obligatorily co-occurs with the enclitic definite article; it cannot be used alone, as in (27c):

(27) a. * *bil-en stora*
car-the big

(27) b. *den stora bil-en*
the big car-the
c. * *den stora bil*
the big car

Interestingly, there is one context where the nonclitic determiner appears alone, and this is when it introduces a restrictive relative structure:

(28) *Jag hörde till den grupp som ledes av NN.*
'I belonged in the group that was led by NN'

The lack of an enclitic definite article in (28) is mysterious from the perspective of the the adjunct analysis, where the "head" is a definite DP and the adjoined relative clause is not expected to interfere with its functional structure. The raising analysis, on the contrary, suggests a possible solution: the nonclitic definite determiner in (28) instantiates the external D° selecting CP; an enclitic definite article should be generated in a lower head Art°, which would be necessarily contained in the raised relative DP. The impossibility of the enclitic definite article then follows from the hypothesis that the relative "head" is not definite:

(29) * [$_{DP}$ *den* [$_{CP}$ [$_{XP}$ [$_{ArtP}$ *-en* [$_{NP}$ *grupp*]]] [$_{CP}$ *som* ...]]]

Another relevant phenomenon is the Spanish dummy preposition *a*. According to Brugé—Brugger (1996), the preposition is obligatorily in front of direct objects bearing the features [+animate] and [+accusative], and it realizes a Case-bearing functional head. The preposition can be omitted if an [+animate] direct object is marked with the Partitive Case, which yields an "existential" interpretation (cf. Belletti 1988). The existential interpretation is incompatible with specificity and definiteness.
From this perspective, it is interesting to note that the dummy preposition can be omitted in the relativization of an animate direct object:

(30) *el señor que María entrevistó ayer* (Toribio 1992: 292)
'the gentleman that Maria interviewed yesterday.'

This suggests that the raised "head" may receive an existential interpretation, i.e. it is non-definite.[26]
To conclude, it has been argued that the presence of a definite morpheme incorporated to the relative determiner does not (necessarily) imply that it is semantically definite. This conclusion will be further justified in

§ 5, where the morphology of relative determiners will be examined from a diachronic perspective.

4. Some issues in the syntax of relative determiners

The data discussed in this section raise substantial problem for the traditional adjunct analysis, and for the pronominal status of the relative morphemes, whereas they receive a more satisfactory account in the raising analysis.

4.1. Correlative clauses

The correlative structure constitutes a normal relativization strategy in various languages, including Mandingo (Bokamba—Dramé 1976), Hindi, Latin, Sanskrit and Old English (cf. Downing 1978: 399–405; Haudry 1973; Hock 1988; 1989; Keenan 1985; Srivastav 1988; 1991). The structure is exemplified by the following Latin example:

(31) $[_{CP}$ $[_{DP}$ *Quibus* $[_{NP}$ *diebus*$]]_i$ *Cumae liberatae sunt obsidione*],
which days$_{ABL}$ Cuma was released from the siege
[[*isdem diebus*]$_i$ *Ti. Sempronius prospere pugnat*]
the same days$_{ABL}$ T. S. wins a victory
'T.S. wins a victory in the days in which Cuma is released from the siege.'
(from Haudry 1973)

Descriptively, the structure is composed of two clauses, the main clause and a dependent clause which appears at the left or right margin of the main clause. The dependent clause and the main clause contain two constituent which are interpreted as coreferent:[27] they will be called the relative and the correlative element respectively. The relative element (*quibus diebus* in (31)) is usually fronted at the beginning of the dependent clause, and the correlative element (*isdem diebus*) may also be fronted in the main clause. The two elements are introduced by specific determiners;[28] their NP complement may be realised in both clauses, but it may also be deleted in either one of them. Assuming that the dependent clause is adjoined to the main clause, the schematic representation of the correlative structure is (32):

(32)

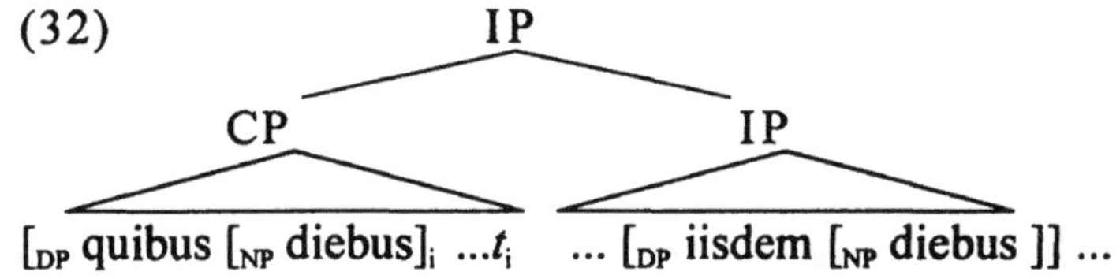

It is interesting to note that in some Indo-European languages both the correlative structure and the headed relative clause are possible, and strikingly, they exploit exactly the same relative morpheme. This is illustrated by the following pairs of examples. (33) is a Latin headed relative featuring the same morpheme *qui* that appears in the dependent clause of (31):

(33) *ex* [$_{DP}$ *iis* [$_{NP}$[$_{NP}$ *rebus*] [$_{CP}$ quas *gerebam*]]] *intellegebatis*
from those$_{ABL}$ things$_{ABL}$ which$_{ACC}$ (I) did (you) understood
'From the things I did you understood...' (Cic. *De Sen.* 22)

The same morphological coincidence is found in Hindi (34)–(35) and in Old English (36)–(37):

(34) [$_{CP}$ [$_{DP}$ jo [$_{NP}$ *larki*]]$_i$ *khari hai*] [*vo*$_i$ *lambi hai*].
which girl standing is, she tall is
(35) [$_{DP}$ *vo* [$_{NP}$[$_{NP}$ *larki*] [$_{CP}$ jo *khari hai*]] *lambi hai.*
that girl who standing is tall is
'The girl who is standing is tall' (Srivastav 1991)
(36) [$_{CP}$[$_{DP}$ ðone *stan*]$_i$ *ðe ða wyrntan awurpon*], [*ðes*$_i$ *is gewerdet*
the$_{ACC}$ stone$_{ACC}$ that the workers rejected, that$_{NOM}$ is become
*on ð*aere *hyrnan heafod*] (Harbert 1983: 552)
on the corner head
'The stone that the workers rejected has become the corner-stone.'
(37) *ure Drihten araerede* [$_{DP}$ *anes ealdormannes* [$_{NP}$[$_{NP}$ *dohtor*] [$_{CP}$ seo *ðe laeg dead*]]
our Lord raised an aldorman$_{GEN}$ daughter$_{ACC}$ who$_{NOM}$ that lay dead (Allen 1977: 87)
'Our Lord raised an aldorman's daughter who lay dead.'

From the perspective of the adjunct analysis, this coincidence is not predicted. Compare the schematic representation of the correlative structure in (32) to that of the adjunct relative clause in (38):

(38)

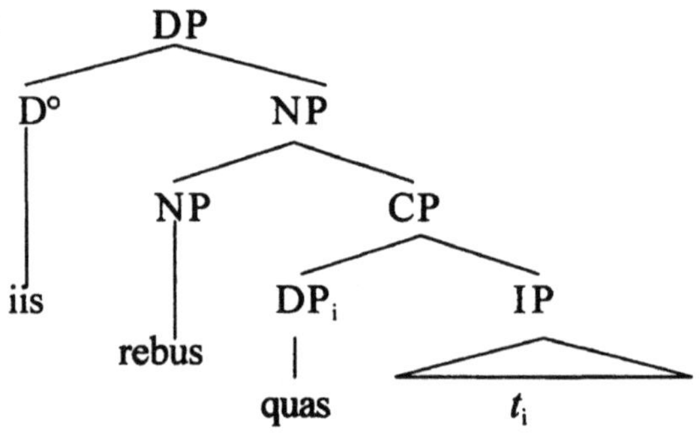

In (32) the relative morpheme is a determiner selecting the noun "head"; in (38), instead, it is an independent pronoun anaphoric to an antecedent NP.

From the perspective of the raising analysis, however, the same element is involved in the two structures. Compare again (32) to the raising relative structure (39):[29]

(39)

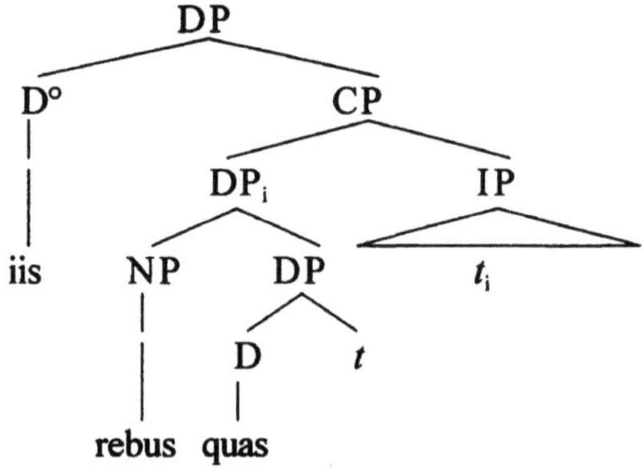

In both structures, the relative morpheme is a determiner selecting the NP "head". The relevant difference is that in the headed relative the "head" moves to the left of the relative D°, in order to enter a checking relation with the external D°, whereas in the correlative structure it stays in the complement position of the relative D°.

4.1.1.

Srivastav (1988; 1991) offers a detailed comparison between the correlative structure and the embedded relative clause of Hindi, exemplified in (34)–(35) above.[30]

Previous analyses tried to reduce the two structures to a single D-structure. Thus, in early matching analyses the correlative pair was derived by extraposing to the left a relative clause generated within the modified NP. The specular approach was to take both the correlative clause and the sentence-internal relative to be adjoined to the matrix clause; but this approach could not explain why the sentence-internal relative appears obligatorily adjacent to the "head".[31]

Srivastav argues that both approaches are flawed in the fundamental assumption that the two structures have a common underlying structure. She points out several systematic differences between the correlative clause and the embedded relative clause:

(1) the embedded relative clause allows internal recursion; namely, the relative clause can contain an NP itself modified by another relative clause. The correlative structure, instead, does not allow internal recursion;

(2) only in the correlative structure can the noun be realised in both members of the correlative pair; in the embedded relative, on the contrary, the "head" is not repeated after the relative determiner:

(40) a. *Jo larki khari hai vo larki lambi hai.*
which girl standing is, that girl tall is
'The girl who is standing is tall.'
b. * *Vo larki jo larki khari hai lambi hai.*
that girl which girl standing is tall is
(Srivastav 1988: 146–147)

(3) in the correlative structure, the correlative element in the main clause must be morphologically definite; it cannot be introduced by an indefinite determiner (e.g. 'two'), unless the latter introduces a partitive structure ('two of them');

(4) the dependent clause of the correlative structure can contain multiple relative NPs, each one resumed by an appropriate correlative element in the main clause, as in (41). The embedded relative, on the contrary, cannot have multiple "heads", as in English or Italian.

(41) *Jis larki-ne$_i$ jis larke-ko$_j$ dekha us-ne$_i$ us-ko$_j$ pasand kiya.*
which girl which boy saw she him liked

Therefore, Srivastav proposes two distinct analyses for the two structures. She analyses the correlative pair as an adjunction configuration, with the dependent clause adjoined to the main clause. The adjoined CP hosts in its Spec the noun phrase headed by the relative determiner *jo*.

The whole dependent CP is interpreted as a universally quantified phrase, and it binds the correlative DP in the main clause.[32] The obligatory definiteness of the correlative DP is determined by an intrinsic quantificational property of the dependent clause: since the latter contains a universal quantification, the set denoted by it is presupposed (cf. Enç 1991), and this is why it must be resumed by a definite DP in the main clause.

As for the headed relative, the author assumes the usual adjunct analysis, with the relative CP adjoined to the NP that it modifies; she then argues that the two analyses correctly determine the properties in (1), (2) and (4). However, it can be shown that the raising analysis of the headed relative clause also yields the correct results.

(1) Recursive embedding is impossible in the correlative structure:

(42) * $[_{IP0}$ $[_{CP1}$ $RelP_j$ $[_{IP1}$ $[_{CP2}$ $RelP_i$ $[_{IP2}$ ... $t_i]]$ $[_{IP1}$ ...DP_i... t_j ...]]] $[_{IP0}$ ... DP_j ...]]

In fact, in (42) the dependent clause CP_1 binds the correlative DP_j within the main clause IP_0; but CP_1 itself contains a correlative DP_i to be bound by a dependent CP_2 adjoined to IP_1. The configuration is excluded because the quantified clause CP_2 adjoined to IP_1 prevents the relative phrase raised to Spec,CP_1 ($RelP_j$) from locally A'-binding its variable t_j.

In the raising relative structure, instead, recursion is unproblematic, since every NP "head" is locally bound by the external D° selecting the relative CP in which the NP originates:

(43) $[_{DP1}$ D_1 $[_{CP1}$ NP_1 $[_{IP1}$ ... t_{NP1} ... $[_{DP2}$ D_2 $[_{CP2}$ NP_2 [... t_{NP2} ...]]]]]]

(2) As for the lexical realization of the noun head in (40a), note that in the correlative structure the relative DP jo larki and the correlative DP vo larki are generated independent of each other and they do not form a chain. In the raising structure (40b), on the contrary, the NP "head" moves from a position following the relative determiner jo to a position immediately adjacent to the external determiner vo:

(44) $[_{DP}$ *vo*$[_{CP}$ $[_{DP}$ *larki* $[_{DP}$ jo $t_{NP}]]$ $[_{IP}$ t_i *khari hai*]]] *lambi hai*
that girl who standing is tall is

Even assuming that the trace following the relative D° is a copy of the NP "head", it is c-commanded by the first link of the chain, and hence it is obligatorily deleted in the phonetic representation by whatever principle prevents the spellout of traces in a chain (cf. the discussion around (63) of chapter II).

(3) The possibility of multiple heads in the correlative structure (41) follows from the hypothesis that the two relative DPs can be interpreted by an operation of quantifier absorption (Srivastav 1991: 667). Without entering into the details of Srivastav's proposal, let us simply assume that the relative DPs are intrinsically non-definite, and they can be both bound by an implicit universal unselective quantifier, which yields a "pair" reading.[33] An unselective binder can bind multiple variables by definition. The raising structure, on the contrary, does not involve unselective binding: the external D° can license only one NP "head", and moreover, the Spec,CP of the relative clause is a unique position, and it can host at most one relative DP. Therefore, multiple "heads" are excluded.

Summarizing, the correlative structure and the internally headed relative differ both syntactically and semantically. From a syntactic viewpoint, in (32) the relative CP is adjoined to the main clause, whereas in (39) it is selected by the external determiner. Nevertheless, the two structures have one point in common: in both cases the CP contains a DP headed by a (non-definite) relative determiner, which takes an NP as its complement; this NP is interpreted as the "head" of the construction.[34]

4.2. From correlatives to externally headed relatives

In a study of Latin subordination, Haudry (1973) proposes a diachronic development in which the headed relative develops from an earlier correlative structure.[35] The starting point is the structure (45a), which Haudry calls the "normal correlative dyptic": the correlative clause precedes the main clause and the relative DP is fronted at the beginning of the former. The second step involves the inversion of the NP "head" and the relative determiner, as in (45b). Once this inversion has taken place, the NP "head" preceding the relative determiner can be reanalysed as a constituent of the main clause, that is, as an external "head" (45c). At this point, the correlative clause can be reanalysed as a modifier of the NP "head"; the string is no longer a subordinate adverbial clause, but a noun phrase embedding a relative clause, and the correlative pronoun in the main clause becomes superfluous.

(45) a. [[*qui vir*] ...] [...*is*...]
which man he

(45) b. [[*vir qui*] ...] [...*is*...]
man which he

c. [[*vir*] [*qui* ...]] ...(*is*) ...
man who (he)

Note that from the perspective of the adjunct analysis, this diachronic evolution requires a radical reanalysis: (i) the NP which was generated within the dependent CP of the correlative structure has to be reanalysed as a constituent of the main clause, after inversion has placed it in front of the relative determiner; (ii) the relative CP originally adjoined to the main clause must be reanalysed as adjoined to the antecedent NP.

With the raising analysis of headed relatives, this evolution receives a more straightforward account. Haudry's proposal (45) can be reconstructed in X-bar theoretical terms along the following lines. The first stage (45a) corresponds to the correlative structure (32): the relative DP is an internal constituent of the dependent CP (46a). The second stage (45b) may be derived by raising the NP "head" to the left of the relative determiner, as represented in (46b):

(46) a. $[_{CP}\ [_{DP}\ \textit{qui}\ [_{NP}\ \textit{vir}]]_i\ [_{CP} \ldots t_i]]$

b. $[_{CP}\ [_{DP}\ [_{NP}\ \textit{vir}]\ [_{DP}\ \textit{qui}\ t_{NP}]]_i\ [_{CP} \ldots t_i]]$

In the resulting structure, the NP "head" occupies the most prominent position within the CP. It suffices to introduce a nominal functional head above this CP to obtain the raising structure (47) (= 39). This external functional head selects the relative CP, turning it into a nominal phrase which can directly occupy an argument position within the main clause (whence the elimination of the correlative pronoun in the latter):

(47) $[_{DP}\ D^{\circ}\ [_{CP} [_{DP} [_{NP}\ \textit{vir}] [_{DP}\ \textit{qui}\ t_{NP}]]_i\ [_{CP} \ldots t_i]]]$

Thus, the crucial step in this diachronic evolution consists in the postulation of a nominal functional head selecting the relative CP. In Latin, this functional head is not always realized as an external determiner; however, it can be assumed to correspond to the Case declension of the noun.[36, 37]

4.3. Inverse attraction

The second step (46b) of the diachronic derivation may seem ad *ad hoc* stipulation; but interestingly, it is attested in a remarkable relative structure, the so called "inverse attraction". In this structure, the "head" of a relative clause bears the same Case as the relative determiner and not the Case corresponding to its syntactic function within the matrix clause

(48) *Urbem quam statuo vestra est*
city$_{ACC}$ which$_{ACC}$ I found yours is
'The city that I found is yours.'
NOM ⇒ ACC (Verg. *Aen.* I, 573)

(49) *Pulli qui nascentur, eos in terra deprimito*
sprouts$_{NOM}$ which$_{NOM}$ germinate, them$_{ACC}$ in the earth (you must) layer
'You must layer in the earth the sprouts that germinate.'
ACC ⇒ NOM (Cato, *Agr.* 51)

(50) *Hunc chlamydatum quem vides, ei Mars iratust*
this$_{ACC}$ soldier$_{ACC}$ whom$_{ACC}$ (you) see, him$_{DAT}$ Mars angry is
'The soldier whom you see, Mars is angry at him'
DAT ⇒ ACC (Plautus, *Poen.* 644)

(51) *Den schilt den er vür bôt der wart schiere zeslagen*
the$_{ACC}$ shield$_{ACC}$ which$_{ACC}$ he held, that$_{NOM}$ was quickly shattered
'The shield that he held was quickly shattered
NOM ⇒ ACC (Old High German; Pittner 1995: 200)

From the perspective of the adjunct analysis, this structure is utterly mysterious (cf. Harbert 1983): in fact, it is unclear how the "head" generated outside the relative CP could come to agree for Case with the relative pronoun in Spec,CP. In the proposed diachronic development, instead, the "inverse attraction" corresponds to the intermediate stage (46b). The structure is still correlative, i.e. the relative CP is not yet introduced by an external D° (cf. the presence of a correlative pronoun in the matrix clause of (49)–(51)); the NP "head" originates in the complement position of the relative D° and shares its Case, though it eventually moves to the left of it.[38]

According to this analysis, the demonstrative morpheme introducing the NP "head" in (50) and (51) is not an external D°: in fact, it bears the Case assigned to the relative DP within the dependent clause. It must therefore be analysed as an internal constituent of the relative DP: [$_{DP}$ *quem* [$_{XP}$ *hunc* [$_{NP}$ *chlamydatum*]]], with leftward movement of XP to the Spec of the relative DP:[39]

(52) [$_{CP}$ [$_{DP}$ [$_{XP}$ *hunc chlamydatum*] [$_{DP}$ *quem* t$_{XP}$]] [$_{CP}$ *vides* t$_{DP}$]]

However, the demonstrative element could have been the "ancestor" of the external determiner: in cases of coincidence of the internal Case assigned to the relative DP with the external Case, its presence in front of the NP "head" could be reinterpreted as evidence for external functional structure introducing the relative clause.

4.4. Case attraction

At this point, it is necessary to explain how the NP "head", which is originally the complement of the relative D°, may come to acquire the Case assigned to the complex DP within the matrix clause, possibly disagreeing with the relative D°, as in (33) above, repeated here:

(53) a. *ex iis rebus quas gerebam*
b. $[_{PP}$ P° $[_{DP}$ $D°_{ABL}$ $[_{CP}$ NP_{ABL} $[D°_{ACC}$ $t_{NP}]$ ...]]]

The solution to this problem rests on a particular conception of morphological Case.

Let us assume that the terminal symbol dominated by a functional head is not a morpheme, but a set of syntactic features. This abstract element is turned into a concrete morpheme in the Morpho-Phonological component (cf. Halle—Marantz 1993).[40] Furthermore, let us assume - following Giusti (1993) - that being Case-marked is a syntactic property of the D° position. The noun (and adjectives) morphologically agrees for Case with the determiner by which it is governed:[41] in other terms, the Case feature is copied into the Agr head which the noun incorporates to; as a result, the noun is spelled out in PF in the form inflected for the corresponding Case.

Consider from this perspective the configuration in (53b). The external D° bears the Ablative Case assigned to it by the preposition, and the relative D° bears the Accusative Case assigned to the relative DP within the relative clause. As discussed in § 2.3, the NP "head" comes to be governed by the external D° by the end of the derivation. Therefore, in the Morpho-Phonological component the Case feature of this D° is copied into it, and it is spelled out in the Ablative form.

Interesting evidence in favour of this hypothesis comes from the phenomenon of Case attraction of the relative determiner:

(54) *notante iudice* quo *nosti*
judging$_{ABL}$ the judge$_{ABL}$ who$_{ABL}$ (you) know
'judging the judge whom you know'
ACC ⇒ ABL (Hor., *Sat* 1,6,15)

(55) ἄνδρες ἄξιοι τῆς ἐλευθερίας ἧς κέκτησθε
men worthy the$_{GEN}$ freedom$_{GEN}$ which$_{GEN}$ (you) possess
'men worthy of the freedom that you enjoy'
ACC ⇒ GEN (Ancient Greek; Harbert 1982: 245)

(56) *syððan hie gefricgeað frean userne ealdorleasne, ðone ðe aer*
when they learn lord$_{ACC}$ our to be dead, who$_{ACC}$ that in the past
geheold wið hettendum hord ond rice...
guarded against the enemies treasure and kingdom
'when they learn that our lord, who in the past defended the treasure and the kingdom from the enemies, is dead...'
NOM ⇒ ACC (Old English; Mitchell—Robinson 1982: 77)

(57) *sie gedâht' ouch maniger leide, der ir dâ héimé geschach.*
she thought also some sufferings$_{GEN}$ which$_{GEN}$ her at home happened
'She thought of some pain that she suffered at home.'
NOM ⇒ GEN (Old High German; Pittner 1995: 198)

Recall that in the proposed raising structure the NP "head" in the Spec of the relative D° is attracted to the Case of the external D° under government:

(58)
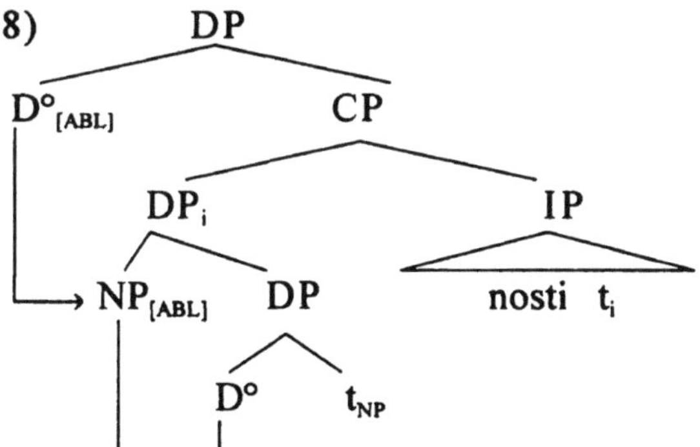

In this configuration, the relative D° too is governed by the external D°. If the relative DP is assigned a structural Case within the relative clause (e.g. the Accusative in (58)), this Case is checked off within IP before the relative DP reaches Spec,CP. Suppose that the checked Case feature is optionally erased (in the sense of Chomsky 1995: 279–282); then the D° can receive another Case feature from the external D° under government.[42]

(59)
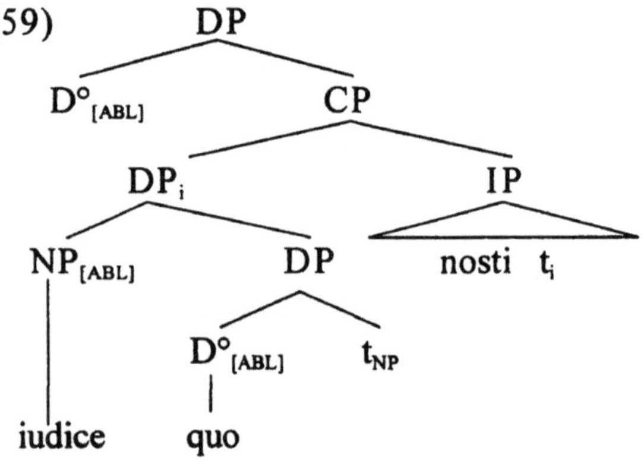

Thus, a relative D° which is assigned a structural Case can be attracted to the Case of the external D° in virtue of a local syntactic relation.

It is not plausible to attribute the same structure to examples like (53) above, where the relative D° is not attracted although it bears a structural Accusative case. The logic of the argument leads to the conclusion that in this case the relative D° is not in a local syntactic relation with the external D° (or with the NP "head" either). Let us assume that the relative determiner can be stranded in a specifier position below Spec,CP, as represented in (60).

(60)
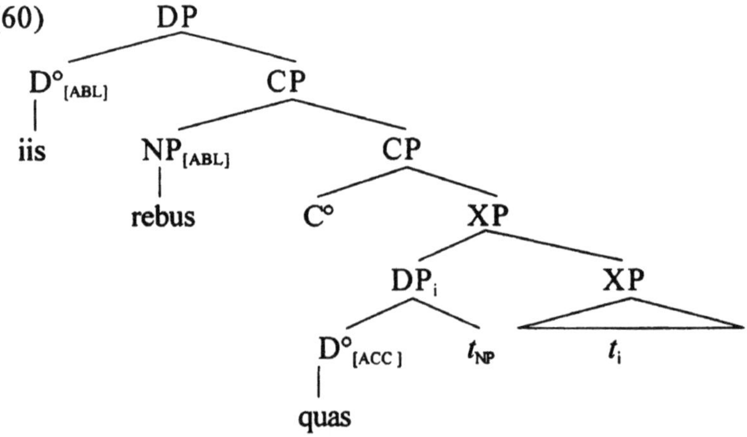

4.5. Postposition of the relative determiner

The proposal in (60) may look implausible; but interestingly, there is evidence to the effect that (in Latin at least) the NP "head" and the relative D° may occupy two distinct specifier positions. This evidence consists in the phenomenon of "postposition" of the relative determiner, exemplified in (61).

(61) *meus vicinus,* [*meo viro*] *qui liberum praehibet locum ...*
(Plautus, *Casina* 536)
my neighbour$_{NOM}$ my husband$_{DAT}$ who$_{NOM}$ offers a free place
'my neighbour, who offers a free place to my husband...'

Here a topicalized dative argument is placed in between the NP "head" and the relative D°; this clearly indicates that the two categories are not in a specifier-head relationship.

Incidentally, note that the example (61) is problematic from the perspective of the adjunct analysis, according to which the relative pronoun is generated independently of the NP "head" and must reach a position close to it in order to be coindexed with it (cf. § II.1). On the contrary, the postposition of the relative determiner is not problematic for the raising analysis: as discussed in § 2.3, it is the NP "head" that must raise to a position locally related to the external D° in order to check the nominal features of the latter; the relative D° is not involved in this checking relation, hence it can be stranded in a lower position.[43]

For the moment, the nature of the XP projection in (60) will not be determined. This problem is deferred to chapter VI, where the distribution of relative pronouns and its interaction with the "complementizer system" will be systematically examined.

4.6. Concluding remarks

The evidence reviewed in this section suggests that the standard analysis of relative morphemes as trace-binding pronouns is not adequate: in particular, it fails to establish a proper relation between the correlative structure and the headed relative clause, and it cannot satisfactorily account for some "unusual" but well attested phenomena such as Case attraction, inverse attraction and postposition.[44] As it can subsume these data, the raising analysis allows for a more unified approach to the various cross-linguistic relativization strategies.

5. The typology of relative determiners in the Indo-European languages

The raising analysis of restrictive relatives implies a nonstandard conception of relative pronouns as stranded determiners of the NP "head";

moreover, it was proposed in § 3 that relative determiners are intrinsically non-definite. These hypotheses seem to be contradicted by the observation that in many Indo-European languages these elements appear to be pronominal and definite.

The issue of the morphological shape of relative "pronouns" and of its diachronic origin has been paid little attention in the generative literature: from a purely synchronic syntactic perspective, it is not obvious that the origin of a given functional element may reveal anything about its present status in the grammar. However, this issue has been extensively discussed by scholars who inquire into the syntax of the ancient Indo-European languages, as a way to shed light on the diachronic development of the syntactic structure of relative clauses.

The main finding of this line of research is that the ancestors of the modern relative "pronouns" were functional elements with a wide distribution in a variety of structures besides relative clauses. The analysis of all these structures leads to the identification of the "original function" of the ancestor elements, of which the relative function constitutes a modification or an extension (cf. Gonda 1954: 2).

This section will summarize some papers representative of this line of research. In the final subsection, the three main types of relative "pronoun" in the Indo-European languages will be examined. Two of them are morphologically definite: the diachronic perspective will support the idea that this definite morphology is actually expletive.

5.1. The correlative origin of relative pronouns

As already mentioned in § 4.2, Haudry (1973) argues against a *communis opinio* about the syntax of proto-Indo-European languages, whereby the subordinate structures of the historically attested dialects developed from the grammaticalization of paratactic structures. Against the myth of the primitive parataxis, Haudry proposes that subordination originated from the correlative structure, a structure "à mi-chemin entre la parataxe et l'hypotaxe". The basic form of this structure is the normal dyptic, consisting of a dependent clause introduced by a relative element and a main clause containing a correlative/resumptive element. The relative function of the correlative dyptic was discussed in § 4.2; however, the dyptic also expressed other subordinate functions, e.g. a temporal clause, as in the following Latin and Old English examples:

(62) *tum denique... nostra intellegimus bona, quom... ea amisimus*
then really our happiness (we) understand, when it (we) lose
'We realize our happiness when we lose it.' (Haudry 1973:159)

(63) *ða* Neron *ða* ðaet geseah, *ða* wende he ðaet hit Godes sunu waere.[45] (Mitchell 1985: § 2576)
then N. then that saw, then thought he that it God's son was
'When Nero saw that, then he thought that it was God's son.'

Subordination developed by embedding the dependent clause into the main clause and eliminating the correlative element in the latter.

Haudry examines the morphology of the relative element in various types of correlative dyptic. In Hittite, Latin and in the ancient Baltic languages the relative element of the dependent clause derived from the stem **kwo-*, which also gave indefinite and interrogative pronouns. Haudry argues convincingly that in the correlative structure the derivative of **kwo-* was an indefinite determiner, and not an interrogative one. In fact, the relative element in the dependent clause constitutes new information, hence it is indefinite, whereas the correlative element in the main clause is resumptive, hence it is anaphoric and definite (derived from the demonstrative theme **to-*). In other terms, the relative use of the derivatives of * *kwo-* in the correlative structure represents a specialized use of the indefinite (Haudry 1973: 166–168; Gonda 1954: 272–273).

In Vedic, relative determiners derived from the theme * *yo-*: there too, the relative function developed from the attested use in the correlative structure, in the way discussed in § 4.2. The descendants of * *yo-* are attested in the Hindi examples (34)–(35), but apart from the Indian branch, they were lost in all the rest of the Indo-European domain (Gonda 1955: 275).

Finally, Haudry considers the purported relative pronouns omophonous to the definite article, derived from the demonstrative theme **to-*, in Homeric Greek. He argues that these structures have been misinterpreted by ancient and modern grammarians: the purported relative clause introduced by a definite article is actually a parenthetic clause containing an anaphoric demonstrative pronoun.

Relative pronouns derived from a demontrastive theme are also found in the Old Germanic languages (e.g. Old English *se, seo, ðaet*), cf. examples (36)–(37) above. Haudry (1973: 169–173) proposes that the relative function of these elements developed under the literary influence of the Greek and Latin model of the Scriptures, where relative pronouns were used as intersentential anaphoric pronouns, in the so called *relatif de liaison*. A demonstrative pronoun was adopted in Old Germanic written languages to translate the *relatif de liaison*, and then it was generalized to all the uses of relative pronouns, including the true embedded relative clauses.[46] This explains the unexpected demonstrative morphology of the relative determiners.

To conclude, Haudry considers the correlative dyptic **kwo- ...*to-...* the prototypical one: the correlative structure consists of an indefinite noun phrase in the dependent clause resumed by an anaphoric element in the main clause. However, independently of their morphological shape, relative pronouns developed their relative function from an original correlative structure.

5.2. Relative pronouns as articles

Benveniste (1966: 208–222) opens his discussion with a methodological warning: there exists no universal relative structure, but there is a universal relative function which can be structurally implemented in various ways. His aim is to identify the nature of the relative function.

Benveniste claims that the relative function is that of an "adjectif syntaxique déterminé", and accordingly, he conceives of the relative pronoun as an "article syntaxique déterminatif". The common property of relative pronouns is their ability to convert a verbal phrase into a nominal modifier: in this, they resemble the articles introducing adjectival and PP modifiers in various languages, such as Ancient and Modern Greek (Adams 1972), Hebrew (Siloni 1991), Arabic (Fassi Fehri 1976), Rumanian (Cornilescu 1993), and Albanian (Morgan 1972; cf. § 3.1).[47] A particularly suggestive case is that of Arabic, where both adjectives and relative clauses have an indefinite and a definite form according to the definiteness of the modified noun phrase. In indefinite noun phrases, adjectives and relative clauses are not introduced by any specific functional element; in definite noun phrases, on the contrary, adjectives are introduced by a definite article, and relative clauses are introduced by the inflected element *allaðii*, composed of the definite article *al*, a demonstrative morpheme, and gender/number agreement morphology.[48] This analogy is attested by the following data (Fassi Fehri 1976: 134–138):

(64) a. *waladun kari:mun*
child generous
'a generous child'

b. *l-waladu l-kari:mu*
the child the generous
'the generous child'

c. *rajulun ʔakala*
man has eaten
'a man who has eaten'

d. *l-rajulu llaði ʔakala*
the man *llaði* has eaten
'the man who has eaten'

Turning to the Ancient Indo-European languages, Benveniste examines the derivatives of the pronominal themes **yo-*, **to-*, **kwo-*. These elements are shown to have a wide distribution in languages like Ancient Greek, Sanskrit, Latin and Hittite: besides introducing relative clauses, they could nominalize adjectives, and they also introduced various nominal modifiers, like adjectives, PPs or DPs in apposition to a noun phrase. In the earlier stages, these nominal modifiers were in the Nominative case, independently of the Case of the modified noun: e.g. Ζῆνα, ὅς τίς τε θεῶν ἄριστος 'Zeus$_{ACC}$, which$_{NOM}$ of the gods the greatest$_{NOM}$'; but they were subsequently attracted by the Case of the latter. According to Benveniste, this use of the relative pronouns is not a secondary development due to the deletion of the copula in the relative clause predicate; rather, it constitutes their primary function, and the relative function is only a special case of it. Therefore, the ancient pronouns were articles introducing a nominal determination.[49]

In a study of Ancient Greek relative clauses, Brunel (1977) further develops Benveniste's analysis. He argues that in Ancient Greek, the pronoun ὅς- ἥ- ὅ, derived from the theme * *yo-*, substantivized the nominal modifiers that it introduced. The resulting phrase allowed both an absolute use and an appositive one; but even in the appositive use, the phrase was syntactically self-sufficient, being independent of the Case of the modified noun phrase. Brunel concludes that these element were originally not articles, but true pronouns, i.e. nominalizers of non-nominal categories. Note that the distinction is obscured in many languages, where the same morpheme can function both as a definite article and as a definite pronoun nominalizing various modifiers: cf. Spanish *el retrato de Rembrandt* 'the portrait of Rembrandt' but also *el interesante* 'the interesting one', *el de Madrid* 'the one from Madrid', *el que tienes en la mano* 'the one that/what you have in your hand' (Schroten 1992).

Thus, according to Brunel (1977: 216), the phrases introduced by the pronoun ὅς- ἥ- ὅ were originally appositive. The restrictive use originated by the gradual weakening of the pronominal force of the pronoun, whereby the originally appositive phrase lost its syntactic independence and was attracted by the Case of the noun phrase that it modified; in the end, having lost its syntactic independence, the phrase allowed a restrictive interpretation. In other terms, the derivative of * *yo-* was originally a pronoun and gradually lost its pronominal force, developing into an adjectival article and a relative determiner.[50, 51]

5.3. The cycle of the definite article

Summarizing, two distinct proposals emerge for the development of relative clauses in the ancient Indo-European languages.

According to Haudry, the relative function developed from a complex *sentential* structure, the correlative one: a dependent clause originally related to the main clause by an anaphoric relation was reanalysed as an embedded modifier of a nominal antecedent (cf. (45) above). The first element of the correlation constituted new information, whence the indefinite morphology of the relative determiner.

According to Benveniste and Brunel, instead, the relative function develops from a complex *nominal* structure, in which a pronoun originally introduced an appositive modifier, but gradually lost its syntactic and referential self-sufficiency, and in the end introduced restrictive modifiers.

The diachronic development proposed by Brunel involves a gradual weakening of a pronoun, which loses its pronominal force and becomes a functional particle introducing restrictive modifiers.

This recalls the so called "cycle of the definite article" proposed by Greenberg (1978): a three stages process which turns an originally deictic pronoun into a pure gender marker.

In the initial stage, the pronoun is a demonstrative endowed with deictic force. Then it loses its deictic value and becomes simply anaphoric, identifying a referent as already mentioned in the previous discourse. This "familiarity" entails definiteness (cf. Heim 1982): this corresponds to the first stage, that of a definite article.

The successive stage involves the loss of definiteness: the article introduces noun phrases with either definite or indefinite specific reference. This is what Greenberg calls a "nongeneric article".

In the final stage, the element becomes a pure nominal marker: the former article occurs obligatorily in any noun phrase, and it is often incorporated to the noun. If in the initial stage the article was inflected for grammatical features, it evolves into a gender marker; if it was uninflected, it becomes a pure marker of nominality. Greenberg exemplifies this process with comparative data from various non-Indo-European languages. In the Indo-European family, the first step of the cycle is exemplified by the Latin demonstrative *ille* becoming the Romance definite article, and by the Old English demonstrative *se, seo, ðaet* becoming the definite article *the* (Traugott 1972: 85–87).

Greenberg's cycle involves a semantic weakening: a deictic element becomes anaphoric, then simply nongeneric, and finally it loses any referential content and becomes a pure agreement marker. However, parallel to this semantic weakening, there is also a syntactic weakening: a pronoun or determiner, which turns its nominal complement into a referential category (cf. (10) of chapter II), becomes an agreement morpheme, whose maximal projection is still predicative and not referential.

Note that Brunel's proposal about the development of the Ancient Greek relative pronoun may be conceived as an instance of this cycle. The varied uses of this morpheme correspond to different stages of this cycle: the nominalizing use corresponds to an earlier stage, and the restrictive relative function corresponds to a later stage in which the element has lost its pronominal force.

5.4. Three types of relative determiner

These remarks suggest a reconsideration of the relative determiners of the modern Indo-European languages. From a morphological viewpoint, these can be grouped in three main types. The first type has an indefinite and/or interrogative theme: e.g. Middle and New English *who* and *which*, Latin *qui-quae-quod*, Rumanian *care,* French *qui* and *quoi*, Italian *cui*, Swedish *vilke*. The second type corresponds to a definite determiner, which is independent of the definiteness of the external determiner: cf. Old English *se-seo-ðaet* and German *der-die-das*.[52] The third type combines the first two: it consists of an interrogative element preceded by a definite article. This type is exemplified by Italian *il quale*, French *lequel*, Spanish *el cual*, Middle English *the which*, Bulgarian *koj-to*, Old Georgian *romel-igi*, Hungarian *amit*.

Recall that the analysis of restrictive relatives proposed by Reinhart (1987) and adopted in § 3 above implies that the relative determiner of restrictive clauses must be nondefinite. The second and third type of relative pronoun are problematic in this respect. However, in the light of the preceding discussion it is possible to assume that the definiteness of these determiners is only apparent: the definite article used as (or incorporated in) the relative determiners actually is in a late stage of Greenberg's cycle, and hence it has lost the feature of definiteness.

This hypothesis is particularly appealing for the third type of relative pronoun. There is clear diachronic evidence that the type *il quale* represents an evolution of an original relative determiner *quale*. This is attested by medieval Spanish *qual* (Rivero 1986) and ancient Italian *quale* (Ageno 1956: 4).[53]

In a diachronic study of French *lequel*, Kunstmann (1991) argues that the definite article was introduced in the relative/interrogative paradigm of *qualis* in order to express the grammatical feature of gender, which had been lost in this paradigm and replaced by an opposition of natural gender (animate versus inanimate). The paradigm of the definite article still preserved the grammatical gender: this is why it was introduced in the relative/interrogative paradigm of the written language under the influence of the original Latin paradigm, which expressed the grammatical gender; later on, it was also adopted by the spoken language.

This diachronic reconstruction suggests that the definite article was introduced before the *wh*-word simply as an agreement marker for the feature of gender. From a semantic viewpoint, it can be considered an expletive: it corresponds to the last stage of Greenberg's cycle.[54]

Summarizing, the morphology of the Indo-European relative determiners can be conceived as the result of three distinct developments. The indefinite/interrogative determiners probably originated in a correlative structure that evolved into an embedded relative clause. The demonstrative determiners may have developed a relative function by a gradual weakening of their referential strenght and syntactic independence along Greenberg's (1978) cycle of the definite article. The mixed type *il quale* seems to have developed from an original indefinite form by the insertion of a definite article with the function of an agreement marker. Therefore, the definite morphology of the second and third type of relative pronoun may be regarded as semantically inert.

6. Summary and open questions

This chapter discussed the status of relative pronouns in the raising analysis of relative clauses. According to Kayne (1994: chapter 8), they must be analysed as the original determiners of the raised "head". It was argued that this proposal does not imply a violation of the Full interpretation principle (Chomsky 1991), given the hypothesis that the relative D° is non-definite and does not saturate the open position of the NP "head".

The proposal has two particularly innovative aspects:

(1) the NP "head" and the relative D° are originally in a complementation relation;

(2) there exists a morphosyntactic interaction between the external D° and the constituent in Spec,CP: Spec,CP is "transparent" (cf. Harbert 1982).

The new analysis allows a more unified approach to two relativization structures - the correlative clause and the headed relative - which were previously considered structurally unrelated, though diachronically related. Furthermore, it opens a new perspective on some "marginal" but problematic aspects of the syntax of relative pronouns.

One problem which was left open in the preceding discussion is the distribution of relative determiners. In English, these can be omitted in some relative structures:

(65) $[_{DP}$ the $[_{CP}$ book$_i$ $[_{CP}$ that I read $t_i]]]$

It is possible to assume that here the raised "head" is a simple NP lacking a relative determiner (Kayne 1994). However, it is in principle possible that the "head" be introduced by a null (or deleted) relative determiner. The analysis of this phenomenon is deferred to chapter VI. The next two chapters focus on the syntactic differences between restrictive and appositive relatives.

Chapter IV
A reconstruction asymmetry

0. Introduction

The preceding discussion has concentrated exclusively on restrictive relative clauses, arguing in favour of the raising analysis proposed by Kayne (1994). At this point, it is necessary to determine whether this analysis can be extended to appositive relatives as well.

A useful criterion is provided by reconstruction effects. As discussed in § II.4.1, in Chomsky's (1993) analysis a syntactic category can only be reconstructed in a position that it has filled at some stage of the derivation. The raising analysis defended in chapter II for restrictive relatives predicts that these should allow the reconstruction of the "head" in the trace position. This criterion can be also applied to appositive relatives.

For reasons that will become clear in the following paragraphs, the argument will be based on Italian data. The possibility of reconstruction will be tested with respect to Principle C, Principle A, scope assignment and quantifier binding. It will be shown that the reconstruction of the relative "head" takes place in restrictive relatives, but not in appositive ones. This raises the problem of how to differentiate the syntax of the two types of relative clause.

Another exceptional property of relative clauses, discussed by Lebeaux (1991), is that when they modify an A' moved phrase, they seem not to be reconstructed along with the latter. Lebeaux proposes a generalized transformation approach crucially based on the adjunct analysis. His empirical generalization, as well as his specific proposal, will be critically reexamined in § 8.

1. Reconstruction as a test for the raising of the "head"

The theory of reconstruction proposed by Chomsky (1993), summarized in § II.4.1, constitutes an important test for the raising analysis of relative clauses. Reconstruction is conceived of as a consequence of the creation of operator-variable pairs in LF: this process consists in the selective deletion of the links of a movement chain, which are copies of the moved category. This implies that reconstruction is only possible in A' structures in which a category has moved from its base position.

As Kayne (1994: fn. 68 to chapter 8) emphasizes, the status of relative operators in this approach to reconstruction is unclear. It is usually assumed that in LF the relative operator inherits the feature specification of its antecedent; but it is unclear whether its traces become "copies" of the antecedent as well. The null hypothesis is that they do not.

This hypothesis is confirmed by the behaviour of parasitic gaps. As discussed by Brody (1995: 91–92) and Munn (1992: 9–13), in a parasitic gap construction it is possible to reconstruct the A' binder in the primary gap position, but not in the parasitic gap position, which is bound by a null operator:

(1) a. [*Which book about himself*$_i$] *did John*$_i$ *file* t [Op *before Mary*$_j$ *read* e]?
 b. * [*Which book about herself*$_j$] *did John*$_i$ *file* t [Op *before Mary*$_j$ *read* e]?

Let us assume the null hypothesis. Then, the reconstruction of a category appearing in an A' position is a diagnostic criterion for the direct movement of that category. Concerning relative clauses, the possibility of reconstructing the "head" within the relative CP becomes an argument for the raising analysis and against the adjunct analysis, which necessarily involves overt or null relative operators.

In fact, reconstruction was already invoked as an argument in favour of the raising analysis by Schachter (1973: 32–34). He noted that pronouns and R-expressions embedded within the "head" behave as if they were in the trace position with respect to binding phenomena:

(2) a. [*The portrait of himself*$_i$ *that John*$_i$ *painted* t] *is extremely flattering.*
 b. [*The interest in each other*$_i$ *that John and Mary*$_i$ *showed* t] *was fleeting.*
 c. [*The opinion of him*$_i$ *that John*$_i$ *thinks that Mary has* t] *is unfavourable.*
 d. * [*The opinion of John*$_i$ *that he*$_i$ *thinks that Mary has* t] *is unfavourable.*
 e. * [*The portrait of John*$_i$ *that he*$_i$ *painted* t] *is extremely flattering.*

In (2a,b) the anaphor embedded in the "head" is bound by the subject of the relative clause: this is only possible if the "head" is reconstructed in the trace position, which is c-commanded by the subject. These data show that the reconstruction of the "head" is possible.

On the other hand, the ungrammaticality of (2d,e) can be reduced to Principle C of the binding theory, if one assumes that the "head" of the relative clause, containing the R-expression *John*, is reconstructed in the trace position, which is c-commanded by the coindexed subject pronoun. This leads to the stronger conclusion that the reconstruction of the "head" is obligatory.[1]

Note that the grammaticality of (2c) is consistent with this conclusion. After reconstruction, the pronoun *him* embedded in the relative "head" is c-commanded by the subject of the relative clause, but the latter is external to the governing category of the pronoun, which is the NP "head" itself. Thus, reconstruction does not lead to a violation of Principle B.

In the following sections, the behaviour of restrictive and appositives relatives vith respect to various reconstruction effects will be systematically compared.

2. Principle C effects

The standard version of Principle C of the binding theory states that an R-expression cannot be A-bound. This principle disallows the reconstruction of an R-expression in a position c-commanded by a coindexed element placed in an A position.

As discussed in § II.4.1, when an A' moved phrase includes an operator and its restrictive term, there are two possible LF derivations. The operator must remain in the A' position in LF; on the other hand, the restrictive term can either be reconstructed in the base position, giving rise to an unrestricted quantification, or it can fail to be reconstructed, yielding a restricted quantification. The LF derivation involves two steps: first, an operation akin to Quantifier Raising separates the operator (including or not the restrictive term) from the material that it has pied piped. Then the remnant constituent is deleted in all the positions of the chain except the lowest one, and vice versa, all the occurrences of the operator except the highest one are deleted (Chomsky 1993: 35–36). The two options are illustrated in (3) and (4):

(3) a. (*I wonder*) [$_{PP}$ *in which house*] *John lived* [$_{PP}$ *in which house*]
b. [*which*] [$_{PP}$ *in* e *house*] *John lived* [*which*] [$_{PP}$ *in* e *house*]
c. [*which*] *John lived* [$_{PP}$ *in* e *house*]
(unrestricted quantification)

(4) a. (*I wonder*) [$_{PP}$ *in which house*] *John lived* [$_{PP}$ *in which house*]
b. [*which house*] [$_{PP}$ *in* e] *John lived* [*which house*] [$_{PP}$ *in* e]
c. [*which house*] *John lived* [$_{PP}$ *in* e] (restricted quantification)

Consider now restrictive relatives. It was argued in § III.3 that the relative determiner is not an operator, but a non-definite determiner: the relative DP provides a free variable to be bound by the external determiner. The relative DP cannot create a restricted quantification structure like (4c) in LF:[2] then, the only possibility is to reconstruct it in the base position. In other terms, since the relative DP cannot be interpreted as a restricted quantifier in Spec,CP, it obligatorily undergoes reconstruction.

This prediction is correct, as shown by the Principle C effects in (2d,e) and in (5) below: an R-expression embedded in a relativized object cannot corefer with the null subject of the restrictive relative clause:

(5) a. * *Quello è l'*[*amico di Gianni*$_i$] *a cui* pro$_i$ *ha offerto un lavoro* t.
this is the friend of Gianni's to whom (he) offered a job
b. * *Questo è il* [*quadro di Gianni*$_i$] *di cui* pro$_i$ *è più orgoglioso* t.
this is the picture by Gianni of which (he) is most proud
c. * *Quello è l'*[*amico di Gianni*$_i$] *a cui* pro$_i$ *ha parlato* t *dei suoi problemi.*
this is the friend of Gianni's to whom (he) spoke about his problems
d. * *La* [*recensione del libro di Gianni*$_i$] *che* pro$_i$ *ha letto* t *sul giornale era negativa.*
the review of Gianni's book that (he) read in the newspaper was negative
e. * *Il* [*libro di Gianni*$_i$] *che* pro$_i$ *ha dedicato ai suoi figli è bellissimo.*
the book by G that (he) dedicated to his children is beautiful

A similar violation of Principle C arises if the R-expression is coindexed with a clitic pronoun that c-commands the relative trace:

(6) a. * *Questo è il* [*quadro di Gianni*$_i$] *che gli*$_i$ *piace* t *di più.*[3]
this is the picture of Gianni's that to-him$_{CL}$ pleases most
b. ?* *Quello è il* [*collega di Gianni*$_i$] *con cui lo*$_i$ *ho riappacificato* t.
this is the colleague of G's with whom I him$_{CL}$ reconciled

On the other hand, the same configurations do not give rise to a violation of Principle C if the relative clause is appositive, as shown in (7):

(7) a. *Il libro di Gianni*$_i$, *che* pro$_i$ *ha sicuramente dedicato* t *ai suoi figli, parla di psicologia infantile.*
the book by Gianni, which (he) surely dedicated to his children, is about children psychology.
'Gianni's book, which he surely dedicated to his children, is about children psychology.'

b. *Quell'amico di Gianni*$_i$, *a cui* pro$_i$ *ha sicuramente parlato* t *dei suoi problemi, è un celebre avvocato.*
that friend of Gianni's, to whom (he) spoke about his problems, is a famous lawyer
'That friend of Gianni's, to whom he surely spoke about his problem, is a famous lawyer.'

c. ? *La recensione del libro di Gianni*$_i$, che pro$_i$ *si è rifiutato di leggere t, era negativa.*
the review of Gianni's book, which (he) refused to read, was negative.
'The review of Gianni's book, which he refused to read, was negative.'

d. *Questi pettegolezzi su Gianni*$_i$, *che* pro$_i$ *probabilmente ha sentito t, devono averlo ferito profondamente.*
these gossips about Gianni, which (he) probably heard, must have hurt his feelings.
'These gossips about Gianni, which he probably heard, must have hurt his feelings.'

e. *L'ultimo quadro di Gianni*$_i$, *che gli*$_i$ *piace* t *moltissimo, non sarà messo in vendita.*
the last picture by Gianni, which to him$_{CL}$ pleases very much, will not be put on sale
'Gianni's last picture, which he likes very much, will not be put on sale.'

f. *Quel collega di Gianni*$_i$, *con cui finalmente lo*$_i$ *ho-riappacificato* t, *è una brava persona.*
that colleague of Gianni's, with whom finally (I) him$_{CL}$ reconciled, is a decent fellow
'That colleague of Gianni's, with whom I have reconciled him at last, is a decent fellow.'

The reconstruction effects in the restrictive relatives (5)–(6) constitute a further argument for the raising analysis. On the other hand, the lack of reconstruction effects in (7) shows that the same analysis is not adequate for appositives.

2.1. The exceptional behaviour of tonic pronouns

The Principle C effects in the restrictive relatives (5)–(6) are considerably weakened if the null subject and the clitic object are replaced by tonic pronouns:

(8) a. *Quello è l'[amico di Gianni$_i$] a cui lui$_i$ ha offerto un lavoro* t
'This is the friend of Gianni's to whom he offered a job.'
b. *Questo è il [quadro di Gianni$_i$] di cui lui$_i$ è più orgoglioso* t.
'This is the picture by Gianni which he is most proud of.'
c. ? *Quello è l'[amico di Gianni $_i$] a cui lui$_i$ ha parlato* t *dei suoi problemi.*
'This is the friend of Gianni's to whom he spoke about his problems.'
d. ? *La [recensione del libro di Gianni$_i$] che lui$_i$ ha letto* t *sul giornale era negativa.*
'The review of Gianni's book that he read in the newspaper was negative.'
e. *Il [libro di Gianni$_i$] che lui stesso$_i$ ha regalato* t *ai suoi$_i$ figli è bellissimo.*
'The book by Gianni that he himself gave to his children is very interesting.'
f. *Questo è il [quadro di Gianni $_i$] che a lui$_i$ piace* t *di più.*
this is the picture by Gianni that to him pleases most
'This is the picture by Gianni that he likes most.'

The contrast between (5)–(6) and (8) must be related to the different types of pronoun involved. The typology of pronouns has been the subject of various recent studies, including Cardinaletti (1994), Cardinaletti—Starke (1994) and Picallo (1994). The intuition shared by these authors is that the different types of pronoun can be characterized by their richness in phi-features and/or in internal functional structure.

The typology proposed by Cardinaletti (1994) and Cardinaletti—Starke (1994) consists of three pronominal classes. One aspect of this classification concerns the licensing position of the pronouns: clitics must be adjoined to a head, whereas other pronouns occupy an XP position. On the other hand, pronouns can be either strong or deficient.

These labels refer to a heterogeneous set of properties: the restriction to [+human] referents, the ability to occur in coordination or in isolation, the possibility of adverbial modifiers and of focalization. All these properties characterize strong pronouns and are not shared by deficient pronouns. Deficient pronouns include clitics, null pronouns, and the so called weak pronouns, which cannot appear in a thematic position in S-Structure. The distinction between strong and deficient pronouns follows from two hypotheses: deficient pronouns have a reduced functional structure; as a consequence of this structural deficiency, they must be licensed in a local relation with a Case-assigning (Agr°) head.

The contrast between (5)–(6) and (8) is apparently sensitive to the strong versus deficient character of the coindexed pronouns. The strong pronouns in (8) appear to be exempt from the Principle C effect: they must be insensitive to the reconstruction of the relative "head", for reasons that remain to be explained.

The behaviour of the strong pronouns in (8) resembles that of focalized R-expressions: these too seem not to be sensitive to Principle C under reconstruction (9b):

(9) a. * *Penso che Gianni stesso*$_i$ *voglia vendere [il ritratto di Gianni*$_i$*].*
(I) think that Gianni himself wants to sell the portrait of Gianni

b. *Questo è il [ritratto di Gianni*$_i$*] che penso che Gianni stesso*$_i$ *voglia vendere* t.
'This is the portrait of Gianni that I believe that Gianni himself wants to sell.'

Another context where a strong pronoun does not yield Principle C effects under reconstruction is the postverbal emphatic use exemplified in (10) (A. Belletti, personal communication):

(10) a. * *Non comprerebbe neanche lui*$_i$ *il quadro di Gianni*$_i$
wouldn't buy even he the picture by Gianni

b. *L'unico [quadro di Gianni*$_i$*] che non comprerebbe neanche lui*$_i$ t *è questo.*
the only picture by Gianni that wouldn't buy even he is this one
'The only picture by Gianni that not even he would buy is this one.'

It seems that in (9b)–(10b) the emphatic function is crucial in licensing coreference of the subject of the relative clause with the R-expression. Suppose that in (8) as well the strong pronouns bear an emphatic function.[4] The generalization would then be that emphatic DPs are insensitive to reconstruction with respect to Principle C.[5]

Adopting the distinction between binding and accidental coreference proposed by Reinhart (1983), the relation between the R-expression and the subject pronoun in (8) is accidental coreference, and it cannot be binding. Binding requires c-command, but in (8) the R-expression does not c-command the pronoun either before or after reconstruction. In fact, replacing the R-expression with a quantified phrase yields an ungrammatical sentence, since quantified phrases disallow accidental coreference.[6] The same point holds for (9b) and (10b): since R-expressions cannot be bound, the relation between the subject of the relative clause and the R-expression embedded in the relative "head" cannot be one of binding, but it must be accidental coreference.

The generalization can then be reformulated as follows: in (8), (9b) and (10b) the emphatic subject of the relative clause can accidentally corefer with the occurrence of the R-expression in Spec,CP, whereas the deficient pronoun in (5)–(6) cannot.

If this generalization is correct, it is necessary to assume that the coreference relation established by deficient pronouns is different from that of emphatic DP, in that the former but not the latter is sensitive to reconstruction. This distinction may be expressed by the following assumptions:

(a) emphatic pronouns may count as R-expressions;[7]

(b) accidental coreference between R-expressions does not "see" reconstruction; on the contrary, binding and coreference of deficient pronouns are sensitive to it.[8]

Hypothesis (b) requires an unorthodox conception of the deletion of chain links. If reconstruction of the relative "head" consists in the deletion of the highest link of its chain, deletion must make this link invisible for certain syntactic relations, but leave it visible for others. Put differently, reconstruction should take place in a syntactic module that also contains binding and coreference of deficient pronouns, whereas accidental coreference between R-expressions should belong in a different LF module. These remarks are completely speculative; for the moment, the problem of (8)–(10) remains open.

The alternative is, of course, to deny that the ungrammaticality of (5)–(6) is a Principle C effect due to the reconstruction of the relative "head". But there is no obvious alternative explanation for the ungrammaticality of these examples, nor for the contrast with the appositive clauses in (7).

3. Principle A effects

Before examining the possibility of anaphor binding under reconstruction of the relative "head", a preliminary counterargument must be considered.

As shown in (2a,b), the relevant configuration involves an anaphor embedded in the NP "head" of the relative clause. Recently, Reinhart—Reuland (1993) have argued that *self*-anaphors embedded within NPs are actually not anaphors, but are interpreted logophorically: namely, they can corefer with an antecedent which is prominent in the discourse, even though they are not c-commanded by it.

According to Reinhart and Reuland, the logophoric reading is possible whenever the *self*-anaphor is not the argument of a syntactic predicate.[9] This is the case when the anaphor is embedded in an adjunct (11a), in a coordination (11b), or within a NP lacking an overt subject (11c). The latter case follows from the hypothesis that the syntactic realization of the external argument is optional for nouns: since lexical heads without an external argument do not qualify as syntactic predicates, a subjectless NP is not a syntactic predicate, and an anaphor embedded in it can be used logophorically.

(11) a. *Clara$_i$ found time to check that [apart from herself$_i$], there was a man from the BBC.*
b. *Max$_i$ boasted that the queen invited [Lucie and himself$_i$] for a drink.*
c. *Lucie$_i$ saw [a picture of herself$_i$].*
(Reinhart—Reuland 1993: 670; 681)

But if anaphors embedded within subjectless NPs allow a logophoric use, an anaphor embedded in a subjectless NP "head" will be able to corefer with an antecedent internal to the relative clause without having recourse to reconstruction. For instance, (2a,b) can be analysed as involving the logophoric use of the anaphors rather than the reconstruction of the relative "head".

Fortunately, Reinhart—Reuland's argument does not carry over to Italian, where logophoric elements are morphologically distinct from true anaphors at least in the third person. Thus in the Italian examples (12), parallel to (11a) and (11b), the logophoric element *lei stessa/lui stesso* is (marginally) allowed, but the anaphor *se stesso/a* is excluded, since it cannot satisfy Principle A:

(12) a. *Clara*$_i$ *si accorse che*, [*a parte*
  lei$_i$], *c'era uno della BBC*
  ? *lei stessa*$_i$
  ?* *se stessa*$_i$

b. *Max*$_j$ *si vantò che la regina aveva invitato a cena* [*Lucia e*
  lui stesso$_j$][10]
  * *se stesso*$_j$

Interestingly, in an NP internal position true anaphors are possible and are actually better than logophoric elements:[11]

(13) a. *Lucia*$_j$ *vide* [*una foto di se stessa*$_j$] *nella vetrina del fotografo*
b. ? *Lucia*$_j$ *vide* [*una foto di lei stessa*$_j$] *nella vetrina del fotografo*
'Lucia saw a picture of herself in the photographer's shop-window.'

Therefore, the fact that in English logophoric elements are not distinguished from true anaphors seems to be accidental.[12] In Italian there is a clear morphological distinction between the two types of element in the third person, and hence the logophoric interpretation cannot be confused with true binding. Thus, in Italian third person *self*-anaphors offer a reliable test for the reconstruction of the relative "head".

The idea that in Italian a *self*-anaphor embedded in a NP is not logophoric is further confirmed by the following example, due to R. Kayne:

(14) * *La* [*descrizione di se stesso* $_i$] *che* [*sua*$_i$ *moglie*] *ha letto* t *nel rapporto della polizia è molto accurata.*
the description of himself that his wife read in the police's report is very accurate

Since logophoric elements need not be c-commanded by the DP they corefer with, if the *self*-anaphor embedded in the "head" were used logophorically, it should be able to corefer with the non-c-commanding possessive *sua* embedded in the subject of the relative clause. But this is impossible.

Moreover, the following example (15) shows that the anaphor *proprio*[13] cannot be used logophorically. The anaphor is contained in a complex noun phrase including a relative clause, but it is not contained in the "head" of the relative clause:

(15) * *Questo è il* [*proprio*$_i$ *contributo* [*al* [$_{CP}$ [*volume*] *di cui Gianni*$_i$ *è il curatore* t]]]].
this is his own contribution to the volume of which G is the editor

The ungrammaticality of (15) shows that the possessive anaphor *proprio* is not used logophorically: in order to corefer with the R-expression *Gianni*, it must be reconstructed in a position c-commanded by it. This is impossible in (15) because the anaphor is not included in the relative "head". (15) contrasts minimally with (16), where the anaphor is contained in the relative "head" and hence undergoes reconstruction:

(16) ? *Questi sono* [*i* [$_{CP}$[*contributi al proprio*$_i$ *volume*] *dei quali Gianni*$_i$ *è soddisfatto* t].
these are the contributions to his own volume with which Gianni is satisfied
'These are the contributions to his volume that Gianni is satisfied with.'

Another potential counterargument concerns the possibility that in examples like (1a,b) the anaphor be bound by a phonetically null category within the NP "head" rather than by the subject of the relative clause under reconstruction.[14] This possibility is illustrated by the examples in (17) (from Vergnaud 1974: 230). In (17a), the anaphor embedded in the NP "head" seems to be bound by the subject of the relative clause. However, in (17b) a pronoun within the "head" cannot be coindexed with the relative clause subject: there is an unexpected Principle B violation. Since the NP "head" is a governing category for the pronoun, there must be within it a phonetically null DP coindexed with the relative clause subject, as represented in (17c):

(17) a. *I hate the story about himself*$_j$ *that John*$_j$ *always tells* t.
b. * *I hate the story about him*$_j$ *that John*$_j$ *always tells* t.
c. * *I hate* [*the* [$_{NP}$ PRO$_i$ *story about him*$_i$] *that John*$_i$ *always tells* t].

But if the PRO within the NP "head" can bind the pronoun in (17b), it can also bind the anaphor in (17a), thus satisfying Principle A without recourse to the reconstruction of the "head".

This factor must be taken into account. Whenever the NP "head" of a relative clause exhibits Principle B effects, it has an obligatory implicit subject, and hence it cannot be used to test anaphor binding under reconstruction. But note that this interfering factor is not excluded even when the NP "head" does not present Principle B effects. In fact, in certain NPs the syntactic realization of a null external argument may be optional: then the same NP "head" could lack an implicit subject when it contains a pronoun, but have one when it contains an anaphor, as represented in (18a) and (18b) respectively:

(18) a. *i [pettegolezzi su di lui$_i$] per cui Gianni$_i$ si è offeso*
the gossips about him at which Gianni took offence

b. ? *i* [PRO$_i$ *pettegolezzi su di sé$_i$*] *per cui Gianni$_i$ si è offeso*
the gossips about himself at which Gianni took offence
'The gossips about himself by which Gianni was offended...'

Note however that in (18b) the implicit subject of the NP "head" controlled by *Gianni* must be interpreted as an abstract possessor rather than as the "agent": in fact, the sentence does not naturally allow an interpretation in which Gianni himself spread gossips about himself. In this respect, the PRO subject would differ from overt subjects, which are necessarily interpreted as agentive (e.g. in *i tuoi pettegolezzi su Gianni* 'your gossips about Gianni' or *i pettegolezzi di Maria su Gianni* 'Maria's gossips about Gianni'). This may be an argument against the presence of an implicit PRO controlled by *Gianni* in (18b). Anyway, in order to completely avoid the possibility of interference of an implicit subject, in the following examples the anaphor will be embedded in NP "heads" that cannot take an external argument, or it will be embedded in the external argument of the NP "head".

Let us now consider the relevant examples.

The possibility of anaphor binding under reconstruction in restrictive relatives is aknowledged by all speakers:

(19) a. ? *Il* [*collega della propria$_i$ moglie*] *del quale Gianni$_i$ è geloso* t *è un noto dongiovanni.*
'The colleague of his wife's who Gianni is jealous of is a notorious lady-killer.'

(19) b. *Questi sono gli [aspetti della propria*$_i$ *personalità] che Gianni*$_i$ *non riesce ad accettare* t.
'These are the aspects of his personality that Gianni is unable to accept.'

c. *L'unica [conseguenza della propria*$_i$ *decisione] che Gianni*$_i$ *non aveva considerato* t *si rivelò disastrosa.*
'The only consequence of his decision that Gianni hadn't taken into consideration turned out to be disastrous.'

d. *I [pochi beni dei propri*$_i$ *genitori] che Gianni*$_i$ *ha rivendicato* t *non erano menzionati nel testamento*
'The few goods of his parents' that Gianni vindicated were not mentioned in the testament.'

e. *L'[impronta della propria*$_i$ *personalità] che l'artista*$_i$ *ha impresso* t *in questa scultura è del tutto originale.*
'The mark of his personality that the artist expressed in this sculpture is absolutely original.'

f. *Le [prove a proprio*$_i$ *carico] che l'imputato*$_i$ *ha distrutto* t *erano decisive.*
'The evidence for his prosecution that the defendant destroyed was of crucial importance.'

g. *La [copia di se stesso*$_i$*] che Bono*$_i$ *ha visto* t *al museo delle cere è perfetta.*
'The copy of himself that Bono saw in the wax-work museum is perfect.'

h. ? *Nel corso dell'analisi è emersa [una parte di se stesso*$_i$*] che Gianni*$_i$ *ancora non conosceva* t.
in the course of the psychoanalysis there emerged a part of himself that Gianni didn't know
'By means of psychoanalysis a part of himself was revealed that Gianni didn't know of.'

i. *Il poeta descrive il [riflesso di se stesso*$_i$*] che Narciso*$_i$ *vide nella fonte.*
'The poet describes the reflection of himself that Narcissus saw in the fountain.'

The corresponding appositive clauses give contradictory results:

(20) a. ?? *Il principale della propria*$_i$ *moglie, del quale Gianni*$_i$ *è molto geloso, è un noto dongiovanni.*
the boss of his own wife, of whom Gianni is very jealous, is a notorious lady-killer

(20) b. ? *Questi aspetti della propria$_i$ personalità, che Gianni$_i$ non riesce ad accettare, sono emersi nel corso dell'analisi.*
these aspects of his own personality, that Gianni is unable to accept, emerged in the course of the psychoanalysis

c. ? *Quella conseguenza della propria$_i$ decisione, che Gianni$_i$ non aveva considerato, si rivelò disastrosa.*
that consequence of his own decision, that Gianni hadn't considered, turned out to be disastrous

d. ?? *Alcuni beni dei propri$_i$ genitori, che Gianni$_i$ ha rivendicato, non erano menzionati nel testamento.*
some goods of his own parents, that Gianni vindicated, were not mentioned in the will

e. * *Un'originale impronta della propria$_i$ personalità, che l'artista$_i$ aveva impresso in quella scultura, passò inosservata.*
an original mark of his own personality, that the artist had imprinted on that sculpture, remained unnoticed.

f. ? *La più importante prova a proprio$_i$ carico, della quale l'imputato$_i$ era a conoscenza, sparì durante il processo.*
the most important evidence for his own prosecution, with which the defendant was acquainted, was stolen during the process

g. * *Una copia perfetta di se stesso$_i$, che Bono$_i$ ha visto al museo delle cere, è stata realizzata l'anno scorso.*
a perfect copy of himself, that Bono saw in the wax-works museum, was made last year

h. * *Nel corso dell'analisi è emersa questa parte di se stesso$_i$, che Gianni$_i$ ancora non conosceva.*
in the course of psychoanalysis there emerged this part of himself, that Gianni didn't know

The possibility of anaphor binding in (20b,c,f) seems to imply that at least some appositive clauses allow the reconstruction of the "head".

Note however that in these examples the anaphor embedded in the relative "head" is not c-commanded by any potential antecedent in S-structure. If one adopts Chomsky's (1986a: 171–172) version of the binding theory, it turns out that the anaphor has no governing category, since there is no binding-theory compatible indexing that could satisfy Principle A. It has been proposed in the literature that when an anaphor lacks a governing category, it is not subject to Principle A and it can co-refer freely (see Manzini—Wexler 1987: 422). This proposal can explain the grammaticality of (20b,c,f), but at the same time, it makes the data in (19a-h) irrelevant for testing the reconstruction of the relative "head": in (19a-h) as well, the anaphor could be taken to corefer freely. However, this approach does not explain the ungrammaticality of (21), where the anaphor *propri* is the possessive subject of a subject DP, hence it lacks a governing category, and it is predicted to corefer freely:[15]

(21) * [*I propri*$_i$ *genitori*] *amano Gianni*$_i$.
his own parents love Gianni

A clear contrast between restrictive and appositive clauses emerges if the anaphor is c-commanded by a potential antecedent in S-structure. In such a configuration, the restrictive relative allows the anaphor to be bound under reconstruction, but the appositive relative does not. Compare (22) to (20b) and the (b) and (c) examples in (23) and (24) below:[16]

(22) * *L'analista*$_j$ *ha scoperto questi aspetti della propria*$_i$ *personalità, che Gianni*$_i$ *non riesce ancora ad accettare* t.
the psychoanalyst discovered these aspects of his own personality, that Gianni is still unable to accept

(23) a. *L'imputato*$_i$ *produsse un'unica prova della propria*$_i$ *innocenza.*
'The defendant produced a single proof of his innocence.'

b. *Il giudice*$_j$ *invalidò l'unica prova della propria*$_i$ *innocenza con cui l'imputato*$_i$ *sperava di scagionarsi* t.
'The judge invalidated the only proof of his innocence by which the defendant hoped that he could exculpated himself.'

c. * *Il giudice*$_j$ *invalidò un'unica prova della propria*$_i$ *innocenza, con la quale l'imputato*$_i$ *era certo di scagionarsi* t.
the judge invalidated a single proof of his own innocence, by which the defendant was able to exculpate himself

(24) a. *Gianni*$_i$ *si è offeso per queste critiche al proprio*$_i$ *operato.*
'Gianni was offended by these criticisms of his work.'

b. ? *Maria*$_j$ *ci riferì le critiche al proprio*$_i$ *operato per cui Gianni*$_i$ *si era offeso* t.
'Maria related to us the criticisms of his work by which Gianni was offended.'

(24) c. * *Maria$_j$ ci ha riferito quelle critiche al proprio$_i$ operato, per le quali, a quanto pare, Gianni si è offeso* t
Maria told us those criticisms of his work, by which, so it seems, Gianni was offended

If reconstruction were possible in appositive as well as in restrictive relatives, it is unclear why the presence of a potential antecedent for the anaphor should block reconstruction in the former but not in the latter.[17] Therefore, it seems reasonable to conclude that reconstruction takes place only in restrictive relatives.[18]

The somewhat surprising pattern attested in (20) and (22)–(24) is also observed in other A' structures. Brody (1995: 92) attributes to Browning (1987) the observation that in the "easy to please" structure reconstruction in the gap position is possible only insofar as there is no antecedent available for the anaphor in its S-structure position. Thus in the ECM structure (25a) the main clause subject is an available antecedent for the anaphor embedded in the ECM subject, and the anaphor cannot be bound under reconstruction. In (25b), on the contrary, the anaphor is separated from the main clause subject by a clausal boundary, so that it has no available antecedent; in this case, it can be bound by the subject of the infinitival *easy* complement

(25) a. * *Mary$_j$ expected [those pictures of himself$_i$] to be easy [to make John$_i$ to buy* e]
b. *Mary$_j$ said [that [those pictures of himself$_i$] are easy [to make John$_i$ to buy* e]

Note that the *easy to please* structure is usually assumed to involve a null operator binding the gap in the infinitival clause; therefore, true reconstruction in the gap position is by hypothesis impossible.

4. Reconstruction for scope assignment

The observation that the "head" of a restrictive relative can be reconstructed for the purposes of scope assignment was anticipated in § II.3.4, in connection with the hypothesis of the external determiner. This section will provide some more data and a comparison with appositive relatives.

Compare (31a) of chapter II, repeated in (26a), with its appositive counterpart (26b):

(26) a. *Ho telefonato ai due pazienti che ogni medico visiterà* t *domani.*
'I called the two patients that each doctor will examine tomorrow.'
b. *Ho telefonato ai due pazienti cardiopatici, che ogni medico visiterà* t *domani*
'I called the two cardiopath patients, who each doctor will examine tomorrow.'

The asymmetry between restrictive and appositive relatives is confirmed: the wide scope reading of the universally quantified subject of the relative clause over the "head" is possible in (26a) and excluded in (26b). In (26b) the "head" of the appositive clause denotes a specific set of two patients: this means that the numeral *due* cannot be reconstructed in the scope of the relative clause subject. The contrast is confirmed by (27) and (28).

(27) a. *Scriverò su fogli diversi l'equazione che ciascuno studente dovrà risolvere* t.
'I will write the equation that each student will have to solve on different sheets.'
b. *Scriverò sulla lavagna l'equazione, che ciascuno studente dovrà risolvere* t *per conto suo.*
'I will write on the blackboard the equation, which each student will have to solve by himself.'
(28) a. *Ora ripeterò la risposta che ciascuno di voi ha dato* t.
'Now I will repeat the answer that each of you gave.'
b. ? *Ora ripeterò la risposta corretta, che ciascuno di voi ha dato* t.
'Now I will repeat the correct answer, which each of you gave.'

The impossibility of scope assignment under reconstruction is in accordance with the referential independence of the "head" from the appositive relative.

5. Quantifier binding

Reinhart (1983: 122) points out that a pronoun cannot accidentally corefer with a quantified expressions, but it must be bound by it under c-command. Let us examine the possibility for a pronoun embedded in the relative "head" to be bound by a quantified phrase contained in the relative clause.[19]

In the restrictive relatives (29), the bound reading of the pronoun (or anaphor) embedded in the NP "head" is very natural:

(29) a. *La fase della sua*$_i$ *carriera che ogni impiegato*$_i$ *vorrebbe dimenticare* t *è quella iniziale.*
'The phase of his career that every clerk would like to forget is the beginning one.'

b. *Il periodo della propria*$_i$ / ? *sua*$_i$ *vita di cui nessun uomo*$_i$ *parla volentieri* t *è l'adolescenza.*
'The period of his life which nobody is willing to speak about is adolescence.'

c. *L'immagine di se stesso*$_i$ *che ogni uomo*$_i$ *cerca di trasmettere* t *agli altri è spesso lontana dalla realtà interiore.*
'The image of himself that every man tries to convey to other people is often very different from his inward reality.'

d. *La parte del suo*$_i$ *stipendio che ho anticipato* t *ad ogni impiegato*$_i$ *verrà sottratta dalla busta paga.*
'The part of his salary that I paid in advance to every clerk will be deducted from the pay-sheet.'

The bound reading is much more difficult to get in appositive clauses:

(30) a. ?* *L'inizio della sua*$_i$ *carriera, del quale nessun impiegato*$_i$ *parla* t *volentieri, è pieno di frustrazioni.*
the beginning of his career, about which no clerk speaks willingly, is full of frustration

b. * *Il primo periodo della sua*$_i$ *vita, che nessuno*$_i$ *può ricordare* t, *è fondamentale per lo sviluppo psichico.*
the first period of his life, that nobody can remember, is of great importance for psychic development

c. ?? *L'immagine esteriore di se stesso*$_i$, *che ogni paziente*$_i$ *cerca di difendere* t, *è molto lontana dalla realtà interiore.*
the exterior image of himself, that every patient tries to protect, is very different from inward reality

d. * *Una parte del suo*$_i$ *stipendio, che ho anticipato* t *ad ogni impiegato*$_i$, *verrà sottratta dalla busta paga.*
a part of his salary, that I paid in advance to every clerk, will be deducted from the pay-sheet

These data confirm the expected asymmetry in reconstruction effects.

6. Idiomatic expressions

In § II.4.1 it was assumed, following Chomsky (1993), that the interpretation of idiomatic expressions requires the adjacency of its syntactic constituents in LF. Thus, if an idiom chunk is susceptible of A' movement, it must nevertheless be reconstructed in its base position in LF. Independently of the syntax of relative clauses, this requirement is proved by the interaction of the idiomatic interpretation with anaphor binding discovered by Chomsky (1993: 38–39).

In the same section it was shown that restrictive relativization of an idiom chunk is compatible with its reconstruction in LF (although not every idiom chunk can be relativized, for reasons discussed in part in § II.3.1). On the other hand, it has been observed by several authors that the relativization of idiom chunks and predicate nominals[20] is completely excluded in appositive clauses (cf. Emonds 1979: 233–234, Fabb 1990: 71, Postal 1994: 168 on English; Huot 1978: 119–120, Vergnaud 1974: 181 on French):

(31) a. * *That headway, which the students made* t *last week, was phenomenal.*
b. * *We were surprised by the inexpected advantage, which he had taken* t *of the situation.*
c. * *That horrible face, which Harry made* t *at Peter, scared him.*
d. * *The fun, which they made* t *of me...*

(32) a. * *Nous avons été surpris par ce parti remarquable, qu'il a tiré* t *de cette situation*
we were surprised by that remarkable advantage, that he took of that situation
b. * *Cette part immense, qu'il a pris* t *au congrès, a bouleversé les données.*
that enormous part, which he took at the conference, has confused the data
c. * *Jean n'est pas ce comédien, que son père était* t *il y a vingt ans.*
Jean is not that comedian, that his father was twenty years ago

Assuming Chomsky's (1993) approach to idiomatic expressions, these data can be conceived as another consequence of the lack of reconstruction in appositive clauses. However, there is also an alternative explanation which does not directly involve reconstruction. As is well known, appositives differ from restrictive relatives in that their "head" is referentially autonomous from the relative clause; it is a referring expression by itself. Moreover, the referential nature of the "head" of appositive clauses is restricted: for instance, it cannot be a quantified phrase (Jackendoff 1977: 175):

(33) * *Any / every man, who drives a Cadillac, is insane.*

Cinque (1990: chapter I) argues that idiomatic expressions and quantified phrases differ from definite-specific arguments in that they are not "Discourse-linked". If there is a requirement to the effect that the "head" of an appositive relative must be a D-linked DP, this suffices to exclude the appositive relatives in (31)–(32), independently of the possibility of interpreting the idiomatic expression under reconstruction.

The fact that this requirement is independent of reconstruction seems to be confirmed by the behaviour of cleft sentences. The clefted constituent cannot be an idiom chunk ((34a,b); cf. Bianchi 1993: 377–381 and Grewendorf—Poletto 1989: 113) nor a quantified phrase (34c); however, cleft sentences allow reconstruction, as shown in (35):

(34) a. * *E' / sono grandi progressi che ha fatto* t.
it is / are much headway that he made
b. * *E' un boccone amaro che ho dovuto ingoiare* t.
it is a bitter pill that I had to swallow
c. * *E' / sono tutti i libri che ho letto* t.
it is / are all the books that I read

(35) *E' [di se stesso$_i$] che Gianni$_i$ vuole parlare* t.
it is about himself that Gianni wants to speak
'It's himself that Gianni wants to talk about.'

Given this alternative analysis, the cause of the ungrammaticality of (31)–(32) remains undetermined.[21] However, the data are at least consistent with the generalization that appositive clauses have no reconstruction of the relative "head".

7. Reconstruction of pied piped material in appositive clauses

In the preceding section it was argued that restrictive and appositive relatives differ with respect to the reconstruction of their "head". One possible approach to this asymmetry is to assume that the two relative structures involve two different A' dependencies, and that the specific dependency involved in appositives disallows reconstruction. For instance, one could speculate that restrictive relatives are derived by A' movement, whereas appositive clauses involve a base generated A' dependency.[22] The latter does not contain a trace of movement, hence it disallows reconstruction..[23]

This approach is excluded by the following observation: although appositive relatives disallow the reconstruction of the relative "head", they allow (and in fact, require) the reconstruction of material pied piped by the "head".

The reconstruction of the pied piped material is attested by the Principle C effect in (36a), where an R-expression contained in the pied piped phrase is coindexed with the null subject of the relative clause; by the binding of the long-distance anaphor *proprio* in (36b); and by quantifier binding in (36c):

(36) a. * *Andrea,* [*le cui insinuazioni su Beatrice*$_i$] pro$_i$ *non è più disposta a sopportare* t...
Andrea, whose insinuations on Beatrice (she) is not willing to stand any more...

b. *Andrea,* [*la cui passione per la propria*$_i$ *moglie*] *Juan*$_i$ *non è disposto a tollerare* t...
'A, whose passion for his wife J is not willing to tolerate...'

c. *Questo farmaco,* [*il cui effetto sui propri*$_i$ *pazienti*] *nessun medico*$_i$ *è in grado di prevedere* t, ...
'This medicine, whose effects on his patients no doctor is able to foresee...'

Within the approach to reconstruction assumed here, these data show that the A' dependency created by the pied piped phrase allows reconstruction; more specifically, it is an instance of movement. It seems implausible to stipulate that appositive clauses with and without pied piping involve two completely different A' dependencies; therefore, appositive clauses must have a movement derivation. The problem is then to explain why the relative "head" cannot be reconstructed as well. This problem will be dealt with in the next chapter.

8. Argument-adjunct asymmetries in reconstruction

Let us now examine a problem concerning the reconstruction of the relative clause itself. Lebeaux (1991) argues that there is a contrast between DP-internal arguments and adjuncts: when an A' moved DP is reconstructed, the internal noun complements are obligatorily reconstructed along with it, whereas for a modifying relative clause reconstruction is optional. In fact, in (37a) the complement clause embedded in a *wh*-phrase gives rise to a Principle C violation under reconstruction, but the relative clause in (37b) does not:

(37) a. * [*Which claim that John*$_i$ *was asleep*] *was he*$_i$ *willing to discuss* t?
b. [*Which claim that John*$_i$ *made* t] *was he*$_i$ *willing to discuss* t?

Lebeaux's solution is based on the assumption of generalized transformations. He argues that noun complements are necessarily inserted in the tree in D-structure, because they are visible for the Theta Criterion.[24] On the contrary, a relative clause is not part of the thematic structure of the sentence, and hence it can fail to be inserted in D-structure; it can be attached to the tree by a generalized transformation at a later stage of the derivation. Thus in (37a) the noun complement clause must be inserted in D-structure in the base position of the complex DP, which is c-commanded by the matrix subject *John*; consequently, the complement clause is reconstructed, and a Principle C violation arises. In (37b), on the contrary, the relative clause need not be inserted in D-structure: it can be attached to the *wh*-phrase directly in Spec,CP, and thus, it is not c-commanded by the matrix subject at any level of representation. This explains the lack of a Principle C effect.[25]

The hypothesis that relative clauses can be inserted at any step of the derivation obviously presupposes the adjunct analysis. Moreover, it presupposes that a relative clause can be analysed as an "argument skeleton" generated independently of the matrix clause: for instance, in Lebeaux's analysis the sentence *John saw the woman whom he knew* is built from two argument skeletons, *John saw the woman* and *whom he knew*. But note that this splitting is impossible in the relativization of an idiom chunk: e.g. the sentence *I was surprised by the advantage that he took of the situation* would have to be built by combining the argument skeletons *I was surprised by the advantage* and *he took* Op *of the situation*, where *Op* is an abstract relative operator. Clearly, neither of the argument skeletons is interpretable by itself, since each one contains only a fragment of the idiomatic expression *to take advantage (of st)*. Even assuming that the two argument skeletons are directly combined in D-structure (which would imply the obligatory reconstruction of a relative clause modifying an idiomatic "head"), it is impossible to build two distinct argument skeletons in the first place. Thus, it is necessary to stipulate a different derivation for relative clauses with an idiomatic "head" - a derivation that does not involve the generation of two argument skeletons.

Besides this conceptual problem, the generalization represented by the contrast in (37) is called in question by some data reported in the literature. In (38), a noun complement fails to be reconstructed and does not manifest the expected Principle C effect;[26] conversely, (39b) is a Principle C violation caused by the obligatory reconstruction of a relative clause.

(38) a. ? [*Which claim that John*$_i$ *was a thief*] *did he*$_i$ *deny* t ? (Huang 1993)
b. [*The rumour that John*$_i$ *stole the money*], *he*$_i$ *has always denied* t. (Guéron 1984: 150–164)
c. [*Which biography of Picasso*$_i$] *do you think he*$_i$ *wants to read* t? (Higginbotham 1983: 407–408)
(39) a. [*Which man who admires his wife*$_i$] *tries to please her*$_i$?
b.* [*Which man who loves his wife*$_i$] *does she*$_i$ *try to please* t ? (Higginbotham 1983: 412)

These data contradict the generalization discussed by Lebeaux, contrasting with (37a) and (37b) respectively. It is likely that in every sentence the possibility of reconstruction be affected by various interpretive factors: for instance, the naturalness of the restricted quantification reading, which corresponds to the lack of reconstruction of the restrictive term (cf. § 2), and perhaps the possibility of emphasizing the pronominal subject (cf. § 2.1). The data seem questionable and require further research; for the moment, they do not constitute decisive evidence in favour of the adjunct analysis.

9. Summary

Chomsky's (1993) analysis of reconstruction, based on the copy theory of traces, constitutes an important test for the raising analysis of relative clauses. The data concerning binding under reconstructions are often obscure, showing that many factors must be at work which the standard theory of binding is still unable to capture. Nevertheless, the plausible generalization that can be drawn from the Italian data discussed in this chapter is that the reconstruction of the relative "head" is possible in restrictive relatives, but impossible in appositive ones. This does not mean that appositive relativization by itself blocks reconstruction: in fact, pied piped material undergoes obligatory reconstruction. It is the "head" that has a special status in this respect.

These data show that the raising analysis cannot directly account for the syntax of appositives, whereas it correctly predicts the possibility of reconstructing the "head" in restrictive relatives. The analysis of appositive relatives will constitute the topic of the next chapter.

Chapter V
Appositive relative clauses

0. Introduction

The reconstruction phenomena illustrated in the preceding chapter suggested two conclusions: relativization in appositive clauses is an instance of A' movement; but the "head" cannot be reconstructed in the trace position. This is not predicted by the raising analysis defended in chapter II for restrictive relatives.

Recall from §§ II.2.1–2.2 that the hypothesis of the external determiner was imposed by the Linear Correspondence Axiom because the relative clause had to be c-commanded by the determiner in order to fall within its restrictive term:

(1) [$_{DP}$ *the* [$_{CP}$ *book*$_i$ *that I read* t_i]]

But this requirement does not hold for appositive relatives: the latter are not included in the restrictive term of any determiner, but they are simply predicated of an autonomous referential expression, the "head". Thus, it is possible to elaborate a completely different syntactic analysis for them. This possibility will be explored in the following discussion.

After a brief summary of previous approaches (§ 1), section 2 reviews some properties that distinguish appositive relatives from restrictive ones. After this, two alternative analyses are discussed. The first one assumes that in appositive clauses it is not the relative "head" that raises to Spec,CP, but rather a relative operator, as in the standard adjunct analysis. The second one, proposed by Kayne (1994: 110–115), assumes that appositive relatives have a raising derivation, and differ from restrictive ones only at the level of Logical Form.

1. Previous approaches

The status of appositive modifiers is problematic in the Principles and Parameters framework: these elements receive a parenthetical intonation, and with respect to various phenomena they seem not to be included in the syntactic structure of the clause in which they occur.

Various approaches try to implement this syntactic independence of appositive relatives by means of exceptional rules of attachment to the tree. There are at least three variants of this general idea: the coordinate analysis, McCawley's (1982) discontinuous constituent structure analysis, and Safir's (1986) LF' analysis.

Other authors instead propose that appositive relatives differ minimally from restrictive ones in that they are not c-commanded by the determiner of the relative "head". This derives from a difference in the level of adjunction of the relative clause. These approaches are briefly summarized in the following subsections.

1.1. The coordinate analysis

According to Emonds (1979), an appositive relative originates as coordinate to the right of the clause in which it occurs,[1] like any other parenthetic expression; however, it contains a pronoun that must be anaphoric to an antecedent contained in that clause.[2]

The surface form of the appositive relative is derived in three steps. Since the second conjunct must be adjacent to its antecedent in order to receive the appositive interpretation, whatever constituent of the first conjunct intervenes between the antecedent and the second conjunct is moved to the right of the latter by the transformation of Parenthetical Formation. As a second step, the rule of S' Attachment deletes the conjunction and attaches the second conjunct directly to the first one. Finally, the rule of *wh*-fronting moves the coreferent pronoun at the beginning of the attached S'. This analysis implies that the appositive relative, although apparently embedded, does not form a constituent with the antecedent of the relative pronoun, but it is attached directly to the root S node.

This approach seems to be supported by the observation that in certain Italian dialects appositive clauses, contrary to restrictive ones, do not show any overt mark of subordination, like e.g. a complementizer or a relative pronoun, but they are simply parenthetical clauses containing a clitic pronoun coreferent to an antecedent in the main clause (cf. Cennamo forthcoming):

(2) *Mario, tu l'avevi visto ieri, è partito.*
Mario, you him had seen yesterday, has left
'Mario, whom you saw yesterday, has left.'
(Modern Tuscan; Benincà 1994)

It is unclear how Emonds' analysis could be implemented in the current syntactic framework. The rule of Parenthetic Formation should probably be formulated as an instance of rightward adjunction, which is incompatible with the Linear Correspondence Axiom. Moreover, the rule of S' Attachment carries out a radical restructuring of the tree, which is hardly compatible with current assumptions about the power of transformations.

1.2. The discontinuous constituent structure analysis

A second possibility is to assume a special phrase structure status for appositive relatives, and more generally, for parenthetical elements. According to McCawley (1982), the peculiarity of parenthetical phrases consists in the fact that their surface position in the string does not correspond to their level of embedding: in other terms, a parenthetical phrase appearing between two sister nodes is not necessarily itself a sister to these nodes. This discontinuous constituent structure is the effect of a special transformation which modifies the linear order of the constituents without modifying the phrase structure of the sentence. A discontinuous constituent structure is exemplified in (3):

(3)

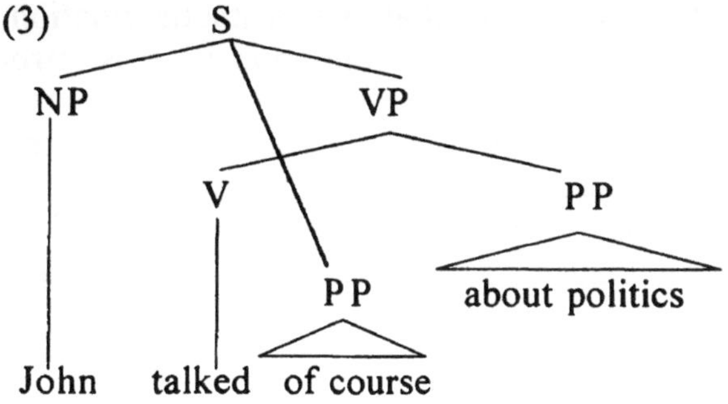

As for appositive relatives, this approach shares with Emonds' the idea that the appositive relative, like parenthetical phrases, is a daughter to S and does not form a constituent with the "head".

The discontinuous constituent structure is inconsistent with the mapping of asymmetric c-command into linear precedence required by the Linear Correspondence Axiom; therefore, McCawley's proposal will not be adopted here.

1.3. The LF' analysis

Contrary to Emonds and McCawley, Safir (1986: 672–677) proposes that appositive relatives form a constituent with their "head", but they do so only at a very late stage of the derivation. Developing a suggestion by Chomsky (1982), he proposes that appositive relatives - and more generally, extrathematic material - are attached to the syntactic tree at a level of representation called LF', which follows LF.[3] Crucially, the principles holding at LF do not apply to this later level of representation. This assumption is meant to explain the "islandhood" of appositive relatives with respect to various binding relations: for instance, an appositive relative cannot contain a long-distance anaphor bound by an external antecedent (4), nor a parasitic gap licensed by an external gap (5), nor a pronoun bound by an external quantifer (6).[4]

(4) *Gianni*$_i$ *ritiene che Giorgio*$_j$, t$_j$ *che ama la propria*$_{j/*i}$ *moglie, sia intelligente.*
'Gianni believes that Giorgio, who loves his wife, is intelligent.' (Giorgi 1984: n. 19; Safir 1986: 673)

(5) * *A man who Bill*$_i$, [*who knows* e$_i$], *admires* t$_i$. (Safir 1986: 673)

(6) * *No one*$_i$ *wanted Sue to leave, which suited him*$_i$. (Jackendoff 1977: 176)

These data follow from the assumption that the syntactic relation of binding only holds at S-structure and LF. Since the appositive is introduced in the tree at a stage later than these, the binding relations represented in (4)–(6) cannot be established; in (5), in particular, the parasitic gap fails to be licensed by the operator binding the real gap.

Another phenomenon that Safir wants to derive is the lack of weak crossover effects in appositive clauses, as opposed to restrictive ones:

(7) a. ?* *A man*$_i$ *who*$_i$ *his*$_i$ *wife loves* t$_i$ *arrived early.*
b. *John*$_i$, *who his*$_i$ *wife loves* t, *arrived early.* (Safir 1986: 667)

Following Chomsky (1982), Safir argues that in the appositive clause (7b) the relative pronoun is not coindexed with the "head" until LF'; therefore, even if at LF the the possessive pronoun bears the same index as the "head", the relative pronoun has a distinct index at this level, and the Parallelism Constraint on Operator Binding (Safir 1984) is not violated (8a). When the relative pronoun is coindexed with the "head" in LF', a potential violation of the PCOB arises: in fact, the relative trace and the pronoun become coindexed, but their values for the features [± pronominal] and [± anaphoric] do not match (8b). The crucial assumption is that the PCOB does not apply at LF', so that no violation arises.

(8) a. LF: *John*$_i$, [*who*$_j$ *his*$_i$ *wife loves* t_j] (PCOB respected)
b. LF': *John*$_i$, [*who*$_i$ *his*$_i$ *wife loves* t_i] (PCOB does not apply)

Lasnik—Stowell (1991) propose a different solution for the contrast in (7). In their analysis, the weak crossover effect is related to the quantificational nature of the binding phrase, which they formalize by the notion of "true quantified phrase":[5] only a trace bound by a true QP qualifies as a variable; the principle responsible for weak crossover only applies to variables. They argue that the interpretation of restrictive relatives involves restricted quantification, whereas in appositive clauses the "head" is a referential expression and the appositive clause is predicated of it. Thus, the different interpretation of the relative clause explains the contrast with respect to weak crossover.[6] The interest of this proposal lies in the fact that it does not only hold for relative clauses, but for A' structures in general, whereas Safir's analysis cannot be easily extended to other cases.

Even with respect to (4)–(6), Safir's solution is problematic. His discussion seem to imply that at LF' binding relations are not visible; but this incorrectly predicts that no Principle C violation should arise in (9), where the R-expression *Gianni* is embedded in an appositive clause c-commanded by the coindexed subject of the matrix clause:[7]

(9) * *Lui*$_i$ *sostiene che Maria, che ama Gianni*$_i$, *è felice.*
he claims that Maria, that loves Gianni, is happy.

Finally, Safir himself (fn. 11) notes a problem with Romance appositive clauses. These can be introduced by the declarative complementizer: in the adjunct analysis, this means that the relative pronoun in Spec,CP has been deleted (10a), or that Spec,CP is filled by a null relative operator (10b):[8]

(10) a. *Gianni*, [$_{CP}$ wh$_i$ [*che* t_i *è un vecchio amico*], *mi ha aiutato.*
↓
∅
b. *Gianni*, [$_{CP}$ Op$_i$ [*che* t_i *è un vecchio amico*], *mi ha aiutato.*
Gianni, that is an old friend, helped me
'Gianni, who is an old friend, helped me.'

As for the representation (10a), Cinque (1978; 1982: 260) argues that the phonetic deletion of the relative pronoun is possible only if the latter is c-commanded by the relative "head". But in Safir's analysis the relative pronoun is not c-commanded by the "head" until LF', a level which is independent of the Phonetic Form; therefore, the PF deletion of the relative operator should not be licensed.

The representation in (10b) is problematic as well. According to Chomsky (1986a: 85; 1995: 152–153), the null operator is equivalent to a free variable, and it must be strongly bound by an antecedent in LF in order to comply with the Full Interpretation Principle. But if the appositive clause in (9b) is attached to the antecedent "head" only at LF', the null operator violates Full Interpretation at LF.

1.4. Asymmetry in the level of adjunction

All the preceding approaches assume that the appositive relative has a special phrase structure status. Other authors instead propose that appositive relatives differ minimally from restrictive ones with respect to the level of adjunction: restrictive relatives are right-adjoined to an internal projection of the "head", which is c-commanded by its determiner; appositive relatives, on the contrary, are right-adjoined to a projection that dominates the determiner (cf. § II.2.1). A recent version of this hypothesis is found in Fabb (1990), who takes restrictives to be adjoined to N', and appositives to NP. Assuming the DP hypothesis, however, restrictive relatives can be adjoined to NP,[9] but appositives must be adjoined to the DP itself (cf. Toribio 1992: 297).

The different level of adjunction explains why in case of stacking an appositive relative necessarily follows any restrictive relative:[10]

(11) *The man that came to dinner, who was drunk, fainted.*
(12) * *The man, who was drunk, that came to dinner fainted.*
(Jackendoff 1977:171)

Note that this approach involves the rightward adjunction of the relative clause to a nominal projection, a configuration which is ruled out by the LCA. In Kayne's (1994) X-bar theory, the hypothesis that the appositive relative is not internal to the DP "head" must be structurally implemented by a syntactic relation different from adjunction. An attempt in this direction will be made in § 3 below.

2. Properties of appositive relatives

Besides the asymmetry in the reconstruction of the "head", shown in chapter III, appositive relatives also systematically differ from restrictive ones with respect to a number of other syntactic properties. These will be reviewed in the following subsections.[11]

2.1. Definiteness of the relative trace

Recall that in the raising structure the determiner introducing the relative "head" is generated outside the relative CP, and the relative DP itself is non-definite. This explains the lack of a definiteness effect in the restrictive relative (26b) of chapter II, repeated here as (13):

(13) [*The* [*men*$_j$ [*that there were* t_j *in the garden*]]] *were all diplomats*

On the contrary, appositive relatives show a clear "definiteness effect":[12]

(14) * *She left the magazines, which there are* t *on the table.* (Postal 1993b: 745)

2.2. Floating quantification

In § II.3.5 it was shown that the raised "head" of a restrictive relative cannot strand a floating quantifier within the relative CP. On the contrary, a floating quantifier can be related to the gap of an appositive clause:

(15) a. *Ho interrogato gli studenti, che avevano superato tutti la prova scritta.*
'I examined the students, who had all passed the written exam.'

b. *I miei amici, che ho invitato tutti alla festa, mi hanno fatto un regalo.*
my friends, that I invited all to the party, gave me a gift
'My friends, all of whom I invited to the party, gave me a gift.'

(16) a. *Les livres de Jean-Paul, qu'elle lira tous, sont très bons.*
the books of J.P., that she is going to read all, are very good
'Jean Paul's books, all of which she will read, are very interesting.'

(16) b. *Les amis de Pierre, que j'ai tous connus à l'âge de sept ans, sont sympathiques.*
the friends of Pierre, that I have all known since the age of seven years, are nice
'Pierre's friends, all of whom I met at the age of seven, are nice.' (Kayne 1975: 6–7)

2.3. *Relatif de liaison*

There is at least one nonrestrictive use of relative pronouns that seems not to be reducible to a raising analysis: the so called *relatif de liaison,* in which a relative pronoun is used to express intersentential anaphora, and can even be followed by a noun different from the one contained in its antecedent:

(17) a. *Ha difeso la sua tesi quasi contro tutti.* La quale *sosteneva la necessità del non intervento.*
he defended his thesis against almost everybody. Which asserted the necessity of non-intervention
'He maintained his thesis against almost everyone. Which thesis asserted the necessity of non-intervention.'
(Cinque 1978: 79)

b. *Rien n'advint de notable jusqu'au lundi de la semaine suivante,* auquel jour *le prince avertit sa femme qu'il allait à Rome.*
nothing remarkable happened until Monday of the following week, in which day the prince informed his wife that he was going to Rome
'Nothing remarkable happened until the next Monday, when the prince informed his wife that he was leaving for Rome.' (Sandfeld 1965: 179)

c. *Amici sunt firmi eligendi,* cuius generis *est magna penuria.*
friends are faithful to be chosen, of which kind is great scarcity
'one must choose faithful friends; but of this kind (of friends) there is great scarcity'
(Cic., *Laelius* 62, from Huot 1978: 120)

(18) ...*whom we name hereafter the Prince of Cumberland:* which honour *must not unaccompanied invest him only...*
(*Macbeth* I.4, 38–40)

2.4. Non-nominal antecedents

As discussed by Jackendoff (1977: 175) and Fabb (1990) among others, appositive relatives differ from restrictive ones in that they can take non-nominal antecedents, e.g. an AP (19a),[13] a PP (19b), a VP (19c) or a clause (19d):

(19) a. *Mary is [courageous], which I will never be.*
b. *John is [in the garden], which is where I should be.*
c. *Mary has [resigned], which John hasn't.*
d. *[John was late], which was unfortunate.*

2.5. Islandhood for binding

Finally, recall from § 1.3 that appositive relatives are islands for binding relations, as opposed to restrictive ones (cf. (4)–(6) above).

3. A non-raising analysis

In § II.2.1, the restrictive term of a determiner was defined as its c-command domain in LF. By this definition, relative clauses receiving an appositive interpretation cannot be c-commanded by the determiner of the "head" in LF. On the other hand, the LCA requires that the "head" asymmetrically c-command the appositive relative in order to yield the correct linear order.

The analyses discussed in § 1 satisfy the first condition, but they are inconsistent with the Linear Correspondence Axiom for various reasons.

There is at least one structure compatible with the LCA that satisfies both conditions: the syntactic connection between the "head" and the relative CP is created by an abstract functional head which takes the "head" as its specifier and the relative CP as its complement:[14]

(20) $[_{XP}$ DP $[_{XP}$ X° CP]]

In this structure, the DP "head" can be generated directly in Spec,XP; then the relative CP must contain an operator binding the relative trace, as in the adjunct analysis:

(21) $[_{XP}$ DP $[_{XP}$ X° $[_{CP}$ OP$_i$ $[_{CP}$ ... t_i ...]]]]

The structural representation in (21) is consistent with the LCA, but the nature of the syntactic relations involved in it remains to be explained. Which relation does X° establish between the DP in its Spec and its CP complement? If X° cannot be assimilated to any independently justified functional head, then (21) is reduced to an *ad hoc* notational variant of rightward adjunction.

The following subsections present a possible interpretation of the abstract structure in (21) and examine its consequences as well as its empirical problems.

3.1. X° as a "predicative" head

Suppose that in (21) the relative CP is a predicative category, with the relative operator in its Spec effecting a sort of lambda abstraction over the trace, much as in the traditional adjunct analysis. Then, the head X° could be analysed as a Predication head, mediating the relation between a predicative category (CP) and a referential one (DP).

The problem with this hypothesis is that, though X° is predicative, its maximal projection XP must be a referential argument; in particular, it cannot be assimilated to a small clause (denoting a proposition, or a situation), but it must be equivalent to a DP, with a nominal denotation, since the constituent formed by the antecedent DP and the appositive relative fills an argument position. Thus, there is an irreducible tension between the function of X° and the function of its maximal projection.[15]

3.2. X° as a "specifying" coordination

Another possibility is to analyse X° as a "specifying coordination"[16] relating two independently referential categories. The appositive structure (22a) is then essentially equivalent to (22b):

(22) a. [$_{XP}$ [$_{DP}$ *this book*], [X° [$_{CP}$ *which I read thoroughly* t]]]...
b. [*this book*], *namely* [*the one that I read thoroughly*]...

This implies that the relative CP in (21) is not a predicative category, but it is essentially similar to a definite free relative.

This analysis seems to be appropriate for a particular relative structure of Basque which de Rijk (1972: 129–131) calls pseudo-extraposition. This structure consists of a DP followed by a free relative clause coreferent with it (23a): the order of the constituents contrasts with normal relative clauses, which are prenominal (cf. (23b)):

(23) a. *Bein ba zan errege bat zazpi seme-alaba zitu-en-a.*
once there was king a seven sons and daughters had+REL+the
'once upon a time there was a king, one who had seven sons and daughters...'
b. *Bein ba zan zazpi seme-alaba zituen errege bat.*
once there was seven sons and daughters had+REL king a
'once upon a time there was a king who had seven sons and daughters'

The free relative is characterized by a verbal form in final position bearing the relative suffix and a suffixed definite article, which is the "head" of the free relative.

Since the free relative precedes this determiner, let us assume, following Kayne (1994: 94–95), that after the incorporation of the verbal form up to the external D°, the remnant IP (or CP) [17] has moved leftwards to Spec,DP: $[_{DP}$ IP $[_{DP}$ V°+I°(+C°)+D° t_{IP}]]

What is relevant for this discussion is the possibility of coordinating a free relative clause to a DP which is interpreted as its antecedent.[18]

A similar analysis has also been proposed for the Old English *se'ðe* relative clause, in which the "head" is followed by the relative pronoun *se* and the clausal particle *ðe*, and the pronoun is attracted by the Case of the antecedent. Since the same sequence of relative pronoun and clausal particle also introduces free relative clauses, the *se'ðe* relative can be analysed as a free relative in apposition to an antecedent noun phrase; this may represent at least the diachronic origin of the structure (Mitchell 1985: §§ 2161, 2175).

(24) *Ic wat wytodlice ðæt ge secað ðone hælend ðone ðe on rode ahangen wæs.*
I know truly that you seek the saviour$_{ACC}$ whom$_{ACC}$ that on cross hung was
'I know for certain that you seek the saviour who was hung on the cross.' (from Allen 1977: 87)

Note that a similar covert coordination seems to be independently required to account for nominal appositions like e.g. *Clinton, the president of the U.S.A.* The coordination in (22a) may seem problematic in that it relates two constituent of a different category, DP and CP. However, Schachter (1977) emphasizes that coordination does not require strict identity of categorial features of the two conjuncts, but rather, identity of function.

The "specifying conjunction" in (21) can never be spelled out. However, the intonation break which typically characterizes appositive relatives is also found in asyndetic coordination. This intonational contour can be conceived of as the phonological manifestation of the conjunction X°.[19]

Note that the "specifying coordination" differs from usual coordination in that the two conjuncts are interpreted as coreferent. Coreference forces the matching of the phi-features of the DP "head" and of the relative pronoun in Spec,CP: this matching is mediated by the head X°, which enters in two distinct agreement relations: it must agree with the DP in its Spec and also with the relative pronoun in Spec,CP:[20]

(25) $[_{XP}$ DP$_i$ $[_{XP}$ X° $[_{CP}$ OP$_i$ $[_{CP}$ ... t ...]]]]

The ability of conjunction to establish an agreement relation with both conjuncts is also observed in normal coordination, where the phi-features of the two conjuncts are combined in a fixed way: for instance, the coordination of two singular DPs gives plural agreement (cf. (26a)); the coordination of a first person DP with a second person DP gives first person agreement (cf. (26b)), and so on:

(26) a. $[_{\&P}$ *Gianni [e Mario]] sono venuti / * è venuto.*
Gianni and Mario have/* has come
'Gianni and Mario came.'

b. $[_{\&P}$ *Io [e te]] saremo / * sarete felici insieme.*
I and you will be$_{1PL./*2PL.}$ happy together
'You and I shall be happy together.'

c. $[_{\&P}$ *Tu [e lui]] sarete / * saranno felici insieme.*
you and he will be$_{2PL./*3PL.}$ happy together
'You and he will be happy together.'

This matching of the two conjuncts is not easy to formulate in the theory of checking proposed by Chomsky (1993; 1995), where agreement is conceived as a biunique relation between a head and a constituent in its checking domain. Anyway, this problem is not peculiar to the appositive structure proposed in (21).

In (25), the morphological agreement of the relative pronoun with the DP "head" is mediated by a local relation between the pronoun and the abstract conjunction X°. But when the relative pronoun is embedded within a pied piped constituent, as in (27), it cannot be governed by the X° external to the relative CP:

(27) a. *Gianni,* [*per incontrare* il quale] *sono venuto a Lucca, è un vecchio amico.*
'Gianni, in order to meet whom I came to Lucca, is an old friend of mine.'
b. *The elegant parties,* [*to be admitted to one of* which] *was a privilege, had been usually held at Delmonico's*
(Ishihara 1984: 398)

The solution to this problem may be Safir's (1986: 678–681) proposal that in pied piping contexts the relative pronoun moves to the highest position in the pied piped constituent in LF.[21] This proposal is supported by the observation that in English this movement is marginally possible in overt syntax (28b):

(28) a. *Those reports,* [*the height of the lettering on* which]$_j$ *the government prescribes* t_j, *are tedious.*
b. ? *Those reports,* [which$_i$ *the height of the lettering on* t_i]$_j$ *the government prescribes* t_j, *are tedious.*

This overt movement of the relative pronoun is impossible in contemporary Italian, but it is found in earlier stages of the language, especially in the literary form of the *relatif de liaison.*[22] In the following examples, a relative pronoun is fronted at the beginning of a clause-initial subordinate clause in which it is generated:

(29) a. *Non si meravigli dunque alcuno se lunga è la digressione della mia scusa, ma, sì come necessaria, la sua lunghezza paziente sostenga.* [La quale$_i$ *proseguendo* t_i], *dico che ...*
(Dante, *Convivio*, I, X)
hence nobody be astonished if the digression of my justification is long, but, as (it is) necessary, its lenght with patience tolerate. Which continuing, I say that ...
'Hence nobody be surprised if my justifying digression is long, but may he tolerate it as it is necessary. In continuing which digression, I say that...'

(29) b. [Alle quali cose$_i$ *qual fine serbi* t$_i$ *il giudicio di Dio*], *coloro il veggiano che governano il timone di questa nave ...*
to which things which end the judgement of God will set, those should see it who hold the steering wheel of this ship...
'Which end God's judgement will set to these things, those should see it who hold the steering-wheel of this ship...'

c. *...il nostro riconoscimento attende*; [il quale$_i$ *se a lungo andare* t$_i$ *non seguirà*], *niuno dubiti che ...*
our gratefulness (he) waits for; which if at lenght will not follow, nobody can doubt that...
'He expects our gratefulness; and if this (gratefulness) will not be expressed in due time, nobody can doub that...'
(Boccaccio, *Vita di Dante* I)

If in (27)- (28a) the relative pronoun moves covertly to the front of the pied piped phrase, it can establish a checking relation with X° at LF.

3.3. Problems with the conjunction analysis

One problem for this approach is the nature of the relative morphemes: in languages like English and Italian, the same relative forms appear both in restrictive and appositive relatives, and this is the most common situation across languages.[23] Therefore, the *wh*-morphemes which were analysed as relative determiners in the raising structure (chapter III) must be analysed as pronouns in the appositive structure (21). Actually, it is possible to analyse them as determiners selecting a *pro* NP category in (21), whereas in restrictive relatives they select a lexical NP, the "head".

The main problem with this analysis is the status of the relative CP in (21). It was suggested above that the CP must be interpreted as a referential expression. This means that the appositive relative is assimilated to a definite free relative coordinated to the antecedent DP. But in Italian and in English this assimilation is problematic, because the two structures differ in many respects:

(a) in Italian, free relatives may be introduced by the interrogative-like *wh*-pronoun *chi*. Appositive relatives instead have the same relative elements as restrictive relatives (*cui* and *il quale*) and disallow *chi*:

(30) [Chi/ *il quale *ha fatto questo*] *verrà punito*.
who did this will be punished
'The one who did this will be punished.'

(31) * *Gianni*, [il quale/ *chi *ha telefonato ieri*], *è un mio vecchio amico.*
'Gianni, who telephoned yesterday, is an old friend of mine.'

(b) Appositive relatives allow the pied piping of heavy constituents (cf. (25)), which is instead excluded in free relatives (this is the so called "matching effect"):

(32) a. *Gianni*, [*per incontrare* il quale] *sono venuto a Lucca, è un vecchio amico.*
'Gianni, in order to meet whom I came to Lucca, is an old friend of mine.'
b. * [[*Per incontrare* chi] *sono venuto a Lucca*] *è un vecchio amico.*
in order to meet whom I came to Lucca is an old friend.

(c) Kayne (1994: 113) observes that French appositive relatives, like restrictive ones, cannot be introduced by a direct object relative pronoun: in case of direct object relativization the relative pronoun is obligatorily omitted (cf. chapter VI). On the contrary, free relatives disallow the omission of the relative pronoun. The same contrast also holds in Italian:

(33) a. * *Le garçon qui Marie préfère s'appelle Georges.*
the boy who Marie prefers is called Georges.
(Kayne 1976: 260)
b. * *Jean, qui je connais bien, est intelligent.*
Jean, who I know well, is intelligent (Kayne 1994: 113)
c. *Qui tu as rencontré est malade.*
who you met is sick
'The one whom you met is sick.'
(Hirschbühler—Rivero 1983: 517)

(d) A related problem, mentioned in § 1.3 above, is that Romance appositive clauses may lack an overt relative pronoun; this requires the postulation of a null operator in the Spec,CP of the appositive relative. However, null operators are absolutely excluded in free relatives. In fact, their function is to turn a CP into a predicate (Browning 1987: 50–63): thus, it is difficult to claim that an appositive relative containing a null operator may count as a referential expression in a structure like (21).

These asymmetries suggest that it is incorrect to assimilate appositive relatives to free relatives.[24] This, in turn, casts doubt on the hypothesis that the X° head in (21) can be assimilated to a "specifying coordination".

3.4. Summary

Summarizing, the preceding discussion has examined an analysis of appositive relatives that is made available by Kayne's (1994) X-bar theory, namely the structure in (21), where the relation between a DP and an appositive CP is mediated by a functional head. This structure implements in a straightforward way two basic properties of the appositive relative: it is not c-commanded by the determiner of the "head" (at any level of representation), and the "head" cannot be reconstructed within the relative CP. Neverheless, the nature of the abstract head X° and the relation that it establishes between DP and CP remain unclear. Unless this problem is adequately solved, the structure (21) is nothing more than a notational variant of a rightward adjunction structure.

4. Kayne's LF movement approach

Kayne (1994: 110–115) argues that the relatively small syntactic differences between restrictive and appositive relatives do not justify the assumption of two completely unrelated structures. In the Romance languages, for instance, the distribution of relative pronouns is subject to similar restrictions in both structures (cf. (33a,b)); moreover, in some languages with prenominal relative clauses, like Japanese or Basque, the restrictive and appositive interpretations are not formally distinguished in any way.

Therefore, Kayne proposes that appositive relatives have the same base structure as restrictive ones and are derived by raising the relative "head" to Spec,CP:

(34) $[_{DP}$ D° $[_{CP}$ $[_{DP}$ NP $[_{DP}$ D_{REL} $t_{NP}]]$ $[_{IP}$ ... t_{DP} ...$]]]$

All the observed differences between the two relative structures are determined by the different derivation in LF. Kayne argues that the appositive interpretation results from the leftward movement of the IP of the relative clause to the Spec of the external determiner, which extracts the relative IP from the c-command domain of the latter:[25]

(35) $[_{DP}$ $[_{IP}$ ... t_j ...$]$ $[_{DP}$ D° $[_{CP}$ $[_{DP}$ NP $[_{DP}$ D_{REL} $t_{NP}]]_j$ $[_{CP}$ C° $t_{IP}]]]]$

As a result, the IP is not included in the determiner's restrictive term (cf. § II.2.1).

The LF movement of the relative IP is triggered by a feature which is manifested in the phonetic form as the intonation break characterizing appositive relatives.[26] This hypothesis receives independent justification from the phenomenon of right dislocation, which Kayne analyses as the covert equivalent of left dislocation: a constituent in its base position is assigned a feature which triggers its LF movement to a topic position; this feature is interpreted in PF as an intonation break preceding the relevant constituent (Kayne 1994: 78–83).

In languages with prenominal relatives, the movement of IP to Spec,DP takes place in overt syntax, and the trigger feature is checked before the Spellout operation. This explains why at least some languages with prenominal relatives do not mark the appositive interpretation by any special intonation.

Furthermore, the assumption of a common base structure for restrictive and appositive relatives allows a unified approach to the distribution of relative pronouns in both relative clause types. The only observed asymmetry is that in Modern English appositive relatives do not allow the omission of the relative determiner, unlike restrictives. This problem will be discussed in § VI.9.

Let us now see how Kayne's analysis can derive the other properties of appositive relatives.

4.1. Lack of reconstruction of the "head"

Kayne argues that appositive relatives show reconstruction effects. But if the arguments of chapter IV are correct, once various interferences are factored out, the "head" of appositive relatives turns out not to be reconstructed within the relative CP, contrary to the material that it pied pipes. This is at first sight problematic for a raising analysis, since both the "head" and the pied piped material originate within CP.

Chomsky's (1993) approach to reconstruction provides a solution to this problem. Recall that Chomsky conceives reconstruction as a part of the LF derivation that separates an operator from its nuclear scope. If the NP complement of a determiner is not reconstructed in the trace position, it becomes the restrictive term of the determiner in LF; if it is reconstructed, it becomes part of the nuclear scope, and the structure is interpreted as unrestricted quantification (cf. (3)–(4) of chapter IV). In other terms, reconstruction is not "free", but it is directly linked to the final interpretation of the sentence.

The appositive interpretation is determined by the LF raising of IP to the Spec of the external determiner, as shown in (35). Suppose that the relative "head" has been reconstructed within IP: the resulting LF representation is (36).

(36) $[_{DP} [_{IP} \ldots [_{DP}\ D_{REL}\ NP] \ldots] [_{DP}\ D^{\circ}\ [_{CP}\ [e]\ [C^{\circ}\ t_{IP}]]]]$

In this representation, the c-command domain of the external determiner is empty in LF: it does not contain any variable to be bound by it. This is an instance of vacuous quantification, and it is ruled out by the Full Interpretation principle. The structure in (35) can only be interpreted if the "head" is *not* recostructed within the relative CP.

The same principle explains the obligatory reconstruction of the material pied piped by the relative "head", attested by the Principle C effect in (36a) of chapter IV, repeated here as (37):

(37) * *Andrea, [le cui insinuazioni su Beatrice*$_i$] pro$_i$ *non è più disposta a sopportare* t
A, whose insinuations on B she is no longer willing to stand

If the pied piped DP is not reconstructed, the LF representation of (37) is (38):[27]

(38) $[_{DP} [_{IP} \ldots t_i \ldots] [_{DP}$ A. $[_{CP}[_{DP}$ le cui t_N insinuazioni su B.$]_i$ $[_{CP}\ C^{\circ}\ t_{IP}]]]]$

After the LF movement of IP, the pied piped material is stranded in Spec,CP, but it cannot receive any interpretation within the restrictive term of the external D°. In order to be correctly interpreted, the pied piped material must be reconstructed within IP.

As mentioned in § IV.6, the lack of reconstruction of the relative "head" is consistent with the fact that idiom chunks cannot undergo appositive relativization, under the assumption that the constituents of the idiomatic expression must be adjacent in the LF representation:[28]

(39) a. * *the headway, which we made*
b. LF: $[_{DP} [_{IP}$ we made $t_i] [_{DP}$ the $[_{CP}[_{DP}[_{NP}$ headway$]$ $[_{DP}$ which $t_{NP}]]_i$ $[_{CP}\ C^{\circ}\ IP]]]]$

4.2. The definiteness of the relative trace

Let us now consider the contrast between restrictives and appositives with respect to the definiteness of the trace, exemplified in (13)–(14), repeated here:

(13) *The men*$_j$ *that there were* t_j *in the garden were all diplomats.*
(14) * *She left the magazines, which*$_i$ *there are* t_i *on the table.*
(Postal 1993b: 745)

An approach to the contrast between (13) and (14) can be built upon Heim's (1987: 22–24) proposal that individual variables count as strong DPs, and as such they cannot occur in the postcopular position of the existential *there* construction at LF. In Heim's analysis, an individual variable is an A' bound DP trace occupying an argument position: an example is the trace of the *wh*-phrase *which actors* in (40):

(40) a. ?? [*Which actors*] *were there* t *in the room*?
b. LF: [*Which actors*]$_i$ *were there* x_i *in the room*?
(Heim 1987: 27)

In the following example (41), instead, the *wh*-phrase *how many soldiers* allows a nonpresuppositional interpretation in which the restrictive term is reconstructed in the trace position; this yields the LF representation in (41b), with the postcopular position filled by a weak DP [x-*many soldiers*]:

(41) a. [*How many soldiers*] *were there* t *in the infirmary*?
b. LF: *How were there* [x-*many soldiers*] *in the infirmary*?
(Heim 1987: 29)[29]

The contrast between (40b) and (41b) closely resembles the restricted and unrestricted quantification structures proposed by Chomsky (1993): unrestricted quantification corresponds to the reconstruction of the restrictive term in (41b), and it yields a LF representation that does not violate the definiteness constraint.[30]

The contrast between (13) and (14) can now be derived as follows. As already discussed in § III.3, in the restrictive relative (13) the "head" [*men*] is reconstructed in the trace position, yielding a representation closely parallel to (41b): since the reconstructed "head" is non-definite, it does not give rise to the definiteness effect.

(42) [$_{DP}$ *the* [$_{CP}$ *there were* [*men*] *in the garden*]]

In the appositive relative (14), instead, the "head" cannot be reconstructed in the postcopular position; the latter necessarily contains an individual variable in LF,[31] which qualifies as a strong DP, and gives rise to the definiteness effect:[32]

(43) [$_{DP}$[$_{IP}$ *there are* t_i *on the table*] [$_{DP}$ *the* [$_{CP}$[$_{DP}$ *magazines* [*which* t]]$_i$ [$_{CP}$ C° t_{IP}]]]]

4.3. Floating quantifiers

The possibility of floating quantification in appositive relatives allows a similar explanation. The example (15a), repeated as (44a), has the LF representation (44b) in the restrictive interpretation and (44c) in the appositive interpretation:

(44) a. *Ho interrogato gli studenti che avevano superato tutti* t *la prova scritta.*
I examined the students that had passed all the written exam
'I examined the students all of whom had passed the written exam.'
b. * [$_{DP}$ *gli* [$_{CP}$ *che avevano superato* [$_{QP}$ *tutti* [*studenti*]] *la prova scritta*]]]
c. [$_{DP}$ [$_{IP}$ ... [$_{QP}$ *tutti* t_j] ...] [$_{DP}$ *gli* [$_{CP}$ [*studenti*]$_j$ [$_{CP}$ C° t_{IP}]]]]

According to Kayne (1994: 87), in relative clauses lacking an overt relative pronoun, like (44), the raised constituent is a NP. However, in the next chapter it will be argued on independent grounds that in these structures the NP "head" is introduced by a phonetically null relative D°; under this assumption, the relative phrase in (44) is a DP:

(45) [$_{DP}$ *gli* [$_{CP}$ [$_{DP}$ D° [$_{NP}$ *studenti*]] [$_{CP}$ *che* ... [$_{QP}$ *tutti* t_{DP}]]]]

In § II.3.5 it was assumed that the Q° *tutti* obligatorily selects a [+definite] DP. However, the grammaticality of the following examples (46) show that the Q° also allows an individual variable in its complement position:

(46) a. ? [*Quali libri*]$_i$ *hai deciso di vendere* [*tutti* t_i] *a Gianni* ?[33]
which books did you decide to sell all to Gianni
'All of which books did you decide to sell to Gianni?'
b. [*Quels enfants*]$_i$ *crois-tu que Pierre a reconnus* [*tous* t_i] ?
which children believe you that Pierre has recognized all
'All of which children do you believe that Pierre recognized?' (Doetjes 1992: 316)

This suggests that *tutti* has a weaker requirement: namely, it disallows in its complement position weak (non-definite) phrases.

Let us now reconsider the contrast in (44). In the restrictive relative (44b), the relative DP is reconstructed in the complement position of the universal Q° *tutti,* but since it is non-definite, the resulting representation is illicit. In the appositive relative (44c), on the contrary, the relative DP is not reconstructed, and it leaves an individual variable in the complement position of Q°, resulting in an interpretable structure.

4.4. *Relatif de liaison* and non-nominal antecedents

In the examples (19), repeated here, the "head" of the appositive clause is a non-nominal category:

(19) a. *Mary is [courageous], which I will never be.*
b. *John is [in the garden], which is where I should be.*
c. *Mary has [resigned], which John hasn't.*
d. *[John was late], which was unfortunate.*

A raising analysis of these examples is problematic in two respects. First, it must be assumed that the AP/PP/VP/IP "head" originates as a complement to the relative determiner *which*; second, it is unclear which external determiner would select the relative CP.

In his monograph, Kayne left open the question whether examples like (19) should receive a raising analysis; some comparative evidence suggests that they probably should not. In the Italian equivalent of (19a) or (19d), the purported relative clause cannot be introduced by the complementizer *che* or by a simple relative determiner, but it must be introduced by the relative-like connectors *cosa che* 'which thing', *il che* 'the that', *la qual cosa* 'the which thing':

(47) a. *Maria è coraggiosa, cosa che io non sarò mai.*
b. *Gianni arrivò tardi, il che / la qual cosa fu controproducente.*

Similarly, in the French equivalent of (18d) the relative clause is introduced by a pronominal "head" *ce* distinct from the purported clausal "head":

(48) *Jean était en retard, ce qui était embarassant.*

If (19) are true appositive relatives, this asymmetry is mysterious.

Recall also from § 2.3 that in many languages relative pronouns can be used for cross-sentential anaphora in the so called *relatif de liaison.* An example is (18), repeated here:

(18) *...whom we name hereafter the Prince of Cumberland:* which honour *must not unaccompanied invest him only...*
(*Macbeth* I.4, 38–40)

These examples cannot be reduced to the raising analysis, because the purported "head" is separated from the relative clause by a sentence boundary. Here the relative morphemes are used as anaphoric determiners or pronouns.[34] But if this possibility is allowed, then nothing in principle excludes that the same anaphoric use is involved in the English examples (19). From this perspective, (19) may not be relative structures at all: the relative pronoun is actually an anaphoric pronoun, and the purported appositive relative is either coordinated to the main clause or parenthetical.

Unless this alternative analysis can be excluded, and it can be convincingly demonstrated that the examples in (19) must be true relative structures, they do not constitute counterexamples to the raising analysis of appositives.

4.5. Islandhood for binding

As noted by Kayne (1994: fn. 72 to chapter 8), the islandhood of appositive relatives for binding (cf. (4)–(6)) does not follow from his analysis, since the appositive CP is internal to the matrix clause. However, he points out that, independently of the appositive interpretation, the possibility of binding from outside could be sensitive to the nature of the external determiner: in fact, definite descriptions in general are opaque for binding.

Intuitively, the islandhood for binding seems to depend on the fact that appositive relatives constitute "backgrounded" information with respect to the matrix clause. It is unclear whether this property is in some way codified in the syntax. It may perhaps be assumed that the appositive IP moves out of Spec,DP to a Topic position of the matrix clause, where it is not c-commanded by any matrix binder. However, in many cases this "long" LF scrambling of IP would be problematic with respect to Subjacency.[35]

Be this as it may, in this respect Kayne's analysis of appositives is simply equivalent to the adjunct analysis which takes them to be adjoined to the antecedent DP (§ 1.4).

4.6. Independent evidence for movement of IP

The hypothesis that the IP category can move out of the complement position of C° may seem problematic, in that IP is not a "perfect" clausal projection in the sense of Grimshaw (1991). However, this hypothesis is independently supported by a focus construction of Caribbean Spanish discussed by Toribio (1992: 299–303):

(49) a. *Mamá tenía que ir era al mercado.*
mummy had to go was to the market
'It was to the market that mummy had to go.'
b. *Ese niño está es enfermizo.*
that child is is sickly
'It is sickly that that child is.'

The presence of a copula makes this structure similar to a pseudocleft; but strikingly, the preposed clause in (49) is not introduced by a relative pronoun.

Toribio proposes that the clause preceding the copula contains a null operator in Spec,CP, but this leaves unexplained why this constituent cannot be introduced by a complementizer:[36]

(50) a. * $[_{CP}$ Op$_i$ $[_{CP}$ *que yo vi* t_i $]$ *fue* [*a Juan*]
that I saw was Juan

Let us consider an alternative derivation.

As discussed in § I.2.1, Rizzi (1995) proposes that Italian has a functional projection FocusP lying between C° and IP. The Spec of this projection is the landing site of focussed and *wh*-phrases. In focus constructions, the head Focus° is empty (its phonetic manifestation may be the stress on the focussed constituent in its Spec).

Let us assume that the Focus projection is present in Spanish as well. Suppose that in this focus construction the copula does not select a full CP, but rather a Focus Phrase (51a). The focussed constituent moves to Spec,FocP, a position immediately following the copula (51b). The surface form of (49) is obtained by further raising the IP complement of Focus° to the left of the copula, as represented in (51c):

(51) a. *era* $[_{FocP}$ Foc° $[_{IP}$ *mamá tenía que ir* $[_{PP}$ *al mercado*$]_i]]$
b. *era* $[_{FocP}$[*al mercado*$]_i$ $[_{FocP}$ Foc° $[_{IP}$ *mamá tenía que ir* $t_i]]$
c. $[_{IP}$ *mamá tenía que ir* $t_i]$ *era* $[_{FocP}$[*al mercado*$]_i$ $[_{FocP}$ Foc° $t_{IP}]]$

If the constituent preceding the copula is IP, the fact that it cannot be introduced by a complementizer or a relative pronoun follows without further assumptions.[37]

5. Summary

Various approaches to the syntax of appositive relatives assume that they do not form a constituent with the noun phrase that they modify, or that they do so only at a very late stage of the derivation. These approaches are in general incompatible with the Linear Correspondence Axiom.

The minimal difference between restrictive and appositive relatives is that the former but not the latter are c-commanded by the determiner of the modified noun phrase. This difference cannot be expressed in terms of the different level of the adjunction of CP, because rightward adjunction is excluded by the LCA. There remain two alternative solutions.

If one assumes that the appositive relative is generated outside the scope of the determiner of the "head", the only possibility consistent with the LCA is to postulate an abstract functional head mediating the relation between an antecedent DP and the relative CP. But the nature of this head is unclear.

The second possibility is to assume that the appositive relative is generated in the c-command domain of the determiner, but it moves out of it in LF (Kayne 1994). In this way, the syntactic difference between restrictive and appositive relatives is reduced to the minimum: the overt syntax is identical, and the only difference lies in the LF derivation.

The hypothesis that appositive relatives are derived by raising the "head" is apparently contradicted by the lack of reconstruction observed in chapter IV. However, this apparent contradiction can be explained on the basis of the theory of reconstruction proposed by Chomsky (1993).

The next chapter is devoted to an apparently superficial, but important aspect of the syntax of relative clauses: the distribution of relative determiners and complementizers.

Chapter VI
Doubly filled Comp effects

0. Introduction

One of the most interesting aspects of the syntax of relative clauses is the distribution of relative determiners and its interaction with the syntax of complementizers. This range of phenomena is intricate and subject to considerable cross-linguistic variation. This chapter will compare the interaction of relative determiners and complementizers in English and Italian, with some occasional references to French and other languages. The following discussion will show that this apparently superficial phenomenon is actually related to some important properties of the functional structure of the clause, especially of the outermost part of it, the "Comp area", the investigation of which is only at the beginning stage.

The main part of the discussion will be devoted to finite restrictive relatives. Some peculiarities of appositives will be discussed in § 9. Infinitival relatives will only be taken into account with respect to the relevant phenomenon; other important aspects of their syntax will be admittedly disregarded. In particular, the differences between the Italian relative determiners *cui* and *il quale*, specifically related to the phenomenon of pied piping (cf. Cinque 1978), are discussed in Bianchi (1995: chapter VI).

Let us now review the basic paradigms of English and Italian finite restrictive relatives.

(1) a. *the book* that *I read*
b. *the book* which *I read*
c. *the book I read*
d. *the way in* which *he acted*

(2) a. *il libro* che *ho letto*
b. * *il libro* il quale *ho letto*
c. * *il libro ho letto*
d. *il modo* in cui *agiva*

The relative clause type exemplified in (1a) and (2a) will be called *that*-relative: in this type, the relative CP is introduced by the declarative complementizer of finite clauses, and it has no overt relative pronoun.[1]

The relative clause exemplified in (1b) will be called *wh*-relative. This structure consists in the relativization of a DP by means of an overt relative pronoun, where the latter does not pied pipe a larger constituent.

In this respect the *wh*-relative is distinguished from the pied piping relative exemplified in (1d)–(2d). As shown in (2b), the *wh*-relative is ungrammatical in contemporary Italian.[2] In English, it is more marked than the other options, at least when it is introduced by the relative determiner *which*.

The third type, exemplified in (1c), will be called the zero relative, because it is not introduced by any overt marker of subordination, either the complementizer or a relative pronoun. This structure too is ungrammatical in contemporary Italian (cf. (2c)).

A further difference between the two languages is that English allows preposition stranding, whereas Italian does not. As this difference is not specific to relative clauses, it will not be discussed here. Pied piping is possible in both languages, as shown in (1d) and (2d), although in English pied piping of a preposition is often more marked than preposition stranding.

1. Approaches to the problem in the adjunct analysis

1.1. The standard analysis

In the adjunct analysis, a relative pronoun moves to Spec of the relative CP, to the left of the declarative complementizer:

(3) *the book* [$_{CP}$ *which* [*that* [$_{IP}$ *I bought* t]]]

The surface string corresponding to (3) is ungrammatical in Modern English: * *the book which that I bought*. Chomsky—Lasnik (1977: 446) propose a specific filter, the Doubly Filled Comp Filter, which rules out the cooccurrence of a *wh*-phrase and an overt complementizer in the same Comp position; an updated version of the filter can be formulated as follows:

(4) Doubly Filled Comp Filter: Spec, CP and C° cannot be both lexically filled.[3]

According to Chomsky—Lasnik, the filter must be satisfied by deleting either one of the two elements in Comp. The deletion of the complementizer in (3) yields the *wh*-relative (1b). Alternatively, the deletion of the relative pronoun yields the *that*-relative (1a). In headed relatives, the relative pronoun can be deleted because its content can be inferred from the antecedent which it is coindexed with (cf. § II.1), and hence its deletion complies with the recoverability condition.

However, when Spec,CP contains lexical material pied piped by the relative pronoun, the recoverability condition blocks its deletion; in this case, the filter can be satisfied only by deleting the complementizer. This gives the pied piping relative (1d), where the complementizer is obligatorily omitted.

Finally, the zero relative in (1c) is taken to derive from the *that*-relative in (1a) by an optional rule of *that* deletion. This rule is not specific to relative clauses, but it also applies to declarative complement clauses, as shown in (5b):

(5) a. *I think* [$_{CP}$ *that you're right*].
  b. *I think* [$_{CP}$ *Ø you're right*].

Thus, the distribution of relative pronouns and complementizers in English is determined by the Doubly Filled Comp Filter plus an optional rule of complementizer deletion.

In Italian the situation is more complex. The Doubly Filled Comp Filter accounts for the obligatory deletion of the complementizer *che* in the pied piping relative (2d), and it is consistent with the grammaticality of the *that*-relative (2a). However, it does not explain the ungrammaticality of the *wh*-relative (2b): in fact, in (2b) it should be possible to delete the complementizer in order to satisfy the filter, in the same way as in (2d). Therefore, the *wh*-relative must be excluded by an independent condition. Kayne (1976) proposed for French a rule of Relative NP Deletion which deletes a "bare" relative NP in Spec,CP. A similar rule was adopted for Italian by Cinque (1978).

The ungrammaticality of the zero relative (2c) follows from the fact that contemporary Italian does not have an optional rule of complementizer deletion: in fact, the equivalent of (5) requires an overt complementizer:[4]

(6) a. Pro *penso* [$_{CP}$ *che hai ragione*].
  b. * Pro *penso* [$_{CP}$ *Ø hai ragione*].

The differences between the English and Italian paradigms are thus reduced to two distinct parametric properties: the rule of Relative NP Deletion, which exists in Italian but not in English,[5] and the optional rule of C° deletion, which exists only in English. The Doubly Filled Comp Filter holds in both languages, but it is not a universal principle: in fact, the string <*wh that*> was grammatical in Middle English, both in interrogative and in relative clauses.

What is very appealing about the deletion analysis is its economy: three distinct relative clause types are reduced to a single underlying structure, and variation is determined by superficial deletion processes. But from the perspective of the current syntactic framework, the deletion rules that this analysis relies on raise two substantial problems:
(1) the constraints on the deletion rules are unclear;
(2) the status of the trigger of these rules is unclear.
Starting from the first problem, consider the recoverability condition, which blocks the deletion of Spec,CP in (1d) and (2d). Though intuitively clear, the notion of recoverability has never been explicitly formulated, and its application is not straightforward in a number of cases. For instance, relative pronouns are taken to be recoverable in restrictive relatives, because their feature content can be inferred from that of the antecedent NP which they are coindexed with; in English appositive relatives, however, the relative pronoun cannot be deleted, and the only grammatical form is the *wh*-relative:

(7) a. * *This book, that I read thoroughly, ...*
b. *This book, which I read thoroughly, ...*
c. ** *This book, I read thoroughly, ...*

Chomsky—Lasnik (1977: fn. 46) stipulate that in appositives the relative pronoun has some semantic import that makes it unrecoverable. But as Maling (1978: 721) points out, this stipulation is flawed because it does not account for Romance languages nor Old/Middle English, in which appositive relatives allow the deletion of the relative pronoun.[6]
A second constraint affects deleted complementizers in that they are phonetically empty, namely, the empty category principle. In fact, optional deletion cannot apply to a complementizer that is immediately dominated by a barrier, as in (8), and thus fails to be properly governed:

(8) a. [*(*That*) *John left*] *is a pity.* (Authier 1992: 334)
b. *the proof* [* (*that*) *this set is recursive*] (Ross 1967: 80)

But at this point, a paradox arises: in the adjunct analysis (3), the relative CP is a strong island: therefore, the deleted complementizer of the zero relative (1c) is wrongly predicted to violate the ECP.
Apart from these specific problems, there is a broader conceptual question related to the independence of the PF and LF components (Chomsky 1993; 1995). From a minimalist perspective, deletion should be conceived of as a process applying in the PF component, which marks a constituent as "invisible at the [PF] interface" (Chomsky 1995: 228). But note that the recoverability condition is not phonological,

nor has it ever been stated in purely configurational terms; on the contrary, it seems to make reference to the semantic content of the items involved, which should be visible at the LF interface only. This seems to imply that PF deletion is sensitive to some properties of the items to be deleted which cannot be reduced to purely formal features accessible throughout the derivation. Or, reversing the point, it must be assumed that the LF component "keeps track" of the fact that a certain lexical item has been marked as invisible at the PF interface.

Similarly, note that if deletion is a PF process, and if the empty category principle is sensitive to the phonetic content of the categories, then it follows that the ECP applies to the output of the PF component to rule out (8). But it is not obvious that a purely configurational principle like the ECP should apply to the PF representation (cf. Chomsky 1995: fn. 19 to chapter 3). Actually, it has been argued by many authors (among them Kayne (1984) and Longobardi (1991)) that the ECP applies to the traces left by LF movement; but note that under the copy theory of traces, LF traces are not "phonetically empty" in any sense.[7] This suggests that the ECP is not sensitive to the phonetic content of the categories, but rather to some other property that is visible in LF.

In conclusion, it appears that the best established constraints on PF deletion face empirical as well as conceptual problems.

But the problem of the trigger is even more serious: it is unclear how the operation of PF deletion contributes to the convergence of the derivation. The question is particularly worrisome with respect to the optional rule of *that-* deletion applying in (1c) and (5b), which is by definition not necessary in order for the derivation to converge.

Intuitively, the Doubly Filled Comp Filter seems to be related to economy considerations: the "cost" of a PF representation may be a function of the number of overt terminal symbols that it contains.[8] Though fairly plausible, a similar economy measure requires a remarkable crosslinguistic parametrization which remains essentially stipulative: in fact, the three relative clause types in (1) are equally costly in English, whereas in Italian the *that*-relative seems to be more economical than the *wh*-relative, and economy cannot force the deletion of both the relative pronoun and the complementizer as in (2c). Thus, it seems that seems that an economy principle of this type cannot correctly trigger the deletion rules to account for the paradigms in (1) and (2).[9]

1.2. Pesetsky's (1994) optimality approach

Within the framework of Optimality Theory, Pesetsky (1994) conceives the deletion of relative pronouns and complementizers as an effect of "telegraphic speech" of the sort that is found in early stages of language acquisition and in agrammatic aphasia. The omission of function words does not imply the absence of the corresponding functional categories in the phrase structure, but rather, it is an effect of a general principle that prevents the pronounciation of function words:

(9) Telegraph: a function morpheme must be unpronounced.
(10) Deletion: Mark α [+silent].

The weight of telegraphic effects in a given language is determined by the relative ranking of Telegraph with respect to other principles in the grammar. In agrammatic aphasia, Telegraph is ranked higher than the recoverability principle, so that even unrecoverable function words fail to be pronounced. Pesetsky proposes that the limited omission of recoverable function words in normal adult grammars is determined by the same principle, ranked after Recoverability. The cross-linguistic variation of deletion phenomena depends on the different ranking of Telegraph with respect to other relevant principles. It is thus possible to subsume under Telegraph the Doubly Filled Comp Filter, the rule of Relative NP Deletion, and the optional *that* deletion of English.

The Doubly Filled Comp Filter derives from the interplay of Telegraph with the following principle:

(11) Left Edge (C): A complementizer must be pronounced, and it must be pronounced at the left edge of CP.

Pesetsky proposes the following hierarchy of principles for French and Italian:

(12) Recoverability >> Left Edge (C) >> Telegraph[10]

By assumption, the declarative complementizer and relative pronouns are recoverable.

Consider first the complementizer deletion in (6b): it satisfies Telegraph, but it also violates Left Edge(C), which is more highly ranked. On the other hand, the realization of the complementizer only violates Telegraph. Therefore, the second option is optimal and excludes the first one.

The hierarchy in (12) also derives the paradigm (2). Pesetsky assumes the equivalent of (3) as the pre-spellout structure of the relative clause:

(13) *il libro* $[_{CP}$ *il quale* $[_{CP}$ *che* [*ho letto* t]]]

The following options constitute the candidate set of output forms:
(a) the pronounciation of both Spec,CP and C° violates Left Edge(C) and Telegraph;
(b) the deletion of the relative pronoun only (= *that*-relative) violates Telegraph;
(c) the deletion of the complementizer (= *wh*-relative) violates Left Edge(C);
(d) the deletion of both Spec,CP and C° (= zero relative) violates Left Edge(C).

The optimal form is (b), since it violates only one principle, Telegraph, which is ranked below the principle Left Edge (C) violated by all the other options. The hierarchy thus subsumes the Doubly Filled Comp Filter, and it determines the effects of the rule Relative NP Deletion and the impossibility of the zero relative (cf. (2a–c)).

As for the pied piping relative (2d), its pre-spellout structure is (14):

(14) *il modo* [$_{CP}$ *in cui* [$_{CP}$ *che* [pro *agiva* t]]]

Here the presence of an unrecoverable preposition in Spec,CP makes the principle Left Edge(C) irrelevant: in fact, in order to place the complementizer at the left edge of CP, it would be necessary to delete the pied piped preposition: but this deletion is prevented by Recoverability, which is ranked higher than Left Edge(C). Thus the only two options to be compared are: (a) no deletion, or (b) deletion of the declarative complementizer. The first option violates Telegraph, the second option does not: hence the latter is optimal. In this way, the hierarchy in (12) derives the whole paradigm in (2).

Turning to the English paradigm in (1), the differences with respect to Italian derive from a different ranking of the same principles; in particular, Left Edge(C) and Telegraph are equally ranked:

(15) Recoverability >> Left Edge(C) <> Telegraph

Consider the underlying relative structure (3), repeated here:

(3) *the book* [$_{CP}$ *which* [*that* [$_{IP}$ *I bought* t]]]

The candidate set for (3) is identical to the one of the Italian equivalent (13), but the different ranking of the principles in English determines a different status of the various options. The *that*-relative violates Telegraph, because a function word is pronounced; the *wh*-relative instead violates Left Edge(C), because the complementizer does not appear at the left edge of the relative clause. The zero relatives also violates Left-Edge(C). Each option violates only one of two equally ranked principles. On the other hand, if both the relative pronoun and the complementizer are pronounced, both principles are violated. Therefore, the last option is ruled out; but the first three options are equivalent, and hence they are all grammatical.

With respect to pied piping, the hierarchy in (15) gives the same results as the one in (12): the pied piped material cannot be deleted, and it necessarily violates Left Edge(C); but pronouncing the complementizer yields an extra violation of Telegraph. Thus, all the paradigm in (1) is correctly derived.

An interesting aspect of Pesetsky's analysis is that it establishes a correlation between the optional complementizer deletion of the zero relative and the grammaticality of the *wh*-relative: both options are allowed by the equal ranking of Left Edge(C) and Telegraph in English. The standard analysis, on the contrary, did not predict any correlation between the two relative clause types, since they were determined by two independent rules (*that*-deletion versus Relative NP Deletion).

On the other hand, one aspect of this analysis is unconvincing. The relative pronoun is by hypothesis recoverable, and intuitively, it should be classified as a function word in the same way as the complementizer. Yet the relative pronoun is not visible for the principle Telegraph: its pronounciation does not give rise to a violation of this principle. If it did, the English *wh*-relative (1b) would violate both Telegraph and Left-Edge(C), and hence it would not be optimal as compared to the *that*-relative and the zero relative, each of which violates only one principle.[11]

Moreover, the obligatoriness of relative pronouns in English appositives seems to require the stipulation that the pronouns are unrecoverable in this syntactic context (whereas they are recoverable in Romance appositives). To this effect, one could invoke Cinque's (1982: 280) proposal that relative pronouns are recoverable only when they are c-commanded by the "head", but in English appositives they are not c-commanded by it. This requires the assumption of a structure in which the appositive clause does not form a constituent with its "head" (cf. §§ V.1.1–1.2).

Note also that in this system a relative clause which is introduced by both a relative pronoun and a complementizer necessarily violates at least one principle more than the simple *that*-relative. This predicts that there cannot exist a language with both *that*- and <*wh that*>-relative clauses: but this possibility is attested in Old[12] and Middle English (see Allen 1977: 239–242; Grimshaw 1975: 216–217, and § VII.2 below).[13]

1.3. Rizzi's (1990) feature compatibility approach

Rizzi (1990: 65–71) reduces the Doubly Filled Comp effects of Modern English to an incompatibility of the feature specifications of the complementizer *that* and of *wh*-pronouns (either interrogative or relative). He proposes a classification of the complementizer system of English based on the features [± *wh*] and [± predicative]. The feature [+predicative] characterizes relative clauses, as opposed to interrogative and declarative clauses:

(16) a. [+*wh* -pred] → ∅ *I wonder* [$_{CP}$ *what* [∅ [*you saw* t]]].
b. [+*wh* +pred] → ∅ *the thing* [$_{CP}$ *which* [∅ [*you saw* t]]].
c. [-*wh* +pred] → *that* *the thing* [$_{CP}$ Op [*that* [*you saw* t]]].
d. [-*wh* -pred] → *that* *I know* [$_{CP}$ *that* [*you saw it*]].

The form of the complementizer agreeing with a [+*wh*] specifier is phonetically null, as shown in (16a–b). On the other hand, the complementizer *that* is intrinsically specified as [-*wh*]: therefore, it appears in declarative complement clauses, where Spec,CP is empty (16d), and in predicative CPs whose Spec contains a null operator, under the assumption that the latter is underspecified for [± *wh*] (16c). The ungrammaticality of the sequence * <*wh that*> in Modern English is thus reduced to feature incompatibility: a [+ *wh*] cannot occupy the Spec of the [- *wh*] C° *that.*

(17) a. * *I wonder* [$_{CP}$ *what*$_{[+WH]}$ [$_{CP}$ *that*$_{[-WH]}$ [*you saw* t]
b. * *the thing* [$_{CP}$ *which*$_{[+WH]}$ [$_{CP}$ *that*$_{[-WH]}$ [*you saw* t]

The grammaticality of (17) in Middle English may be derived from the assumption that the complementizer *that* was underspecified for [± *wh*] in that stage of the language, so that no feature clash arose.

In current terms, the configuration does not satisfy the Wh-criterion (Rizzi 1991); the consequences for Middle English are unclear. (See § VII.2 for a different approach to the problem.)

Note that Rizzi's approach does not account for the Italian paradigm (2). Italian has both the null operator (cf. (2a)) and the [+wh] relativization strategy (cf. the pied piping example (2d)); it is unclear why the latter strategy cannot be exploited for the relativization of a noun phrase, as in (2b) above.

A development of Rizzi's approach is proposed by Toribio (1992), who tries to derive from the typology of complementizers in (14) the effects of the rule of Relative NP Deletion in Spanish.

According to Rizzi (1990: chapter II), subject relativization in English leaves a trace in Spec,IP that must be properly head-governed by the immediately c-commanding head C° in order to satisfy the ECP (Rizzi 1990: 87). C° is by itself inert for government, but it can be turned into a proper governor by means of an agreement relation which coindexes it with the trace in Spec,IP. In the following abstract configuration:

(18) [NP_i [$_{CP}$ OP_i [$_{CP}$ C° [$_{IP}$ t_i ...]]]

C° can get coindexed with the trace by means of two distinct agreement patterns: it can be coindexed with the relative operator in its Spec by Spec/head agreement, or it can agree directly with the NP "head" of the relative clause, in virtue of the relation of predication (cf. § II.1). The first relation is called A'-agreement, the second one A-agreement.

These two agreement relations are exploited by the complementizers of *wh*-relatives and *that*-relatives respectively: the null C° [+*wh*, +pred] agrees with the relative pronoun in its Spec, which is overtly specified for phi-features. The [-*wh* +pred] complementizer *that*, on the contrary, cannot agree with the null operator in its Spec, because the latter is intrinsically anaphoric, and it is incompatible with agreement; therefore, *that* must agree directly with the "head" of the relative clause.

Toribio assumes that in Spanish too subject relativization leaves a trace in Spec,IP which is subject to the Empty Category Principle. As in English, there are two distinct relative complementizers: the complementizer *que* hosts a null operator in its Spec and exploits A- agreement with the "head"; the zero complementizer instead agrees with an overt relative pronoun in its Spec:

(19) a. *el dueño* [Op_i [*que* [$_{IP}$ t_i *vende la casa*]]]
A-agr
the owner that sells the house

b. *el dueño* [*quien*$_i$ [Ø [$_{IP}$ t_i *vende la casa*]]]
A'-agr
the owner who sells the house
'The owner who is selling the house...'

Toribio notes that in Spanish the *wh*-relative (19b) is only possible as an appositive: in other terms, restrictive relativization of the subject shows the effects of Relative NP Deletion, whereas appositive relativization does not. She relates this contrast to a difference in the level of adjunction of the relative clause (cf. § V.1.4). The restrictive relative is adjoined to the NP level, and it is in the domain of A-agreement with the head N°: therefore, A-agreement is obligatory, yielding the *that*-relative (19a). The appositive relative, on the contrary, is adjoined to DP: the relative C° cannot establish a direct A-agreement with the head N°, but only an indirect relation mediated by the functional head D°. In this configuration, A-agreement is possible, but it is not mandatory: therefore, both of the variants in (19) are possible as appositive relatives.

Note that this analysis derives the ungrammaticality of the *wh*-restrictive relative only for the case of local subject relativization: in fact, the analysis relies on the presence of agreement in C°, which is only required when C° must properly govern a trace in the Spec,IP. However, the *wh*-relative is ungrammatical also for the restrictive relativization of direct objects and of other "bare" DPs, and even for non-local subject relativization (Brucart 1992: 120–124; Donati 1995; cf. also Cinque 1981 on Italian).[14] In order to cover these cases, Toribio has to assume that the relative C° must always be licensed by some agreement relation, even in configurations that do not involve a potential ECP violation.

As a final point, note that, though Rizzi's analysis correctly derives the Doubly Filled Comp effects in English and Italian headed relative clauses, it cannot be easily extended to other languages. Consider for instance the following Swedish paradigms: among various *wh*-structures, the Doubly Filled Comp effect emerges only in headed relative clauses:

(20) a. *Jag undrar* vem som *Jan mötte på puben*
I wonder who C° Jan met at the pub
'I wonder who Jan met at the pub.'

b. *Jag undrar* vilket lag som *är bäst*
I wonder which team C° is the best
'I wonder which team is the best.'

c. *Vi åt* vad som *Kalle bjöd oss*
we ate what C° Kalle offered us
'We ate what Kalle offered us.'

d. Vilken idiot som *var här!*
what an idiot C° was here
'What an idiot was here!'

(21) a. *Den hund* som *vi såg heter Rasmus*
the dog C° we saw is called Rasmus

b. *Den hund* vilken *vi såg heter Rasmus*
the dog which we saw is called Rasmus
'The dog that we saw is called Rasmus.'

(21) c. * *De hundar* vilka som *var här är farliga*
the dogs which C° were here are dangerous
(Swedish; Andersson 1975:156–166)

The complementizer *som* cooccurs with *wh*-phrases in indirect questions (20a–b), free relatives (20c) and exclamative clauses (20d); therefore, it seems to be compatible with a [+*wh*] specifier. However, in the headed relative clauses (21) it is in complementary distribution with the [+*wh*] relative pronoun *vilken,* showing a Doubly Filled Comp effect.

Of course, within Rizzi's (1990) approach it is possible to assume that the *wh*-pronouns in (20) have a different feature specification than the relative *vilken* in (21). In fact, in headed relatives the *wh*-pronoun can be considered semantically vacuous, whereas in interrogatives, free relatives and exclamative clauses it is endowed with quantificational force. The complementizer *som* could then be sensitive to some specific feature which sets apart the relative *vilken* in (21) from the other *wh*-forms in (20). But note that this solution is stipulative: there is no independent evidence that can prove the relevance of a given feature for the agreement with the complementizer *som.*

1.4. Consequences for the raising analysis

The previously discussed approaches were formulated within the adjunct analysis of relative clauses. Let us now consider their consequences for the raising analysis. The structural representations proposed by Kayne (1994: ch. 8) for the relative clauses in (1a–d) are reproduced in (22).

(22) a. $[_{DP}$ *the* $[_{CP}$ *book*$_i$ [*that* $[_{IP}$ *I read* t_i]]]] (*that*-relative)
b. $[_{DP}$ *the* $[_{CP}[_{DP}[_{NP}$ *book*] $[_{DP}$ *which* t_{NP}]]$_j$ [C° $[_{IP}$ *I read* t_j]]] (*wh*-relative)
c. $[_{DP}$ *the* $[_{CP}$ *book*$_i$ $[_{CP}$ C° $[_{IP}$ *I read* t_i]]]] (zero relative)
d. $[_{DP}$ *the* $[_{CP}[_{PP}[_{NP}$ *way*] $[_{PP}$ *in* $[_{DP}$ *which* t_{NP}]]] [C° $[_{IP}$ *he acted* t_{PP}]]]] (pied piping relative)

The Doubly Filled Comp Filter, as formulated in (4), predicts that the *that*-relative in (22a) should be ungrammatical: in fact, contrary to the adjunct analysis, in (22a) the Spec of *that* is filled by the relative "head". Similarly, Pesetsky's (1994) approach predicts that (22a) should have the same nonoptimal status as the ungrammatical * *the book which that*

I read: in both cases, the presence of lexical material in Spec,CP causes a violation of the principle Left Edge(C), in addition to a violation of Telegraph. With respect to Italian, this violation of Left Edge(C) in the *that*-relative, added to the violation of Telegraph, would imply the optimality of the *wh*-relative and of the zero relative, which only violate Left Edge(C); but these relative clause types are actually ungrammatical in Italian.

Only Rizzi's approach is compatible with the structures in (22), precisely because it does not make reference to the overt realization of the material in Spec,CP, but rather to its feature specification. Thus in (22a) the complementizer *that* agrees with the relative "head", which can be classified as [-*wh*]; in (22b) and (22d), on the contrary, C° cannot be instantiated as *that*, because it agrees with a [+*wh*] phrase in Spec. A similar approach is entertained by Kayne (1994: 91).

Note that this perspective dissociates the Doubly Filled Comp effects from the cross-linguistic distribution of *wh*-relatives ((1b) versus (2b)). The ungrammaticality of the *wh*-relative in Italian will have to be derived from an updated equivalent of the rule of Relative NP Deletion. However, in order to derive the *that*-relative from the pre-Spellout structure:

(23) $[_{DP}$ *the* $[_{CP}[_{DP}[_{NP}$ *book*$]$ $[_{DP}$ *which* $t_{NP}]]_j$ $[_{CP}$ *that* $[_{IP}$ *I read* $t_j]]]]$

the rule would have to selectively delete the relative determiner in Spec,CP, while leaving intact the NP "head" in its Spec.

2. The omission of the relative D°

Within the raising analysis, there is an intuitive difference between the *wh*-relative and the *that*-relative. In the *wh*-relative two elements must occur in Spec,CP: the NP "head" and the relative determiner. In the *that*-relative, instead, a single constituent seems to be fronted to Spec,CP. Informally, the ungrammaticality of the *wh*-relative in Italian can be reduced to the hypothesis that there isn't simply "enough room" for both the relative determiner and the NP "head" in Spec,CP.

2.1. Base generation of NP

Kayne (1994: 87–90) proposes a version of this idea, starting from the hypothesis that in *that*- and zero relatives the relativized phrase is generated as a bare NP lacking a relative determiner. The NP category raises to Spec,CP, where it is bound by the external determiner:

(24) a. $[_{DP}$ *the* $[_{CP}[_{NP}$ *book*$]$ $[_{CP}$ (*that*) $[_{IP}$ *I read* $t_{NP}]]]]$
b. $[_{DP}$ *il* $[_{CP}[_{NP}$ *libro*$]$ $[_{CP}[_{IP}$ pro *ho letto* $t_{NP}]]]]$

When the NP "head" is instead introduced by a relative D°, the whole DP moves to Spec,CP, and the NP then raises to Spec,DP, as represented in (22b) above.

The possibility of the *wh*-relative in a given language depends on the ability of the relative determiner to host the NP in its Spec. In English Spec,DP can be used as a landing site for the NP "head", so that there is enough room for both the "head" and the relative D° in Spec,CP.

In Italian (and French), on the contrary, the configuration in (22b) is impossible: the Spec of the relative D° is not available for NP. On the other hand, NP cannot remain in the complement position of the relative D°, since it would fail to be related to the external D°. Therefore, the only possibility is to generate the relative phrase as a simple NP, as in (24b). This explains the impossibility of the *wh*-relative.

The pied piping configuration (22d) is instead possible both in English and in Italian (cf. (1d) and (2d)): in this case, the NP "head" moves to the Spec of the pied piped PP. This means that the specifier of the preposition is an available landing site in both languages.

Kayne leaves open the problem of what difference between the relative D° of the two languages determines the availability of Spec,DP as a landing site for the NP "head".[15]

However simple and appealing, this analysis is inconsistent with the general principle (10) of chapter II, whereby only a DP projection, but not an NP, can be an argument. By this principle, in (1a), (1c) and (2a) the relative "head" must be a DP, since it originates in an argument position.

Moreover, there is also English-internal evidence against Kayne's proposal. Consider the paradigm of English infinitival relatives:

(25) a. *I found a topic to work on.*
b. * *I found a topic which to work on.*
c. *I found a topic on which to work.*
(Chomsky—Lasnik 1977: 460–470)

This paradigm shows a significant gap with respect to that of finite restrictive relatives in (1): namely, the infinitival *wh*-relative (25b) is impossible. In this respect, (25) is closer to the Italian restrictive paradigm in (2). This raises a problem for Kayne's analysis: though the possibility of NP raising to Spec,DP may be cross-linguistically parametrized, it is hardly plausible that it may be sensitive to the tensedness of the relative clause.[16] The relevance of tensedness suggests instead that what is involved is the internal functional structure of finite vs. infinitival clauses.

2.2. A null relative determiner

In Kayne's approach, a relativized argument can belong to two different syntactic categories: in *that*-relatives it is a NP, whereas in *wh*-relatives it is a DP. It is necessary to explain why pied piping relatives only allow the second option, requiring an overt relative determiner:

(26) a. *the man with whom you're sure to have a good time*
b. * *the man with (that) you're sure to have a good time*
(Kayne 1984: 65)

In (26a), the NP "head" raises to the Spec of the pied piped PP, stranding the relative determiner in the complement position of P°, as represented in (27a) below. (26b) should allow a similar derivation: if a relative NP can be directly generated in an argument position, then the NP "head" can originate as a complement to the preposition and raise to Spec,PP, as represented in (27b):[17]

(27) a. [$_{DP}$ *the* [$_{CP}$[$_{PP}$[$_{NP}$ *man*] [*with* [$_{DP}$ *whom* t_{NP}]]] [C° [$_{IP}$ ...t_{PP}...]]]]
b. * [$_{DP}$ *the* [$_{CP}$[$_{PP}$[$_{NP}$ *man*] [$_{PP}$ *with* t_{NP}]] [$_{CP}$ C° [$_{IP}$ ... t_{PP} ...]]]]

The contrast between (27a) and (27b) can be reduced to the Empty Category Principle under the assumption that in (27a) the relative determiner *whom* is able to properly govern the trace of its complement NP, whereas the preposition in (27b) is unable to do so. In fact, note that P-stranding in English is only possible in argument positions, but not in Spec,CP (Kayne 1984: 60; Postal 1972: 213).[18]

Consider however the examples (28a,b), featuring the pied piping of a larger phrase:

(28) a. *the* [*author* [*recent books by* [*whom* t_{NP}]]]$_j$ *you should read* t_j
b. * *the* [*author* [*recent books by* t_{NP}]]$_j$ *you should read* t_j
c. [*Which author*]$_j$ *did you read* [*recent books by* t_j]?

Here too, pied piping is only possible in the presence of a relative determiner. But in this case, the ungrammaticality of (28b) cannot be derived from the Empty Category Principle: in fact, (28c) shows that the preposition *by* is able to properly govern the trace of its own complement. This suggests that an ECP approach is inadequate. Let us then consider an alternative solution.

Suppose that the structure in (24) is impossible: by principle (10) of chapter II, the relative phrase cannot be generated as a bare NP. It follows that in *that-* and zero relatives, the NP "head" is selected by a relative D°:[19]

(29) $[_{DP}$ *the* $[_{CP}[_{DP}$ D_{REL} *book*$]$ $[_{CP}$ *that* $[_{IP}$ *I read* $t_{DP}]]]]$

If the relative determiner is syntactically realized, the problem is to explain why and how it fails to be spelled out in this structure.

An interesting clue emerges from the comparison of (29) with the pied piping structure in (27a), where the NP "head" raises to the Spec of the pied piped PP, stranding the relative determiner in the complement position of P°.

The impossibility of deleting the relative determiner in (27a), contrary to (29), suggests the following generalization: the relative determiner can be null only if it is governed by the external D°. This is the case in (29), where the relative DP in Spec,CP falls in the minimal domain of the external D° (cf. § II.5.2), but not in (27b), where the relative DP is in the minimal domain of the pied piped preposition. This suggests that the external D° plays a crucial role in licensing the PF deletion of the relative D° in *that-* and zero relatives.[20]

2.3. PF deletion as incorporation

The preceding discussion has led us to the hypothesis that PF deletion is licensed by a sort of head-to-head dependency. An interesting implementation of this idea becomes possible if one conceives of the terminal symbol dominated by a functional head as a set of syntactic features. This abstract element is turned into a concrete morpheme in the Morpho-Phonological component (cf. Halle—Marantz 1993).

Consider from this perspective the incorporation of a functional head X° to another functional head Y°:

(30) a. $[_{YP}$ Y° $[_{XP}$ X° $]]$
F_1 F_3
F_2 F_4

(30) b. $[_{YP}\ X°+Y°\ [_{XP}\ t_X\ ]]$
F_1 F_3
F_2 F_4

Suppose that in the morpho-phonological component the complex head X°+Y° undergoes an operation of "unification" of the two sets of features (31a).[21] The resulting set is spelled out in PF as a single morpheme, whereas the trace of the incorporated head X° is deleted: thus, the position originally occupied by X° remains phonetically empty, and the apparent result is that an independent morpheme has been deleted (31b): [22,23]

(31) a. $[_{YP}\ X°+Y°\ [_{XP}\ t_X\ ]]$
F_1
F_2
F_3
F_4

b. $[_{YP}$ *morpheme* $[_{XP}$ t]]

Of course, in order for unification to take place the two sets of features X° and Y° must be compatible with each other, namely they cannot include inconsistent feature values.

Consider now the relative structure (29). The relative D° and the external D° both agree with the NP "head", and hence they share the same phi-feature specification. Furthermore, it was argued in § III.3 that the relative D° is not specified for the feature of definiteness.[24] It follows that the two sets of features are fully compatible for the purposes of unification. If the relative D° incorporates to the external D°, the resulting head can be spelled out as a single determiner in the external position; the apparent effect is the PF deletion of the relative D°:

(32) $[_{DP}\ D_{REL}$+*the* $[_{CP}[_{DP}\ t_D\ [_{NP}$ *book*$]]_i\ [_{CP}$ *that I read* $t_i]]]$

Note also that in (32) incorporation creates an extended minimal domain for D_{REL}+D° (cf. Chomsky 1995: 180) which includes the NP complement of the incorporated D_{REL}; or in other terms, the complex D_{REL}+D° head-governs the NP "head": thus it can check its features against the NP (§ III.2).

Consider then the impossibility of PF deletion in the pied piping relative (27a): as discussed above, the relative D° is separated from the external D° by the intervening preposition, and incorporation is blocked by relativized minimality; as a result, the relative D° must be spelled out as an independent morpheme.

This analysis has two general consequences of some interest. The first one is that deletion in PF is reduced to a general syntactic process - incorporation - which is subject to well defined constraints. The second one concerns the proper notion of recoverability. In this approach, recoverability is reduced to the condition of feature compatibility on unification: the head to be incorporated must consist of a set of features compatible with those of the host head.

2.3.1.

The hypothesis of abstract incorporation may account for a restriction on the omission of the relative determiner in Bavarian, discussed by Bayer (1984: 215–225). Bayer divides the paradigm of the relative pronoun *der-die-des* in two subclasses on the basis of the value of the feature [± oblique]. The value [- oblique] identifies the nominative forms and the forms that are morphologically identical to a nominative one: these are the accusative of the feminine and neuter singular, and the accusative of the plural. The value [+oblique] is attributed to the other forms, namely the accusative of the masculine singular and the dative forms. Bayer's typology is reproduced in (33):

(33)

	NOM	ACC	DAT
masc	*der* -	*den* +	*dem* +
fem	*die* -	*die* -	*der(a)* +
neut	*des* -	*des* -	*dem* +
plur	*die* -	*die* -	*dene(n)* +

This typology allows a generalization on the omission of a relative pronoun: a [- oblique] relative D° can always be omitted, whatever the value of the feature [± oblique] of the external D°; on the contrary, a

[+oblique] relative D° can be omitted only if it is morphologically identical to a [+oblique] external D°.

This generalization is illustrated by the data in (34)–(37). In (34), a [-oblique] relative D° is optionally omitted under government by a [+oblique] external D°:

(34) a. *I sog's* dem *Mõ* (der) *wo im Gartn arwat.*[25]
I say-it the_{DAT} man who_{NOM} C° in the garden works
'I said it to the man who works in the garden.'
b. *Mir song-s* dera *Frau* (die) *wo da Hund bissn hod.*
we tell-it the_{DAT} woman who_{ACC} C° the dog bitten has
'I told it to the woman who the dog bit.'
c. *Mir song-s* dene *Menner* (die) *wo da Hund bissn hod.*
we tell-it the_{DAT} nem who_{ACC} C° the dog bitten has
'I said it to the men who the dog bit.'

In (35a), the [+oblique] relative D° is omitted under government by an identical [+oblique] external D°; in (35b), instead, it cannot be omitted, because the external D° is [- oblique]:

(35) a. *Sie gem's* dem *Mõ* (dem) *wo mir g'hoifa hom*
they give-it the_{DAT} man who_{DAT} C° we helped have
'They gave it to the man whom we helped.'
b. Der *Mõ* *(dem) *wo mir g'hoifa hom*
the_{NOM} man who_{DAT} C° we helped have
'The man whom we helped...'

Finally, in (36a–b) a [- oblique] relative D° in the Accusative case is omitted under government by a [-oblique] nominative D°. In (37), instead, the masculine singular form is [+ oblique] in the Accusative, and it cannot be omitted under government by a [- oblique] external D°:

(36) a. *Die Lampn* (die) *wo i g'seng hob wor greißlich*
the_{NOM} lamp $which_{ACC}$ C° I seen have was ugly
'The lamp that I saw was unsightly.'
b. *Des Audo* (des) *wo i mecht is z'teia*
the_{NOM} car $which_{ACC}$ C° I like is expensive
'The car I like is expensive.'

(37) *Der Mantl* *(den) *wo i kaffd hob wor z'rissn*
the_{NOM} coat $which_{ACC}$ C° I bought have was torn
'The coat that I bought was torn.'

The hypothesis of deletion through incorporation can account for this paradigm. Recall that the determiners in (33) are abstractly characterized as sets of features. Suppose that every [+oblique] form bears a specific Case feature (+gen, +dat...), while the [-oblique] forms are actually not specified for the same features; the other features are common. Then the licensing condition for the omission of the relative determiner in Bavarian follows from the operation of unification proposed in (31): in order for unification to be possible, the features of the relative D° and those of the external D° must be compatible.

Thus, the presence of morphological Case features on the Bavarian determiners constrains the possibility of unification, and hence it constrains the deletion of the relative determiner.

2.3.2.

Van Riemsdijk (1989) discusses the phenomenon of "split topicalization" in German, whereby a nominal constituent is topicalized and a weak determiner is left floating in the base position. The problem with this structure is that when the fronted constituent contains a count singular noun, as in (38), an indefinite determiner appears in front of it, besides the floating determiner:

(38) [Einen *Wagen*] *kann er sich noch* keinen *leisten.*
a car_{ACC} can he to himself still $none_{ACC}$ afford
'A car, he cannot afford yet.'

The appearance of two determiners seems to be inconsistent with a movement analysis of this structure along the lines of Sportiche's (1988) approach to floating quantification: in fact, the source of the split topicalization in (38) should be a DP with two determiners, which is ungrammatical:

(39) *Er kann sich noch* [*keinen* [(* *einen*) *Wagen*]] *leisten.*
he can himself yet no car afford
'He cannot afford a car yet.'

This paradox leads Van Riemsdijk to propose a process of "regeneration": after a determinerless nominal constituent N' has been topicalized, it is "completed" by adding to it another determiner, which turns it into a maximal projection NP.

The hypothesis of deletion through incorporation offers an alternative solution. It has been argued by many authors[26] that the functional structure of noun phrases includes a head Num° below D° which hosts numeral determiners It is possible to assume that the indefinite determiner contained in the fronted constituent in (38) is not a D°, but rather a Num°. Then the difference between (38) and (39) concerns the spelling out of Num°. In (38), NumP has been fronted and Num° is not governed by D°, so that it cannot be deleted through incorporation to D°. In (39), instead, Num° is governed by D° and it can (in fact, must)[27] incorporate to it, so that it fails to be spelled out as an independent morpheme.

(40) a. $[_{CP}[_{NumP}$ *einen Wagen*] $[_{CP}$ *kann* $[_{IP}$ *er sich noch* $[_{DP}$ *keinen* $t_{NumP}]$ *leisten*

b. $[_{DP}$ Num°+ *keinen* $[_{NumP}$ t_{NUM} $[_{NP}$ *Wagen*]]]

2.3.3.

The preceding discussion has shown that deletion of the relative determiner is possible in the syntactic structure (32), but it has not explained why and how it is triggered.[28] This problem is deferred to § 8, after a discussion of the "deletion" phenomena affecting the declarative complementizer.

3. Two types of null complementizer

Both the *wh*-relative (1b) and the zero relative (1c), repeated below, lack an overt complementizer:

(1) b. *the book* which Ø *I read*

c. *the book* Ø *I read*

At first sight, a unified analysis of the two cases seems to be desirable for the economy of the grammar. But on closer inspection, it turns out that the missing complementizers of the two relative clause types have distinct properties. This is shown in the following subsections.

3.1. Subject relativization

As is well known, the English paradigm of restrictive relatives in (1) is not completely reproduced in case of local subject relativization. The latter is impossible in the zero relative:[29]

(41) a. *the thing*$_i$ [*that* t_i *happened*
b. *which*
c. * Ø

The contrast between (41a,b) and (41c) is often attributed to the ECP. Consider for instance the analysis proposed by Rizzi (1990: chapter II), briefly summarized in § 1.3 above. According to Rizzi, the C° immediately c-commanding the trace in Spec,IP becomes a proper governor for it by means of an agreement relation that coindexes it with the trace. The complementizer *that* in (41a) agrees with the "head" of the relative clause, whereas the null complementizer in (41b) agrees with the relative pronoun in its Spec. In (41c), no agreement relation is possible, because the null complementizer can only agree with its own Spec, but the latter contains a null operator, which cannot license an agreement relation. Thus the trace in Spec,IP cannot be properly head governed, in violation of the ECP.

This approach to (41) can hardly be maintained in the raising analysis: the distinction between A-agreement of C° with the relative "head" and A'-agreement with Spec,CP no longer holds, since the "head" itself is in Spec,CP. In particular, assuming the hypothesis of the deleted relative determiner, the examples in (41) have the following representations (we keep for the time being Kayne's analysis (22b) of the *wh*-relative):

(42) a. [$_{DP}$ *the* [$_{CP}$[$_{DP}$ D$_{REL}$ *thing*] [$_{CP}$ *that* [$_{IP}$ t_i *happened*]]]]
b. [$_{DP}$ *the* [$_{CP}$[$_{DP}$ *thing* [$_{DP}$ *which* t_{NP}]] [$_{CP}$ Ø [$_{IP}$ t_i *happened*]]]]
c. * [$_{DP}$ *the* [$_{CP}$[$_{DP}$ D$_{REL}$ *thing*] [$_{CP}$ Ø [$_{IP}$ t_i *happened*]]]]

Note that both the *wh*-relative (42b) and the zero relative (42c) involve a null complementizer. Within Rizzi's approach, the contrast between the two examples implies that the possibility of agreement between the null C° and the relative DP in its Spec would be affected by the lexical nature of the relative D° .

It is possible to argue that the null complementizer can only agree with a [+wh] constituent: then, Spec/head agreement is possible in (42b), but not in (42c), assuming that the deleted relative D° is not specified as [+wh]. The problem with this assumption is that in long distance subject relativization the null C° is actually able to agree with (the trace of) a relative DP with a deleted D$_{REL}$:

(43) [$_{DP}$ *the* [$_{CP1}$[$_{DP}$ D$_{REL}$ *thing*]$_i$ [$_{CP1}$ *that I think* [$_{CP2}$ t_i [$_{CP2}$ Ø [$_{IP}$ t_i *will happen*]]]]]]

This makes the feature-based approach problematic.

From a purely descriptive viewpoint, the null C° of the *wh*-relative patterns with the overt C° of the *that*-relative, rather than with the null C° of the zero relative. Put differently, the null complementizer of the *wh*-relative (42b) and the one of the zero relative (42c) are not equivalent with respect to proper head government. This asymmetry suggests that they are not exactly the same element.

3.2. Embedded topicalization and negative preposing

A similar point can be made with respect to the phenomena of embedded topicalization and negative preposing. Kayne (1994: 28–29) observes that the optional deletion of *that* in complement clauses, exemplified in (5b) above, is blocked if the complementizer is followed by a topicalized or a preposed negative constituent:[30]

(44) a. *I didn't know [that yesterday Peter danced].*
b. * *I didn't know [Ø yesterday Peter danced].*

(45) a. *I didn't know [that never had Peter danced so well].*
b. * *I didn't know [Ø never had Peter danced so well].*

A similar pattern is observed in relative clauses: embedded topicalization and negative preposing are possible in *that*-relatives but impossible in zero relatives:

(46) a. *This is the kind of car that [for my son] I wouldn't even have considered buying.*
b. * *This is the kind of car Ø [for my son] I wouldn't even have considered buying.*

(47) a. ? *I saw a dress that* [*under no circumstances*] *would I have considered buying for my daughter.*
b. * *I saw a dress Ø* [*under no circumstances*] *would I have considered buying for my daughter.*

The data in (44)–(47) suggest the naive generalization that the preposing structures are only licensed in embedded clauses whose C° is overt. But the generalization is falsified by *wh*-relatives: these have a null complementizer, like zero relatives, yet they pattern with *that*-relatives, allowing both embedded topicalization and negative preposing:

(48) *a man whom [for his brutal insolence and cruelty] Robin had long hated*

(49) *John is the kind of person who [under no circumstances] would I be willing to talk to.*

This further confirms the idea that that the null complementizer of *wh*-relatives is syntactically distinct from that of zero relatives.[31]

In the following sections, this idea will be formulated within a theory of the clausal functional structure recently elaborated by Müller—Sternefeld (1993) and especially Rizzi (1995), who identify different types of A' position above the inflectional functional structure. Such A' positions correspond to the specifiers of a number of functional projections organized in a fixed hierarchical order. These projections constitute the complementizer system of the clause.

4. A characterization of the complementizer system

The argument by Authier (1992) reported in § I.2.1 showed that in English embedded clauses the subject/auxiliary inversion triggered by negative preposing forces the postulation of a functional head between C° and IP. Briefly put, in an example like (45a) the auxiliary incorporates to a head which asymmetrically c-commands the subject position, but which cannot be identified with C°, as it follows the declarative C° *that*.

Authier (1992) suggested that topicalization in (44a) also involves an abstract functional head with the same licensing requirements as the head involved in negative preposing: in fact, certain ungoverned contexts exclude both these types of preposing. The parallelism is confirmed by Kayne's observation that both of them are incompatible with the deletion of the complementizer, as shown in (44b) and (45b).

This argument for a "split CP" converges with an internal requirement of Kayne's (1994) X-bar theory. In the latter there is exactly one adjunct position per maximal projection: thus, a topicalization example like (44a) cannot have both the topicalized adverb *yesterday* and the preverbal subject adjoined to IP; it is necessary to postulate an abstract head between C° and IP, whose Spec hosts the topicalized phrase.[32]

This section will review some further evidence in favour of the splitting of CP into different functional projections. These data are accommodated by the Comp system proposed by Rizzi (1995).

4.1. Relative ordering of preposed constituents

Rizzi (1995) notes that topicalization and negative preposing cannot be taken to involve exactly the same head, because the two structures can cooccur, with the topic to the left of the preposed negative phrase:

(50) *He said [that [beans*$_i$, *[[never in his life] [had [he* t_{AUX} *been able to stand* t_i t_{ADV}]]]]][33]

The obligatory relative ordering of the preposed constituents further confirms the hypothesis that each of them moves to the Spec of a specific functional head. The ordering in (50) is determined by the following hierarchical array of functional heads:

(51) [$_{CP}$ *that* [$_{TopP}$ Topic° [$_{FocusP}$ Focus° [$_{AgrSP}$ ...]]]]

Spec,TopicP is the landing site of the topicalized phrase and Spec,FocusP is the landing site of the focalized negative phrase.[34]
This characterization of the Comp area can be extended to Italian:

(52) *Disse che* [$_{TopP}$ *quel libro*$_j$ [$_{FocP}$ *SOLTANTO A GIANNI*$_i$ [$_{IP}$ pro *avrei dovuto darlo*$_j$ t_i]]]
'He said that that book, only to Gianni I should have given it.'

Here Spec,TopicP is the landing site of Clitic Left Dislocation, whereas Spec,FocusP is the landing site of the focussing structure lacking a resumptive clitic, traditionally called topicalization. To avoid confusion, the latter structure will be henceforth called "focalization", and the term "topicalization" will exclusively refer to the English topic construction lacking a resumptive pronoun.[35]
Rizzi also notes that relative pronouns move to a higher position than interrogative phrases in direct questions. This can be shown by the relative ordering of these elements with respect to topicalized (in English) or left dislocated (in Italian) constituents:

(53) a. *He's a man to whom liberty we could never grant.*
b. * *He's a man liberty, to whom we could never grant.*
(Baltin 1982: 17–19)
c. *In the living room, what did Mary find ?*
d. * *What in the living room did Mary find ?*
(Rochemont 1989: 147)
e. *Tom, why would anyone want to meet ?*

(54) a. *un uomo a cui, il premio Nobel, lo daranno senz'altro*
'A man to whom, the N prize, (they) will surely award it.'
b. * *un uomo il premio Nobel, a cui lo daranno senz'altro*
a man, the Nobel prize, to whom (they) will surely award it
c. *Contatterò chi, le tue idee, le conosce bene.*
I will contact who, your ideas, knows them well
'I will ask someone who knows well your ideas'

(54) d. * *Contatterò le tue idee, chi le conosce davvero.*
I will ask your ideas, who knows them well
e. *Il premio Nobel, a chi lo hanno dato?*
the nobel Prize, to whom did (they) award it?
'As for the Nobel prize, to whom did they award it?'
f. * *A chi, il premio Nobel, lo hanno dato?*
to whom, the Nobel prize, did (they) award it?
(Rizzi 1995)

In English, relative pronouns obligatorily precede topicalized phrases (53a,b), whereas in direct questions interrogative phrases obligatorily follow them (53c,d).[36]

Similarly, in Italian relative pronouns precede left dislocated phrases both in headed relatives (54a–b) and in free relatives (54c–d). Note that the latter contain a relative pronoun *chi* omophonous to an interrogative pronoun. However, despite this morphological similarity, interrogative phrases occupy a different position than relative pronouns: in fact, they follow a dislocated constituent, as shown in (54e–f). These data show that relative and interrogative phrases of direct questions move to two distinct positions.[37]

In indirect questions, the situation is different. In English, the interrogative phrase cannot follow the topic (55), contrary to direct questions (cf. (53c)). In Italian, on the contrary, the interrogative phrase may either precede or follow a topic, as shown in (56), althogh the first option is marginal:

(55) * *I wonder Tom, why anyone would want to meet.*
(56) a. ? *Mi domando a chi, il premio Nobel, lo potrebbero dare*
I wonder to whom, the Nobel Prize, they could give it
b. *Mi domando, il premio Nobel, a chi lo potrebbero dare.*
I wonder, the Nobel prize, to whom they could give it.
'As for the Nobel prize, I wonder to whom they could award it.'

Rizzi (1995) derives the different position of relative and interrogative phrases from the following assumptions:

(a) a relative pronoun must be immediately adjacent to its antecedent, the relative "head" (which is external to the relative CP in the adjunct analysis), in order to be antecedent-governed by it.

(b) In direct questions, Infl bears the feature [+wh], and it must enter in a Spec/head agreement relation with a *wh*-phrase in order to satisfy the Wh Criterion.[38] This requirement is met by incorporating Infl to Focus[39] and by moving the interrogative phrase to Spec,FocusP (cf. Rizzi 1991: § 2).[40]

(c) In English indirect questions, the feature [+wh] is directly selected by the governing verb on the head of its complement (cf. also Rizzi 1991: § 2). In Italian embedded questions, on the contrary, the feature [+wh] is realized on Infl°, which moves to Focus° giving rise to subject inversion; therefore, the Wh-Criterion needs not be satisfied in the highest projection of the embedded question.

From hypothesis (a), it follows that the relative pronoun must always move to the highest position in the Comp area, namely Spec,CP, in order to be adjacent to the external "head".[41] This is why relative pronouns precede any other phrase in the Comp area (cf. (53a–b) and (54a–d)).

According to hypotheses (b) and (c), in English the feature [+wh] has different realizations in main and embedded questions. In the former, the feature is realized on Infl incorporated to Focus°; FocusP can be dominated by a Topic projection, yielding the relative order <topic, *wh*-phrase> in (53e). In embedded questions, instead, the feature [+wh] is selected by the governing verb in the head of its complement. Since selection is a strictly local head-to-head relation, a TopicPhrase cannot intervene between the verb and the [+wh] head whose Spec hosts the *wh*-phrase; this explains the impossibility of (55).[42]

As for Italian, in constituent questions the feature [+wh] is always realized on Infl° which raises to Focus°, and it is not directly selected by the matrix verb. Therefore, Focus° can be dominated by a Topic projection both in direct and indirect constituent questions ((54e) and (56b)).[43]

On the other hand, the order <*wh*-phrase, topic> in (56a) is not predicted by the structure in (51). For the latter case, Rizzi assumes a further Topic position below Focus°. The latter usually hosts adverbial phrases,[44] but in Italian, it may also host dislocated arguments, yielding the order < *wh*-phrase, dislocated phrase>.[45]

The complete structure of the complementizer system is thus (57):

(57) [$_{FP}$ Force° [$_{TopP*}$ Top° [$_{FocP}$ Focus/Wh° [$_{TopP*}$ Top° [$_{FinP}$ Fin° IP]]]]]
      | |
      che   *se*
      that

The highest head Force°, whose spellout is the declarative complementizer, expresses the declarative force of the clause and its Spec provides the landing site for relativization. The Topic Phrases above and below the Focus/Wh Phrase host topicalized or left-dislocated phrases in their Spec. These projections are recursive, as shown by the possibility of multiple topicalization/left dislocation:[46]

(58) a. *To John, this kind of article, I would gladly give.*
b. *On this table, that heavy a book you should never put.*
c. *A Gianni, questo libro, lo regalerei volentieri.*
to G., this book, (I) it_{CL} would give gladly
'To Gianni, this book, I would be glad to give it.'

Focus/wh° carries the interrogative force of the clause and hosts interrogative or focalized phrases in its Spec. Finally, the Finite Phrase immediately above IP constitutes the interface between the inflectional and the complementizer systems.[47]

4.2. Multiple complementizers

The postulation of various functional heads to the left of Infl is supported by the possibility of multiple complementizers in various Germanic dialects: cf. Hoekstra (1992b) and Haegeman (1992: 46–47) on Dutch dialects; Alber (1994) on the Tyrol dialect; Vikner (1991) on Danish; Platzack (1994) on Mainland Scandinavian. The phenomenon is also found in Old and Middle English (Allen 1977: 244–255). Some representative examples are reported in (59)–(62):[48]

(59) *Dat is niet zo gek* als of dat *hij gedacht had*
this is not as crazy as if that he thought had
'This is not so crazy as he had thought it.'
(Dutch dialect; Hoekstra 1992: 191)

(60) *Des isch komischer* als wia wos *I mr gedenkt hat*
this is stranger than C° C° I myself thought have
'This is more strange than I had thought.'
(Tyrol dialect; Alber 1994: 11)

(61) ? *Vi kender mange lingvister* som at der *vil laese denne bog*
we know many linguists C° C° C° will read this book
'We know many linguists who will read this book.'
(Danish; Vikner 1991: 112)

(62) *Loc nu* giff ðatt *tu narrt rihht wod.*
look now if that you not are right crazy
'Now look if you aren't really crazy.'
(Middle English; Allen 1977: 251)

Some of the multiple complementizers in these examples cannot be accommodated by the structure in (57), which suggests that the complementizer system may be even more complex than that.[49] However, the phenomenon clearly supports Rizzi's general approach, which correctly predicts that in every language the complementizers occur in a fixed relative order. Every complementizer is thus conceived as the instantiation of a different head of the Comp area, bearing a different functional feature.

4.3. Omission of the functional heads

The adoption of the complementizer system raises two general questions. The first one concerns the specificity of the landing site for various types of A' movement: every type of movement targets a different Spec in (57). This state of affairs can be easily captured by generalizing Rizzi's (1991) *Wh*-Criterion, so that every constituent raised to the Comp system carries a functional feature (e.g. [+wh], [+topic], [+rel]) to be matched under Spec/head agreement with the identical feature specification on a head of the Comp system:

(63) F- Criterion:
  a. Every [+F] head must agree with a [+F] constituent in its Spec.
  b. Every [+F] constituent must agree with a [+F] head.

On the other hand, Müller—Sternefeld (1993) have argued on independent grounds that the specificity of the A' positions does not only concern the final landing site where the [+interpretable] feature is matched, but also the intermediate links of the A' chain: for instance, German scrambling targets a specific type of A' position and cannot use Spec,CP as an "escape hatch" for extraction from a clause. Thus, the authors distinguish various types of A' position and propose a Principle of Unambiguous Binding whereby the A' positions included in a chain must be of a uniform type:

(64) Principle of Unambiguous Binding
  A variable that is α-bound must be β-free in the domain of the head of its chain (where α, β refer to different types of position).

(63) and (64) restrict the possible cyclic movement paths across the Comp system.

The second question is whether the rich structure of the Comp system in (57) is uniformly projected in all clauses. Note that in Italian, Force° is not realized in an indirect question introduced by the Focus/Wh° *se* or by a *wh*-phrase:[50]

(65) a. *Mi chiese (* che) se avevo letto il libro.*
he asked me that if I had read the book
'He asked me if I had read the book.'
b. *Mi chiese(* che) a chi avevo dato il libro.*
(he) asked me that to whom (I) had given the book
'He asked me whom I had given the book to.'

Conversely, note that in declarative complements introduced by Force° (*che*) the head Focus/Wh° cannot be spelled out as the interrogative particle *se*, although the latter is lower than Force°:

(66) *Ha detto che (* se) verrà domani.*
(he) said that if (he) will come tomorrow
'He said that he will come tomorrow.'

The examples (65) and (66) raise the problem of the syntactic realization of the Comp heads: in English and Italian, only one functional head is overt in every embedded clause, namely the one that is relevant for its interpretation. The problem is how to interpret the omission of the other functional heads of the complementizer system.

A first possibility is to assume that the rich structure of the complementizer system in (57) is uniformly projected in every clause up to the level Force Phrase, but only the functional heads endowed with a semantically relevant functional feature are spelled out. Thus, in a *wh*-complement clause Force° is phonetically null, and conversely, in a declarative complement Focus° is null.

If this is so, in many cases the instantiation of some of the projections will be vacuous: for instance, in a simple declarative clause like (5a) only the [+declarative] head Force° has a clear syntactic function. But natural languages in general seem not to allow the vacuous instantiation of syntactic structures (cf. the principle of Full Interpretation proposed by Chomsky 1995: chapter 2).[51] Let us instead assume that the syntactic realization of a functional projection cannot be vacuous, but it must be structurally licensed according to the following principle:

(67) Every head in the complementizer system must be *structurally licensed* by either (1) or (2):
(1) selection by a head external to the Comp system;

(67) (2) specification of a feature satisfying the F-Criterion (63).

According to (1), a head is syntactically projected when it is necessary in order to mediate a relation of selection. Let us assume with Szabolsci (1994: 214) that a theta assigning head cannot directly select NP or IP, but it must select DP or CP. In our terms, the heads D° and Force° "close" an extended projection (in the sense of Grimshaw 1991), enabling it to act as an argument; this is why Szabolsci defines them "subordinators".

According to (2), a head can be realized when it bears a feature that triggers the movement of a constituent to its Spec (either overtly or covertly).

Thus, for instance, in (5a) only Force° is licensed under selection by the matrix verb; in (45a) too, Force° is licensed under selection, whereas FocusP is licensed by the functional feature that triggers negative preposing (cf. Rizzi 1991).

Principle (67) implies that a functional projection can be realized only when it is structurally licensed (in minimalist terms, when its presence is required for the convergence of the derivation). This flexibility in the internal structure of the complementizer system will be crucially exploited in the analysis of "complementizer deletion" to be proposed in the next section.[52]

5. Optional *that* deletion

Let us consider first the *that*-Ø alternation that is found in the *that*- and zero relatives (1a) and (1c), and in the two complement clauses in (5). This phenomenon is usually labelled "optional *that*-deletion"; but the opposite properties displayed by the two alternating forms (§§ 3.1–3.2) are unexpected if what is involved is simply PF deletion.

The assumption of Rizzi's (1995) complementizer system opens a new perspective on the problem. The alternation can be thought of as the alternative selection of two distinct functional heads of the Comp area: *that* is the spellout of Force°, whereas the zero form corresponds to a lower functional head that intrinsically lacks any phonetic content. The availability of this lower head must be crosslinguistically parametrized so as to account for the lack of "optional deletion" in Italian relative clauses (cf. (2c)) and complement clauses (cf. (6b)).

The essential property of the lower head is that it can "substitute for" Force° in two respects:

(1) it can support declarative force, satisfying the selectional requirements of the selecting head;
(2) it can host the relative DP in its Spec, satisfying the F-Criterion.

It is of course possible to postulate an extra functional head with just these properties, whose availability in the complementizer system (57) is parametrized. But as we have seen, the array of functional projections in (57) is supported by much comparative evidence, and the null hypothesis is that it is fixed by Universal Grammar. A more promising move is to identify the null complementizer with one of the heads in (57) and to parametrize its properties.

Let us assume that the null complementizer corresponds to the phonetically null head Topic°.[53] This assumption will be justified in the next chapter on the basis of some diachronic evidence: it will be shown that in earlier stages of English the complementizer system contained a Topic head that licensed verb second topicalization in main clauses, whereas in embedded clauses it was filled by a complementizer-like particle; the latter could receive the [+declarative] value by a default rule. In the evolution of the complementizer system, this head lost the ability to attract the verb and was reanalysed as a null Topic°, but it retained the ability to introduce declarative clauses.[54]

We therefore propose the following parameter:

(68) ± (Topic° optionally supports the features [+declarative] and [+relative])

5.1. Zero complement clauses

Let us assume that English has a positive setting for (68). We can then propose the following representations for the complement clauses in (5). In (5a), the matrix verb selects a Force Phrase headed by *that*:

(69) *I know* $[_{FP}$ *that* $[_{IP}$ *you're right*$]$.

In (5b), instead, Topic° is selected and carries the feature [+declarative]:

(70) *I know* $[_{TopP}$ *Ø* $[_{IP}$ *you're right*$]$.

Note that the selection of either a Force Phrase or a Topic Phrase leads to a convergent and equally costly derivation; therefore, optionality is guaranteed.

Italian differs minimally from English in that it has a negative setting of parameter (68): only Force° (*che*) carries the features [+declarative]. This explains the ungrammaticality of zero complement clauses (cf. (6b)):

(71) a. Pro *so* [$_{FP}$ *che hai ragione*].
b. * Pro *so* [$_{TopP}$ *Ø hai ragione*].

Recall that according to Szabolsci's (1994) hypothesis, a sentential argument cannot be simply IP, but it must be introduced by a subordinator functional head.[55] The proposal amounts to the claim that in English both Force° and Topic° can accomplish the subordinating function: both projections can be structurally licensed in virtue of selection by a lexical head. In Italian, on the contrary, the lexical verb only selects CP as a declarative complement, and FocusP as an interrogative complement: in other terms, only Force° and a [+wh] Focus° can act as subordinators, whereas Topic° by itself cannot.[56]

5.2. Zero relatives

The proposed analysis can be extended to "complementizer deletion" in zero relatives. Here the external determiner must select a projection of the Comp area whose Spec will host the relativized phrase. The zero relative clause (1c) differs from the *that*-relative (1a) in that its complementizer system contains a Topic Phrase instead of a Force Phrase:

(72) [$_{DP}$ *the* [$_{ForceP}$ [$_{DP}$ *book*]$_i$ [*that* [$_{IP}$ *I read* t_i]]]] (*that*-relative)
(73) [$_{DP}$ *the* [$_{TopP}$ [$_{DP}$ *book*]$_i$ [Topic° [$_{IP}$ *I read* t_i]]]] (zero relative)

The representations in (72) and (73) mirror those in (70): in English, the external determiner may select either a ForceP or a TopicP. By the positive setting of (68), the Topic head carries both the features [+declarative] and [+relative], so that its Spec can be used as a landing site for relativization.

In Italian, instead, the negative setting of parameter (68) makes the equivalent of (73) impossible. As Topic° cannot carry the features [+declarative] and [+relative], it cannot host the relative phrase in its Spec. This explains the ungrammaticality of the zero relative (2c):

(74) a. [$_{DP}$ *il* [$_{ForceP}$[$_{DP}$ *libro*]$_i$ [*che* [$_{IP}$ pro *ho letto* t_i]]]] (*that*-relative)

(74) b. * $[_{DP}$ *il* $[_{TopP}[_{DP}$ *libro*$]_i$ [Topic° $[_{IP}$ pro *ho letto* $t_i]]]]]$
(zero relative)

Note that parameter (68) interacts with the Principle of Unambiguous Binding (64). In a language with positive setting, both Force° and Topic° may carry the feature [+relative]; since relativization is a unitary process targeting a specific type of A' position, it must be assumed that the specifiers of the two heads are positions of the same type. In Italian, instead, Spec,TopicP is not an available landing site for relativization, and therefore it is not necessary to assume that it is a position of the same type as Spec,ForceP.[57]

Note that in Italian a Spec,TopicP filled by a left dislocated phrase does not create an island;[58] in English, on the contrary, topicalized constituents induce an island effect (Lasnik—Saito 1992: 85–86; Rochemont 1989):

(75) *Lo studente a cui*$_i$ $[_{TopP}$ *un libro del genere*$_j$, $[_{IP}$ pro *non lo*$_i$ *consiglierei* $t_i]]$, *è Gianni.*
the student to whom a book like that I wouldn't recommend is Gianni
'The student to whom I wouldn't recommend a book like that is Gianni.'

(76) ?? *He's a man from whom*$_j$ $[_{TopP}$ *money*$_i$ $[_{IP}$ *we could never take* t_i $t_j]]$.

It is unclear whether this can be classified as a weak (relativized minimality) island. Assuming so, the contrast between (75) and (76) may follow from the hypothesis that the island effect only emerges when Spec,TopP is a position of the same type as the landing site of relativization: this is the case in English, but not in Italian.

6. The *wh*-relative

6.1. A two steps relativization

In the analyses proposed in (72) and (73), the relativized phrase may target the Spec of two distinct functional heads in the Comp system. This aspect of the proposal is supported by some evidence from Danish, showing that a relativized subject may move to the Spec of two distinct complementizers (Vikner 1991):

(77) a. *Vi kender de lingvister*$_i$ som t_i *vil laese denne bog.*
b. *Vi kender de lingvister*$_i$ der t_i *vil laese denne bog.*
we know the linguists C° will read this book
'We know the linguists who will read this book.'

In (77a) the relative DP *lingvister* targets the Spec of the relative complementizer *som*, whereas in (77b) it moves to the Spec of the complementizer *der*.[59]

The latter resembles French *qui* in that it is only licensed in case of local A' movement of the subject that it immediately c-commands, as in (77b), and it is excluded in any other context, e.g. in the relativization of a direct object:

(78) * *Vi kender en bog*$_i$ der *denne lingvist har laest* t_i.
we know a book C° this linguist has read

Compare (77b) and (78) to the French equivalents (79a–b):

(79) a. *Nous connaissons le linguiste*$_i$ qui t_i *lira ce livre.*
'We know the linguist who will read this book.'
b. * *Nous connaissons un livre*$_i$ qui *ce linguiste a lu* t_i.
we know a book C° this linguist has read

Vikner assumes for Danish *der* an analysis similar to that proposed by Rizzi (1990: chapter II) for French *qui*: like *qui*, *der* is a functional head endowed with agreement features, which are licensed by Spec/head agreement with a DP moved to its Spec. This allows *der* to act as a proper governor for the trace in Spec,IP in (77b):[60]

(80) $[_{DP}$ *de* $[_{CP}$ *lingvister*$_i$ $[_{CP}$ *der*$_i$ $[_{IP}$ t_i *vil laese denne bog*]]]]

The two particles *som* and *der* cannot be two alternative realizations of the same functional head, since they can cooccur in a fixed order, with the former preceding the latter, as in (61) above or in (81):

(81) *Vi kender de lingvister*$_i$ som der t_i *vil laese denne bog.*
we know the linguists C° C° will read this book
'We know the linguists who will read this book.'

Interestingly, if *der* is an agreeing C° licensed by coindexing with its Spec, it follows that in (81) the relativization of the subject must take place in two steps: first, the relativized subject moves to the Spec of *der*, licensing the latter by Spec/head agreement; then it proceeds to the Spec of the higher complementizer *som*, which is adjacent to the external determiner. The derivation is illustrated in (82):

(82) $[_{DP}$ *de* $[_{CP1}$ *lingvister*$_i$ $[_{CP1}$ *som* $[_{CP2}$ t_i $[_{CP2}$ *der*$_i$ $[_{IP}$ t_i *vil laese*...]]]]]]

Thus, Vikner's discussion implies not only that relativization can target the Spec of two distinct functional heads in Danish, but also that the two positions can be involved in one and the same derivation. This possibility will be crucially exploited in our analysis of the English *wh*-relative.

6.2. Kayne's (1994) analysis

Recall that according to Kayne (1994), the possibility of the *wh*-relative depends on the availability of the Spec of the relative D° for the NP "head", as represented in (22b), repeated here as (83):

(83) $[_{DP}$ *the* $[_{CP}$ $[_{DP}$ $[_{NP}$ *book*] $[_{DP}$ *which* t_{NP}]]$_j$ $[_{CP}$ C° $[_{IP}$ *I read* t_j]]]

But as discussed in § 2.1, this proposal cannot account for the impossibility of the infinitival *wh*-relative (25b) in English. If in English Spec,DP is available for the raising of NP, it is unclear why this possibility can only be exploited in a finite relative clause, but not in an infinitival one.

Thus, it seems that the derivation in (83) should be excluded on general grounds. Interestingly, Szabolsci (1994: 202–203) and Kayne (1994: 26–27) put forth independent evidence to the effect that Spec,DP is only accessible to operator phrases.[61] For instance, in Hungarian quantificational possessor phrases must raise to Spec,DP, where they are assigned the Dative case, and cannot fill the lower possessor position where they receive the Nominative case:

(84) a. *ki-nek a t kalap-ja*
who-DAT the hat-POSS.3SG
'whose hat'
b. * *ki kalap-ja*
who-NOM hat-POSS.3SG (Szabolsci 1994: 202–203)

Let us therefore assume the following hypothesis:

(85) Spec,DP is an operator position.

Note that NP is a predicative category (cf. (10) of chapter II), and therefore its raising to Spec,DP in (83) constitutes an improper movement.

It follows that if there is only one available Spec in the Comp area, the *wh*-relative cannot be derived, because there isn't enough room for both the relative determiner and the NP "head" in a single Spec.[62] By this hypothesis, the derivation of the *wh*-relative requires the availability of two distinct Spec positions for relativization.

6.3. An alternative proposal

Under our assumptions, in the English complementizer system there are two positions available for relativization, namely Spec,ForceP and Spec,TopicP: the former is involved in the derivation of the *that*-relative, and the latter in the derivation of the zero relative. It is then possible to assume a two steps derivation involving both position. First, the relative DP moves to the Spec of TopicP;[63] then the NP "head" is extracted and raises to Spec,ForceP, stranding the relative determiner in the lower Spec. The derivation is illustrated in (86):

(86)

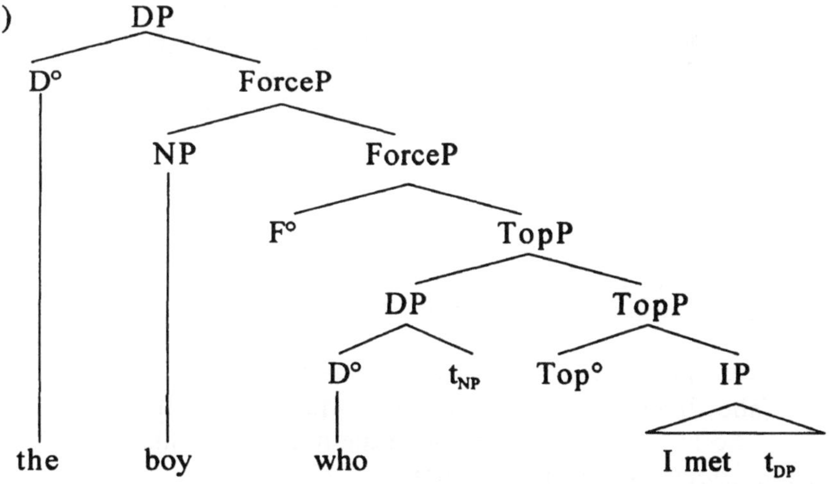

Note that contrary to the Danish example (82), the two steps of the derivation in (86) involve two distinct constituents: in the first step, DP moves to the topic position, and in the second step, the NP "head" is extracted and raises to Spec,ForceP. However, the first step requires the movement of the relative DP to a Spec position below Spec,ForceP.

Assuming with Müller—Sternefeld (1993) that every type of A' movement targets a specific type of position (cf. (64)), the derivation is possible only if there is a position available for the relative DP below Spec,ForceP. This condition is satisfied in English, where Spec,ForceP and Spec,TopicP belong to the same type and Top° bears the feature [+relative].

In Italian, by the negative setting of parameter (68) Topic° cannot host the relative DP in its Spec. As this is a crucial step in the derivation of (86), the *wh*-relative cannot be derived in Italian. (cf. the ungrammaticality of (2b)).

6.4. Evidence for the two steps derivation

The derivation in (86) implies that the relative determiner need not occur in the highest specifier of the relative clause. If the relative determiner occurs in a topic position, it should be possible to find overt material between the NP "head" and it: in particular, a lexical Force° or a preposed constituent.

Neither of these possibilities is realized in the English *wh*-relative. In the latter, the Force° head is obligatorily null, for reasons to be discussed in § 8; and moreover, the relative pronoun obligatorily precedes any topicalized constituent. This may be conceived as an effect of relativized minimality, assuming that the NP "head" in Spec,ForceP must antecedent-govern its trace in Spec,TopicP, and antecedent-government would be blocked by an intermediate Spec of the same type filled by a topicalized phrase.

Nevertheless, the two predictions are fulfilled in other languages. According to Rudin (1990–1991: 439), in Bulgarian a subject topic may intervene between the NP "head" and the relative determiner, as in (87). The same possibility is attested in Hungarian, as shown in (88) (cf. Haberland—van der Auwera 1990: 136).

(87) *živee na ulicata, muzeja* kŭdeto *se namira.*
(she) lives on street-the museum-the where self-finds
'She lives on the street where the museum is situated.'

(88) *a könyv, János* amit *említtet*
the book J. which mentioned
'the book which J. mentioned' [64]

Moreover, Kenesei (1994: 303) mentions that in XVI century Hungarian the complementizer *hogy* could occur between the "head" and the relative determiner:[65]

(89) ... *hogy a városba jónne a farkassal,* hogy ki *igen fene vala.*
that the town would-come the wolf, that who very wild was
'that he would come to town with the wolf which was very wild'

R. Kayne (personal communication) points out another prediction: if the relative D° can be stranded in Spec,TopicP below the landing site of the NP "head", leftward movement of TopicP to the Spec of the external D° should give a prenominal relative clause introduced by a relative determiner. But prenominal relative clauses generally lack relative determiners (cf. Kayne 1994: 93–95).

However, some Latin and Ancient Greek examples appear to be instances of prenominal relative clauses with a fronted relative D°:

(90) a. τούτους καὶ ἄρκοντας ἐποίει ἧς κατεστρέφετο χώρας
these ones$_{ACC}$ and governors (he) made which$_{GEN}$ he conquered country$_{GEN}$
'and he made these ones governors of the country which he conquered' (Xenophon, *Anabasis* 1, 9, 14)

b. ἀδικεῖ Σωκράτης οὓς μὲν ἡ πόλις νομίζει θεοὺς οὐ νομίζων
is wrong Socrates which$_{ACC}$ the city adores gods$_{ACC}$ not adoring
'Socrates is wrong in that he does not adores the gods whom the city adores'
(Xenophon, *Comm.* 1,1,1: from Gonda 1954: 24)

c. ...ἀποπέμψαι πρὸς ἑαυτὸν ὃ εἶχε στράτευμα
to send to him which$_{ACC}$ (he) had army$_{ACC}$
'to send to him all the soldiers he had' (Xen., *Anab.* I,2,6)

d. *agedum odorare hanc quam ego habeo pallam*
come on, sniff this$_{ACC}$ which$_{ACC}$ I have mantle$_{ACC}$
'come on, smell this mantle that I have'
(Plautus, *Maenechni* 166)

e. *illa quae iniuria depulsa fuerat ancilla totam faciem eius fuligine longa perfricuit*
that$_{NOM}$ who$_{NOM}$ rudely repulsed had been maid$_{NOM}$ all face$_{ACC}$ his long signs of soot covered
'that girl who had been rudely repulsed covered all his face with long signs of soot' (Petronius, *Satyricon* 22,1)

f. *mandatae quae sunt, volo deferre epistulas*
entrusted which$_{NOM}$ were (to me) I want to deliver letters$_{ACC}$
'I want to deliver some letters which were entrusted to me'
(Plautus, *Persae* 694)

In these examples the relative determiner precedes the "head". In (b), (c), (d) and (e) the external and internal Case coincide; in (a) and (f), the two Cases do not coincide, and relative "head" is in the external Case; in (a) the relative determiner is attracted by the external Genitive Case of the "head".

The (b)–(e) examples admit of two analyses:

(1) they can be true prenominal relative clauses preposed to the Spec of the external determiner. Since they contain a relative D° fronted at the beginning of the relative clause, it must be assumed that this D° has been stranded in a Spec position below the landing site of the NP "head", and the preposed relative clause is a category larger than IP, including this Spec. Thus, the representation of e.g. (90d) would be (91): [*quam ego habeo* t] is a clausal constituent not including the NP "head", preposed to a Spec of the complex DP:

(91) $[_{DP1}$ *hanc* $[_{XP}$ $[[_{DP2}$ *quam* $t_{NP}]$ [*ego habeo* $t_{DP2}]]_j$ [X° $[_{CP}$ $[_{NP}$ *pallam*] $[_{CP}$ C° $t_j]]]]$

(2) Alternatively, the (b)–(d) examples may be internally headed relative clauses: namely, the NP "head" occupies its base position within the relative CP, and the relative determiner is fronted alone at the beginning of the CP (R. Kayne, personal communication). The possibility of scrambling determiners is typical of Latin, as attested by the following interrogative clause:[66]

(92) *interrogare ergo atriensem coepi*, quas *in medio* picturas *haberent*
to ask then the porter I began, which$_{ACC}$ in the middle picture-s$_{ACC}$ there were
'then I began to ask the porter which pictures there were in the middle' (Petronius, *Satyricon* 29, 9)

Given the coincidence of the external and internal Case, both analyses are in principle possible for the (b)–(d) examples.[67] Note however that the (a) example only allows the first analysis: in fact, the NP "head" χώρας 'country' is in the genitive Case required by the external context, being a modifier of the noun ἄρκοντας 'governors'; by hypothesis, it must have reached Spec,ForceP in order to be attracted by the external Case (§ III.4.4). Moreover, the sentence cannot be analysed as a correlative structure, since the relative clause [ἧς κατεστρέφετο] is clearly embedded in the main clause. Thus, (90a) cannot be an internally headed relative, but it must be a true prenominal relative clause.[68]

If this is the correct analysis, this evidence provides further support for the two-steps derivation (86).

7. Pied piping relatives

In pied piping relatives, the NP "head" moves to the Spec of the pied piped constituent, stranding the relative determiner (Kayne 1994: 89–91; cf. (22d) above). However, given the assumption that in English the relativized phrase may move to either Spec,ForceP or Spec,TopP, it is necessary to determine which position the pied piped phrase moves to.

Pied piping phenomena are usually classified in two types: "light" pied piping involves a preposition selecting the relative DP, whereas "heavy" pied piping involves larger constituents, like e.g. a complex DP or a whole subordinate clause. It is often assumed that heavy pied piping is only possible in appositive relatives, whereas light pied piping is possible in restrictive relatives as well (cf. for instance Cinque 1978 on Italian, Fabb 1990 and Emonds 1979 on English). Actually, heavy pied piping is sometimes possible in restrictive relatives, but it is often less felicitous than in appositive ones.

In his coordinate analysis of appositives (cf. § V.1.1), Emonds (1979: 224–225) suggests that heavy pied piping is actually an instance of topicalization into the second of two coordinated clauses; the topicalized phrase contains a pronoun coreferent to a constituent of the first conjunct. The transformations of Parenthetical Formation and S' Attachment place the second conjunct in a position adjacent to the pronoun's antecedent, and the pronoun takes the *wh*-morphology, yielding an appositive relative. For instance, the pied piping appositive in (93b) is derived from the conjunction (93a):

(93) a. *He should explain how to open both the doors, (for)* [*the keys to them*]$_i$ *he has ruined* t_i.

b. *He should explain how to open both the doors,* [*the keys to which*]$_i$ *he has ruined* t_i.

Emond's hypothesis may be reinterpreted in the following way: whereas light pied piping of a preposition does not require any special licensing mechanism, pied piping of a heavy constituent must be licensed by a topicalized interpretation of that constituent. Then, light pied piping could target either Spec,ForceP or Spec,TopicP, but heavy pied piping would necessarily target Spec,TopicP. Hooper—Thompson (1973: 489–491) observe that topicalization is often more felicitous in appositive than in restrictive relatives: this would explain the marginal status of heavy pied piping in the latter.

Apart from these considerations, nothing in principle prevents a pied piped phrase from moving to either Spec,ForceP or to Spec,TopicP in English. Note that Force° cannot be spelled out either to the left or to the right of the pied piped phrase; but this does not necessarily imply that the ForceP level is missing. In fact, in the next section it will be argued that Force° is subject to PF deletion under specific syntactic conditions. Therefore, the landing site of (light) pied piping in English remains undetermined.

8. Economy and PF deletion

At this point, it is necessary to thoroughly reexamine the interaction of relative determiners with the complementizer *that*, now analysed as Force°. In the preceding discussion, two distinct mechanisms were proposed to account for "deletion" phenomena:

(1) the omission of the relative determiner in *that*- and zero relatives consists in its abstract incorporation to the external determiner;

(2) the optional omission of *that* in zero relatives and complement clauses consists in the selection of a phonetically null head Topic° instead of Force°.

Note that in (86) the head Force° is syntactically present, but it cannot be spelled out (* *the book that which I read*). Therefore, it is necessary to explain how it comes to be deleted in the PF representation.

It is possible to extend to this case the incorporation approach elaborated in § 2.3. In fact, in (86) Force° is directly selected by the external determiner, and the feature contents of the two heads may be assumed to be compatible.[69] The incorporation of Force° yields the final representation (94) for the *wh*-relative:

(94) $[_{DP}$ Force°+D° $[_{ForceP}$ NP $[$ t_F $[_{TopP}$ $[_{DP}$ D_{REL} $t_{NP}]$ [Top° IP]]]]]

But the core of the problem is yet unsolved: what triggers the obligatory incorporation of Force° in (94)? The same question arose with respect to the incorporation of the relative determiner in (32). The answer to this question is the last step of our argumentation.

Note that the effect of abstract incorporation is to fuse two terminal symbols into one without loss of features: in this respect, it yields a more economical representation (at least with respect to the Morpho-Phonological component). Suppose now that we adopt a conception of economy whereby economy principles do not only select the optimal step in a derivation (or the optimal member in a set of derivations), but they may also apply within the computation, triggering a step that leads to a more optimal representation. Then we can assume that abstract incorporation in (32) and (94) is triggered by the following economy principle:[70]

(95) Economy of representation
Unify two terminal symbols into one whenever possible
(where unification is achieved through abstract incorporation)

The qualification "whenever possible" refers to the independent syntactic constraints discussed in § 2.3: relativized minimality and the condition of feature compatibility for unification. There is, however, one further syntactic constraint which immediately derives from Kayne's (1994) Linear Correspondence Axiom. The latter rules out double adjunction to a head, since this would create too symmetric a configuration (Kayne 1994: 19–21). This implies that in every relative structure it is possible to incorporate to the external D° either the relative D° or Force°, but not both of them:

(96) * $[_{DP}$ Force°+D_{REL}+D° $[_{ForceP}$ $[_{DP}$ t_D NP] [t_{Force} IP]]]

We are now in a position to show how the economy principle (95) applies in the derivation of the various relative clause types, subsuming the effects of the Doubly Filled Comp Filter.[71]

Consider first the *that*-relative (33), repeated here as (97):

(97) $[_{DP}$ D_{REL}+*the* $[_{ForceP}$ $[_{DP}$ t_D $[_{NP}$ *book*]]$_i$ $[_{ForceP}$ *that I read* t_i]]]

In this structure, the incorporation of the relative determiner to the external one is possible and, by (95), it is obligatory. On the other hand, this excludes the incorporation of Force°, since the resulting double adjunction structure (96) would violate the Linear Correspondence Axiom; thus, Force° must be spelled out as an independent morpheme.

Note that if Force° were to incorporate instead of the relative D°, the resulting structure would be illformed:

(98) * $[_{DP}$ Force°+D° $[_{ForceP}$$[_{DP}$ D_{REL} NP] [$t_{Force°}$ IP]]]

In fact, the NP "head" is in the minimal domain of the relative determiner, which remains distinct from the minimal domain of the external determiner; in other terms, the external D° cannot govern NP by relativized minimality, and its nominal features remain unchecked.

Let us then consider the zero relative:

(99) $[_{DP}$ D_{REL}+*the* $[_{TopP}$ $[_{DP}$ t_D $[_{NP}$ *book*]]$_i$ $[_{TopP}$ Top° *I read* t_i]]]

Principle (95) applies in the same way as before, triggering the incorporation of the relative determiner. Recall that in this case the omission of *that* is not brought about by the abstract incorporation of Force°, but it consists in the selection of Topic°, which is intrinsically null and need not be licensed through incorporation.

The situation is reversed in the *wh*-relative, repeated here:

(100) $[_{DP}$ F°+*the* $[_{ForceP}[_{NP}$ *book*] t_F $[_{TopP}[_{DP}$ *which* $t_{NP}]_i$ [Top° *I read* t_i]]]]

In this structure, Force° intervenes between the relative D° stranded in Spec,TopP and the external D°. By relativized minimality, the relative D° cannot incorporate and it is obligatorily spelled out. On the other hand, nothing prevents the incorporation of Force° itself, and the economy principle (95) makes it obligatory.

The same holds in the pied piping relative, repeated here as (101):

(101) $[_{DP}$ F°+*a* $[_{FP}$ $[_{PP}[_{NP}$ *man*] $[_{PP}$ *with* $[_{DP}$ *whom* t_{NP}]]]$_i$ $[t_F$ $[_{IP}$ ...t_i...]]]]

As discussed in § 2.3, the relative determiner is prevented from incorporating to the external determiner by the intervening preposition, and consequently it cannot be deleted; however, the incorporation of Force° is possible and hence obligatory.[72]

In conclusion, the "doubly filled Comp effect", i.e. the obligatory deletion of either the relative determiner or *that*, has been reduced to the economy principle (95). This economy principle is "local" in the sense of Chomsky (1995: 225–229), namely it applies at each step of a derivation triggering the next step, but it does not require the comparison of different derivations; in other terms, it has no transderivational power.

8.1. An alternative: the extraction of the "head"

In § III.2, an alternative hypothesis on the raising of the NP "head" was considered, whereby the NP targets the Spec of an Agr_D head external to the relative FP:

(102) $[_{DP}\ D° [_{AgrP}\ NP\ [\ Agr_D\ [_{FP} \ldots t_{NP}]]]$

Agr_D itself is affixal: it must incorporate to D° and get "unified" with it; the result is spelled out as an inflected determiner (cf. § 2.3).

Let us consider whether the analysis of the "doubly filled Comp effect" proposed above is compatible with this alternative.

Starting from the derivation of the *that*-relative, the raising of the relative DP to Spec,FP yields the derived structure (103):

(103) $[_{DP}\ D° [_{AgrP}\ Agr_D\ [_{FP}\ [_{DP}\ D°\ NP]\ [Force°\ IP]]]]$

At this point, two distinct requirements must be met. On the one hand, the strong features of the external Agr_D must be checked. On the other hand, the economy principle (95) must be satisfied by the abstract incorporation of either the relative D° or Force°.

Suppose that the first requirement is met by the raising of NP to Spec,Agr_D, and the second requirement is met by abstract incorporation of the relative D° (104a) or Force° (104b). This predicts that both the *that*-relative and the *wh*-relative can be derived from the underlying structure (103).

(104) a. $[[_{DP}\ D° [_{AgrP}\ NP\ [\ D_{REL}{+}Agr_D\ [_{FP}\ [_{DP}\ t_D\ t_{NP}]\ [Force°\ IP]]]]]$
b. $[[_{DP}\ D° [_{AgrP}\ NP\ [\ F°{+}Agr_D\ [_{FP}\ [_{DP}\ D°\ t_{NP}]\ [\ t_F\ IP]]]]]$

This is, however, an unwanted result, because we would lose the account for the ungrammaticality of the *wh*-relative in Italian (cf. (2b)) and the correlation between the possibility of the *wh*-relative and of the zero relative, previously reduced to the availability of a low landing site for the relative DP in the complementizer system of English.

Keeping to minimalist assumptions, the steps in (104), though legitimate, may be excluded by the existence of a more economical alternative derivation that satisfies both the checking requirement and the economy requirement by a single operation.

Note that from a derivational perspective, in (103) the nominal features of the relative D° must have been checked by the movement of NP to Spec,Agr_D, which then incorporates to D_{REL}. Thus, more accurately, the substructure of the relative DP in (103) is (105):

(105) $[_{DP}\ Agr_D + D_{REL}\ [_{AgrP}\ NP\ [t_{Agr}\ t_{NP}]]]$

Consider now the continuation from (103) to (106):

(106) $[_{DP}$ D° $[_{AgrP}$ $[Agr+D_{REL}]+Agr_D$ $[_{FP}$ $[_{DP}$ t_D ... NP] [Force° IP]]]]]

Here $Agr+D_{REL}$, whose strong features have been checked, abstractly incorporates to the external Agr_D. (The latter will further incorporate to the external D°). As a result, the checked features of the relative D° are "unified" with the unchecked features of the external Agr_D. Suppose that this unification effectively eliminates these unchecked strong features. Then, the incorporation in (106) can satisfy at once both the checking requirement and principle (95).

Thus, considerations of economy of the derivation select (106) as the optimal convergent derivation starting from (103); this yields the *that*-relative. If Force° were instead to incorporate, as in (104b), the checking requirement would have to be satisfied by the independent raising of NP to Spec, Agr_D, resulting in a nonoptimal derivation.

As for the *wh*-relative, let us keep the hypothesis that the relative DP moves to a low Spec position of the complementizer system. This step can be explicitly forced by the hypothesis that in the *wh*-relative the relative DP bears a [topic] feature, corresponding to the somewhat "emphatic" status of this relativization strategy in contemporary English. Thus, the relative DP must stay in Spec,TopP in order to satisfy the Topic Criterion (cf. (63) above).

Furthermore, it will be proposed in § VII.6 that the feature [topic] on Top° must be licensed by a c-commanding Force° in the same complementizer system. This requirement imposes the projection of the Force Phrase above the Topic Phrase:

(107) $[_{DP}$ D° $[_{ForceP}$ Force° $[_{TopP}$ $[_{DP}$ D_{REL} NP$]_i$ Top° $[_{IP}$ ... t_i ...]]]]
[topic] [topic]

In this configuration, as already discussed in connection with (100), the incorporation of the relative D° is excluded by relativized minimality; the only way to satisfy the checking requirement is to raise the NP "head" to the Spec of the external Agr_D. Principle (95) is instead satisfied by the deletion through abstract incorporation of Force°:

(108) [D° $[_{AgrP}$ NP $[F°+Agr_D$ $[_{FP}$ t_F $[_{TopP}[_{DP}$ D_{REL} $t_{NP}]_i$ Top° $[_{IP}$ ...t_i...]]]]]]

The same holds, *mutatis mutandis,* for the pied piping relative, where the relative D° is stranded in the complement position of the preposition. Thus, the proposed analysis of the "doubly filled Comp effects" can be essentially maintained, though in this version it crucially relies on the principle of economy of derivations.

9. Appositive relatives

The analysis developed in the preceding sections accounted for the distribution of relative pronouns and of *that* in restrictive relatives. This section deals with the paradigm of English appositive relatives, which differs sharply from the restrictive paradigm (1).

Recall from § V.4 that in Kayne's (1994: 110–115) approach, appositive relatives have the same derivation as restrictive ones in overt syntax; the appositive interpretation is obtained by moving the IP constituent of the relative clause out of the c-command domain of the external determiner in LF:

(109) IP... $[_{DP}$ D° $[_{FP}$ NP $[_{FP}$ F° $t_{IP}]]]$

Kayne proposes that IP moves to the Spec of the external D°.[73] Note however that the appositive relative receives a "backgrounded" interpretation: for instance, it is not affected by sentential negation in the matrix clause. Let us tentatively assume that the appositive IP moves to a Topic position internal to the functional structure of the DP.

Although the appositive interpretation is brought about in the LF branch, it affects the visible shape of the relative clause: in fact, in Modern English only *wh-* (and pied piping) relatives allow this interpretation, as shown in paradigm (7), repeated here as (110):

(110) a.* *This book,* that *I read thoroughly,* ...
b. *This book,* which *I read thoroughly,* ...
c.** *This book, Ø I read thoroughly,* ...

In Italian, instead, *that*-relatives allow the appositive interpretation:

(111) *questo libro,* che *ho letto attentamente,* ...
this book, that I read thoroughly
'This book, which I read thoroughly,...'

It would be interesting to find a structural reason for the asymmetry between the appositive (11a) and the restrictive (1a) in English, as well as for the lack of a similar asymmetry in Italian.

A closer look at the derivation in (109) reveals a possible line of analysis. Note that in its LF scrambling the IP constituent always has to skip a c-commanding A' specifier, namely the highest Spec of the relative clause, which is invariably filled. Recall that by the positive setting of parameter (68), in the English complementizer system both Spec,ForceP and Spec,TopicP are available for relativization and, with respect to the Principle of Unambiguous Binding, they are positions of the same type. We suggested above that relativized minimality is sensitive to the same typology of A' positions: it follows that in its movement toward a Topic position, the scrambled IP in (109) necessarily crosses an A' position of the same type as its landing site. It may be assumed that this position qualifies as a potential antecedent governor for the trace of IP and blocks IP scrambling under relativized minimality:

(112) a. * IP... [$_{DP}$ D° [$_{FP}$ DP$_{REL}$ [$_{FP}$ F° t$_{IP}$]]] (*that*-relative)
b. * IP... [$_{DP}$ D° [$_{TopP}$ DP$_{REL}$ [$_{TopP}$ Top° t$_{IP}$]]] (zero relative)

On the other hand, Italian has a negative setting of parameter (68), and nothing requires the assumption that Spec, Force Phrase and Spec, Topic Phrase are positions of the same type; the evidence in (75) above suggests that they are not. Then, the LF scrambling of IP past Spec, Force Phrase is unproblematic, and the *that*-relative (111) can receive an appositive interpretation.

It remains to be explained why the English *wh*-relative (110b) allows an appositive interpretation after all. Recall that in the analysis proposed in the preceding section, the *wh*-relative differs crucially from the *that*- and zero relatives in that its head Force° incorporates to the external D°, so that it fails to be spelled out:

(113) [$_{DP}$ *this* Force°+D° [$_{FP}$[$_{NP}$ *book*] t$_{Force}$ [$_{TopP}$[$_{DP}$ *which* t$_{NP}$]$_i$ Top° [$_{IP}$ *I read* t$_i$ *thoroughly*]]]]]

Adopting an idea of Kayne's (1993: 7), let us assume that after the incorporation of Force° Spec,ForceP becomes a "derived" specifier of the complex head Force°+D° and changes its status in the typology of A' positions. Then Spec,ForceP does not qualify as a potential antecedent for LF scrambling and the TopP complement of Force° can be moved past Spec,ForceP without violating relativized minimality:[74]

(114) TopP... [$_{DP}$ Force°+D° [$_{FP}$ NP [$_{FP}$ t$_{Force°}$ t$_{TopP}$]]]

10. Summary

This chapter proposed an analysis of the the "doubly filled Comp" effects observed in headed relative clauses, based on two main hypotheses:

(1) the PF deletion of the relative determiner and of Force° consists in the abstract incorporation of these heads to the external D° of the relative structure; this incorporation process is triggered by a principle of economy of representation, and it is subject to well known syntactic constraints, e.g. relativized minimality, which accounts for the impossibility of deletion of the relative determiner in pied piping structures. The complementarity of deletion of the two elements (apart from zero relatives in English) follows from the Linear Correspondence Axiom, whereby the external D° can license the incorporation of one head at most. The notion of recoverability is reduced to the condition of feature compatibility for the "fusion" of two incorporated functional he-ads.

(2) The possibility of the zero relative (1c) and of the *wh*-relative (1b) in English is determined by the availability of the specifier of the null head Topic° for relativization. This position is instead not available in Italian, accounting for the contrast between the English and Italian paradigms (1) and (2).

The difference between the two languages depends on the setting of parameter (68), which determines some properties of the head Topic°. However, in the preceding discussion the parameter has been assumed without empirical justification. The next chapter will examine various phenomena concerning the complementizer system of headed relative clauses, in order to motivate the proposed parameter and to further explore its consequences.

Chapter VII
The complementizer system of headed relative clauses

0. Introduction

This chapter focusses on the internal structure of the complementizer system in the various relative clause types proposed in the foregoing discussion. The relevant structures are repeated in (1)–(3):[1]

(1) $[_{DP}$ *the* $[_{ForceP}$ $[_{DP}$ *book*$]_i$ $[_{ForceP}$ *that* [*I read* t_i]]]]

(2) $[_{DP}$ *the* $[_{TopP}$ $[_{DP}$ *book*$]_i$ $[_{TopP}$ Top° [*I read* t_i]]]]

(3) $[_{DP}$ *the* $[_{FP}[_{NP}$ *book*] [Force° $[_{TopP}[_{DP}$ *which* $t_{NP}]_i$ [Top° *I read* t_i]]]]

The three relative structures differ in the composition of their complementizer system: the *that*-relative (1) has the highest layer of the system, the Force Phrase, whereas the zero relative (2) has a lower layer headed by a phonetically null head, provisionally identified with Top°; finally, the *wh*-relative has both layers.

The structures (2) and (3) are possible in English, but not in Italian. The availability of the lower structural layer for relativization depends on the setting of parameter (68) of chapter VI, repeated here:

(4) ± (Topic° optionally supports the features [+ declarative] and [+ relative])

In English the parameter has a positive setting: this implies two consequences. On the one hand, Top° can bear the feature [+ declarative], substituting Force° in declarative complement clauses, and it can also bear [+ relative], hosting a relativized phrase in its Spec. On the other hand, since both Force° and Top° can bear the feature [+relative], they provide a landing site for the same type of A' structure and, by the Principle of Unambiguous Binding, their specifiers must be classified as positions of the same type.[2]

In Italian, on the contrary, the parameter has a negative setting. It follows that Top° cannot host a relative phrase in its Spec, nor can it substitute for Force° in any declarative clause: this is why Italian does not allow the optional omission of the declarative complementizer in a way comparable to English. Furthermore, the specifiers of the two heads need not be classified as positions of the same type for the purposes of the Principle of Unambiguous Binding (cf. § VI.9).

The preceding discussion left open two important questions:
(1) why should the null head be identified as Top°?
(2) what determines the setting of parameter (4)?
These questions are addressed in the next section.

1. The Topic parameter

1.1. Finiteness

Consider again the internal structure of the complementizer system, repeated here in its most expanded form:

(5) [$_{FP}$ Force° [$_{TopP*}$ Top° [$_{FocP}$ Focus/Wh° [$_{TopP*}$ Top° [$_{FinP}$ Fin° IP]]]]]

In English, an interesting asymmetry in the composition of this system emerges in infinitival vs. finite clauses. In fact, in infinitival clauses topicalization seems to be generally disallowed:

(6) a. * *My friends tend* [$_{TopP}$[*the more liberal candidates*]$_i$ [$_{IP}$ PRO *to support* t$_i$]] (Hooper—Thompson 1973: 485)
b. ?* *I wonder whether* [$_{TopP}$ *next year* [$_{IP}$ PRO *to go home or not* t]]
c. * *I would like* [*for* [$_{TopP}$ [*my brother*]$_i$ [$_{IP}$ *you to meet* t$_i$]]] (Iwakura 1980: 57)

An explanation for the ungrammaticality of (6) is offered by Rizzi (1995: 18–22), who analyses it as an adjacency effect on Case assignment. Specifically, he argues that in infinitival clauses, contrary to finite ones, the subject must be Case-licensed by a head of the Comp system rather than by Inflection. In (6c), the overt subject in Spec,IP must be licensed under government by the prepositional complementizer *for*; but the intervening head Top° blocks head government by relativized minimality. The same analysis can be extended to (6a,b), under the assumption that PRO must be Case-licensed by an infinitival null C° lying above Top°.

The ungrammaticality of (6a,b) can also be related to a contrast discussed by Kayne (1991: 665):

(7) a. *He doesn't know* [*whether* [PRO *to go to the movies*]].
b. * *He doesn't know* [*if* [PRO *to go to the movies*]].

In infinitival clauses, contrary to finite ones, a contrast emerges between the interrogative particles *whether* and *if.* Kayne argues that (7b) is ruled out by the PRO theorem, because the lexical C° *if* governs PRO in Spec,IP. In (7a), on the contrary, *whether* is a *wh*-phrase in Spec,CP and PRO is ungoverned.[3]

Assuming the complementizer system, a slightly different approach becomes possible: the two particles can be conceived of as two distinct functional heads. Specifically, let us assume that *whether* instantiates an interrogative Force°, whereas *if* instantiates the lower head Focus/Wh°.

This hypothesis is supported by diachronic as well as synchronic evidence. First, *whether* derives from the neuter singular form of the Old English interrogative pronoun *hwæðer* 'which of two': in Old English, this particle could occur in direct questions and its presence blocked verb second (Mitchell 1985: § 1643). This blocking effect can be more naturally attributed to a head than to a *wh*-phrase.[4]

Furthermore, in the following English/French contrasts, discussed by Kayne (1991: fn. 68), *whether* patterns with the French declarative complementizer *que* whereas *if* patterns with the interrogative particle *si* filling Focus°:

(8) a. *Whether they give him a seat or not, he'll be happy.*
b. * *If they give him a seat or not, he'll be happy.*
c. *Qu'on lui donne une place ou non, il sera heureux.*
that one to-him give a seat or not, he will be happy
'Whether they give him a seat or not, he will be happy.'
d. * *Si on lui donne une place ou non, il sera heureux.*
if one to-him give a seat or not, he will be happy

If we assume that the two interrogative particles instantiate two different functional heads, then the ungrammaticality of *if* in (7b) can be compared to the impossibility of topicalization in (6): in both cases, a low projection of the complementizer system is unavailable - the Focus/Wh Phrase and the Topic Phrase respectively. This means that the complementizer system of English infinitival clauses is significantly less rich than that of finite clauses.

Note that no comparable asymmetry is found in Italian: topicalization is allowed in infinitival clauses, and moreover, the particle *se*, which realizes the head Focus/Wh° (§§ I.2.1, VI.4.1) and corresponds to English *if*, occurs in infinitival questions:

(9) a. *Non so*, [$_{TopP}$ *questo libro*, [$_{FocP}$ *se darlo a te o a Gianni*]].
I don't know, this book, if to give it$_{CL}$ to you or to Gianni
'As for this book, I wonder whether to give it to you or to Gianni.'

b. *Pensavo*, [$_{TopP}$ *a Gianni*, [$_{FinP}$ *di regalargli un libro*]].
I was planning to Gianni to give him$_{CL}$ a book
'As for Gianni, I was planning to present him with a book.'

(10) *Non so* [$_{FocP}$ *se partire o restare a casa*].
I don't know if to leave or stay at home
'I wonder whether to leave or to stay at home.'

The comparison of (6) and (7b) to (9) and (10) shows that in English the Topic Phrase and the Focus/Wh Phrase are restricted to finite clauses, whereas in Italian they may be realized both in finite and infinitival clauses. This state of affairs may be determined by the intrinsic feature specification of the corresponding functional heads in the two languages: in English Top° and Focus° bear the feature [+finite], whereas in Italian they are underspecified for this feature.

1.2. Licensing Topic°

Let us now reconsider the principle of structural licensing (67) of chapter VI, reproduced here:

(11) Every head in the complementizer system must be *structurally licensed* by either (1) or (2):
(1) selection by a head external to the Comp system;
(2) specification of a feature satisfying the F-Criterion

Though the principle establishes two different ways of structurally licensing a head, these have one element in common: in both cases structural licensing depends on the presence of some feature on the head. By clause (1), a functional head α is structurally licensed if it meets the selectional requirements of some higher head β; but this means that α contains some feature that satisfies the selectional requirements of β. For instance, if β is a verb selecting a [+wh] clause, it structurally licenses a head α containing the feature [+wh]. By clause (2), a functional head is structurally licensed if it is specified for a feature to be matched by an appropriate phrase in its Spec.

The essential condition stated by principle (11) is, then, that a functional head must contain at least one feature; otherwise, it is not structurally licensed. This condition makes sense in the approach to functional heads proposed in § VI.2.3, whereby a functional head consists of a set of features. Principle (11) states that this set cannot be empty.

This principle seems to be complementary to the Feature Scattering Principle proposed by Giorgi—Pianesi (1996: § V.2):

(12) Feature scattering principle
Each feature can head a projection

According to this principle, each feature can be realised as an independent functional head (although it is also possible to realize two or more (sets of) features in a single syncretic functional head). Conversely, by (11) each functional head must consist of at least one feature.

The option of scattering[5] is restricted by two conditions. The first one is a principle of economy, whereby scattering a feature is only possible if the initial numeration contains some element to be merged in the Spec position of the scattered feature. The second condition is a universal ordering which restricts the hierarchical structure in which the scattered features are realized:[6]

(13) Universal ordering constraint
The features are ordered so that given $F_1 > F_2$, the checking of F_1 does not follow the checking of F_2.

The structure of the complementizer system (5) can then be restated as an ordering of the relevant features:

(14) [±decl] > [topic] >[focus] > [topic] > [±finite]
[presupposed] [±wh]

The order in (14) is by hypothesis universal. The language-specific properties of the complementizer system can then be reduced to the different possibilities of realizing various features syncretically on a single head.

The features in (14) can be distinguished in two subtypes. On the one hand, the features [± declarative][7] and [± finite] presumably characterize every clause and therefore they must always be realized in the complementizer system. On the other hand, the features [topic] and [focus] seem to establish a privative opposition: since topicalization and focalization are marked syntactic processes, it is plausible to assume that a sentence without topicalization or focalization does not contain a negatively specified feature in the complementizer system, but it is simply underspecified for these features.[8]

This corresponds to a hypothesis independently put forth by Rizzi (1995). According to Rizzi, when there is no topicalization or focalization the features [+declarative] and [+finite] are syncretically realized by the head Force°, which directly selects the Inflectional Phrase (15a). When the clause contains a topicalization or focalization, instead, the Topic or Focus Phrase must be projected below the Force Phrase; in this case, the feature [± finite] must be realized by an independent head Finite° below Topic/FocusP, which constitutes the "interface" with the inflectional system (15b):

(15) a. [$_{FP}$ Force° IP]
  Force°: [±decl] [±finite] [±presupp.]

b. [$_{FP}$ Force° [$_{TopP}$ Top° [$_{FinP}$ Finite° IP]
  Force°: [±decl] [±presupp.]
  Top°: [topic]
  Finite°: [±finite]

This proposal can be easily recast in Giorgi—Pianesi's framework: in (15a) three features are syncretically realized on a single head Force°; in (15b), instead, the feature [±finite] is scattered and, given the ordering principle (14), it is realized as the lowest functional head of the complementizer system.

However, the universal ordering (14) can also be satisfied in another way, namely by syncretically realizing the feature [±finite] on the Topic/Focus heads. The evidence reviewed in § 1.1 suggests that this is in fact the option realised in English: the Topic/Focus heads have an intrinsic positive specification for this feature, so that they are unavailable in [-finite] clauses. Under this hypothesis, English topicalization structures would have the complementizer system (16), with a hybrid category Top/Fin°, rather than (15b):

(16) [$_{FP}$ Force° [$_{TopP}$ Top/Fin° IP]
  Force°: [±decl] [±presupp.]
  Top/Fin°: [topic] [+finite]

Let us now reconsider the status of simple declarative clauses without internal topicalization (or focalization): here the feature [topic] is simply not specified, the opposition being privative. One possibility is to have recourse to a syncretic Force°, as in (15a), which will be spelled out as the morpheme *that*. However, our hypotheses also allow another option. When the head Top/Fin° is underspecified for the feature [topic], its content is reduced to the feature [+finite]; but this single feature can be realised as a functional head by itself, according to the scattering principle (12). Now suppose that this underspecified head is optionally assigned the default feature value [+declarative]:[9] this hypothesis is supported by the behaviour of some complementizer-like particles like Middle English *that* (cf. § 2) and Quebec French *que* (cf. § 3.1), which act as generic "subordinators", but also as "complementizers" in declarative complement clauses. As a result, Top/Fin° can be structurally licensed as the unique head in the complementizer system of a declarative clause - in fact, as a "defective" declarative complementizer.[10]

(17) ... [$_{\text{TopP}}$ Top/Fin° IP]
                    |
              [+finite]
              [+decl]

Consider now the situation of Italian. The data in (9) and (10) show that here Top° and Focus° are not intrinsically specified as [+finite].[11] Thus, in Italian the only possible content of Topic° is the feature [topic], to be matched by a topicalized constituent in the Spec. When the clausal structure does not contain a topicalization, Top° cannot be structurally licensed - in fact, it would be a radically empty head. Consequently, Top° cannot be structurally licensed as a "defective" complementizer.

The Topic parameter (4) is thus reduced to the different ways of structurally licensing Top°. In English, an "inactive" Top°, i.e. one that lacks the feature [topic], can be structurally licensed because it contains a "residue", the feature [+finite]; in Italian, on the contrary, Top° only consists of the feature [topic]; an "inactive" Top° simply doesn't exist, because it would be a radically empty head.

Let us also consider the status of (15a) and (17) with respect to the Principle of Unambiguous Binding. Note that the feature content of Top/Fin° in (17) is a proper subset of that of Force° in (15a). The corresponding specifier positions can be assumed to be nondistinct, and they are possible targets for the same type of A' movement. In Italian, on the contrary, the unique feature [topic] of Top° licenses a specifier position of a different type than Force°.[12]

If the setting of the Topic parameter depends on the presence of the feature [+finite] on Top°, the question is why English and Italian differ with respect to this feature specification. In the next section, this difference is tentatively put in correlation with the diachronic development of "verb second topicalization" in the two languages.

1.3. Verb second

Drawing on previous work by Holmberg—Platzack (1991), Platzack (1995) assumes that the phenomenon of verb second, i.e. the obligatory movement of the verb to the highest head position in main clauses, is triggered by the feature [+finite]. Platzack argues that finiteness is a prerequisite for both mood and tense specification, because it indicates the existence of predication at the time of utterance; this relation must be expressed in order to further identify the order of the situation/event time with respect to the time of utterance (tense) and the speaker's attitude with respect to the situation/event (mood). The V/2 parameter is related to the location of the feature [+finite] in the clausal structure:

(18) The V/2 parameter
± ([+finite] is located in C°)

In V/2 languages, the CP layer is obligatorily realised in root clauses, because it hosts the feature [+finite]; this feature can be assumed to be strong, so that it triggers the overt raising of the verb to C° (cf. Wilder—Cavar 1994: 67–72). In non-V/2 languages like Italian, instead, the feature [+finite] is located in the inflectional system.

Platzack assumes that verb second always involves the CP structural layer: in interrogative clauses Spec,CP hosts a *wh*-phrase, whereas in main declarative clauses it hosts either a topic or the subject (which can be considered the "unmarked" topic: Wilder—Cavar 1994: 72). But the assumption of the complementizer system (5) calls into question the uniqueness of the target position of V/2. In fact, the internal composition of the system varies in different clause types: in a direct question, the system contains the head Focus/Wh°,[13] whereas in main declarative clauses with V/2 topicalization it contains the head Top°.[14] Therefore, it must be assumed that the feature [+finite] can be syncretically realized either on Top° or on Focus°. The superficial uniformity of the V/2 phenomenon is due to the fact that the target is always the highest head of a root clause, which is independently licensed by agreement with a filled Spec (cf. clause (2) of principle (11)).[15]

With this assumption, the V/2 parameter can be tentatively restated in the following way:

(19) The V/2 parameter
The highest head[16] of a root clause is assigned a strong [+finite] feature

where the feature is strong in that it triggers the obligatory raising of the verb.[17]

Both English and Italian were almost consistent V/2 languages in earlier stages. Platzack (1995) suggests that the loss of V/2 in English and French is due to a "regression" of the feature [+finite] towards the inflectional system. However, the regression of the [+finite] feature can be thought of as a gradual process. In English, the lower complementizer heads have lost the specification for a strong [+finite] feature;[18] nevertheless, they seem to have preserved a weak [+finite] feature, which makes them incompatible with an infinitival clausal structure. In Italian, instead, this regression has gone one step farther: the feature [+finite] has been completely dissociated from the heads Top° and Focus/Wh°, and it is uniformly realised in the inflectional system. Thus, the different setting of the Topic parameter can be tentatively related to the stage that the language has reached in the gradual process of loss of verb second:

$$\text{(20)}\quad \begin{array}{ccccc} X^\circ & \rightarrow & X^\circ & \rightarrow & X^\circ \\ | & & | & & \\ [+\text{finite}]_{\text{STRONG}} & & [+\text{finite}]_{\text{WEAK}} & & \end{array}$$

The next section discusses in greater detail the diachronic evolution of the complementizer system in English.

2. The complementizer system in Old and Middle English

In its earliest stages, Old English seems to have expressed adverbial subordination by means of a correlative structure with undifferentiated members: the correlative clause is either coordinated or adjoined to the main clause, and a particle derived from the demonstrative theme occurs in both clauses, setting up a correlative pair (Haudry 1973: 173–175; Kiparsky 1995). The main clause is usually marked by V/2:

(21) *ðonne he geseah ða hearpan him nealecan, ðonne aras he for scome from ðæm symble...*
then he saw the harp to him be brought, then raised he for shame from the party
'when he saw that the harp was brought to him, then he left the party full of shame'
(Bede, from Mitchell—Robinson 1982: 206)

There are also some ways to formally mark the dependent clause: the correlative element in the subordinate clause can be doubled, as in (22a); or else, the particle *ðe* can be introduced after the correlative element of the subordinate clause, as in (22b) (Stockwell—Minkova 1991: 374–381; Mitchell 1985: § 2424):

(22) a. *ða Neron ða ðæt geseah, ða wende he ðæt hit Godes sunu waere* (Bl.Hom. 175.4; from Mitchell 1985: § 2576)
then N then that saw,then thought he that it God's son was
'when Nero saw that, he thought that it was God's son'
b. *mid ðam ðe he wolde ðaet weorc beginnan, ða clipode Godes engel of heafenum: Abraham!*
while that he would that deed begin, then called God's angel from the heavens: Abraham!
'when he was about to begin that deed, God's angel called him from heaven: Abraham!'
(Aelfric, from Mitchell—Robinson 1982: 172)

In his extensive study of Old English syntax, Mitchell (1985: §§ 1956–1957) argues that the particle *ðe* accomplishes the function of a subordinator marker, and not that of a true complementizer: in particular, it cannot introduce an argument clause. This function is instead accomplished by the "nominalizing" particle *ðæt*, corresponding to the neuter singular form of the demonstrative pronoun/determiner *se, seo, ðæt.* This particle is optionally followed by *ðe*:[19]

(23) *...sagað Matheus se godspellere ðætte (=ðæt ðe) Hælend wære læded on westen & ðæt he wære costod from deofle*
says Matheus the Evangelist that the Lord was led into the desert and that He was tempted by the devil
'Matheus the Evangelist says that our lord was led into the desert and that He was tempted by the devil'
(Bl. Homilies 27.1, from Mitchell 1985: §§ 1956–1957)

Interestingly, the complementizer of relative clauses differs from that of argument clauses: it is not *ðæt* but *ðe*:[20]

(24) *gemyne he ðæs yfeles ðe he worhte* (Allen 1977: 76)
remember he the$_{GEN}$ evil$_{GEN}$ that he wrought
'let him remember the evil that he wrought'

In this respect Old English *ðæt* differs from Modern English *that*, which occurs in both clause types.

A plausible assumption is that the two particles *ðæt* and *ðe* instantiate two hierarchically ordered heads in the complementizer system of Old English. As for the subordination marker *ðe*, for the time being we shall not try to determine its precise syntactic category. As for the particle *ðæt,* its distribution suggests that it is not a pure complementizer, but it is an intrinsically argumental head, which must be licensed under theta-assignment by a lexical head. In other terms, the particle seems to have an intermediate status between a Determiner and Force. Let us tentatively label this hybrid category D/F°.[21]

The following partial representation of the Old English complementizer system emerges:

(25) [$_{D/FP}$ *ðæt* [$_{XP}$ *ðe* IP]]

If D/F° has to be licensed by theta assignment, its absence in the complementizer system of relative clauses follows directly: in fact, the relative clause is selected by an external determiner which takes up the Case and theta role assigned to the complex DP. Therefore, D/F° is not licensed and the external D° directly selects the lower layer of the complementizer system headed by the particle *ðe*:

(26) [$_{DP}$ D° [$_{XP}$ DP$_{REL}$ [$_{XP}$ *ðe* IP]]]

The complementizer system underwent considerable changes in Middle English. The correlative elements from the demonstrative theme were replaced by the corresponding elements from the indefinite-interrogative theme *wh-* in the subordinate clause of the correlative pair; the subordinate clause was thus unambiguously marked:

(27) *And* whan *that he wel dronked hadde the wyn, thanne wolde he speke no word but Latyn*
(*Canterbury Tales*, *Prologue* 636–637)

A similar process also affected relative determiners: the demonstrative *se* was replaced by *which*, taking both animate and inanimate antecedents:

(28) *he* which *that hath the shorteste shal bigynne*
(*Canterbury Tales, Prologue* 836)

A second important phenomenon was the spread of the particle *that* in all subordinate clause types. In late Old English, the particles *ðæt* and *ðe* were confused with each other and with *ðætte*, possibly because of phonological weakening (Mitchell 1985: § 2430). In early Middle English, the particle *that* was generalized and *ðe* disappeared (Allen 1977: 240–245). Starting from the beginning of the XIV century, *that* began to spread in all subordinate clauses. On the one hand, it continued to introduce declarative complement clauses:

(29) *and whan this goode man saugh* [*that it was so*]...
(*Canterbury Tales, Prologue* 850)

On the other hand, it optionally followed the *wh-* subordinating conjunctions (cf. (27) above), the relative determiners (cf. (28) above), and also the interrogative phrases and the hypothetical/interrogative particle *if* (cf. (30)):

(30) a. *she wolde wepe, if that she saugh a mous kaught in a trappe, if it were deed or bledde*
(*Canterbury Tales* A 144–145)
b. *What that he mente therby, I kan nat seyn.*
(*Canterbury Tales, Wife of Bath* 20)

A possible interpretation of this evolution is that in this stage the Old English D/F° is reanalysed as the lower head of the Comp system corresponding to the Old English subordination marker *ðe*. In other terms, *that* is no longer the highest head of the complementizer system, and it is not intrinsically specified for declarative force, since it also occurs in interrogative clauses like (30b). On the other hand, the particle *if* in (30a) plausibly heads a functional projection Focus/Wh Phrase above *that*. Let us assume that the Spec of this head is the landing site of [+wh] phrases. This gives us the following partial representation of the complementizer system of Middle English:[22]

(31) $[_{FocP}$ whP $[_{FocP}$ *if* $[_{XP}$ *that* IP]]]

Since there is no clear evidence to postulate two distinct particles *that* in Middle English,[23] the null hypothesis is that complement clauses like (29) are introduced by the same subordination marker *that* that apperas in *wh*-clauses. Since this head is not intrinsically specified as [+declarative], it must receive this feature by a default rule:

(32) ...*saugh* [$_{XP}$ *that* [*it was so*]]

[+declarative]

By the end of the XV century, the sequence *wh-that* is substantially eliminated. This change can be tentatively related to a reanalysis whereby *that* ceases being a subordination marker and is reinterpreted as Force°, acquiring the syntactic properties and the position that it has in Modern English.

At this point it is possible to make some speculative remarks on the status of the "subordinator marker" X°. Recall that in Old English *ðe* marked the subordinate clause in a correlative structure. In this respect, it was symmetric to verb second, which marked the main clause (although not quite systematically). Suppose that the two phenomena involved the same functional head: in main clauses the head X° is the target of verb raising, and its Spec hosts a topicalized constituent. The hypothesis that the verb raises to X° is supported by Kiparsky's (1995: 145) observation that verb second topicalization was possible in complement clauses introduced by *ðæt:*

(33) *Be ðam is awriten ðæt betera beo se gelðlega wer ðonne se gielpna*
therefore (it) is written that better is the patient man than that boastful
'therefore it is written that the patient man is better then the boastful one'

Recall that the main clause expresses the speaker's attitude by its mood, and its event time is directly related to the time of utterance: in other terms, the main clause cannot depend on a higher "anchor" for tense[24] and mood, but it constitutes itself an anchor for the subordinate clauses. According to Platzack (1995), the specification for finiteness is a prerequisite for this tense and mood independence.

But the correlative pair is an essentially symmetric structure: the order of the main and dependent clause can be freely inverted, and the subordination relation is expressed by means of two correlative elements belonging to the same syntactic category (e.g. a temporal adverb in (21)).[25] In this symmetric structure, the independent clause can be formally distinguished from the dependent one by means of a strong [+finite] feature assigned to the highest head in it; this feature triggers the raising of the verb. The [+finite] head is normally associated with an assertive declarative force, corresponding to the speaker's "default" attitude.

In other terms, the positive setting of the V/2 parameter (19) originates in the "correlative" stage of English syntax proposed by Haudry (1973), as a way to formally distinguish the independent clause in the correlative pair.

From this perspective, the low head X° of the complementizer systems (25) and (31) can be identified as a syncretic head Top/Fin°. In main declarative clauses, this head bears the strong [+finite] feature and hosts the finite verb. In dependent clauses, instead, Top/Fin° does not carry a strong feature, but it is filled by the "subordination marker" *đe* (later on *that*); the presence of this particle blocks verb second. Plausibly, in this case Top/Fin° bears a weak [+finite] feature which needs not be checked by the overt raising of the verb (cf. Wilder—Cavar 1994: 70–71). The particle can be thought of as a sort of expletive filling the head position (cf. Law 1991 on the expletive status of overt complementizers).[26]

The reanalysis of *that* as Force° and the consequent loss of the sequence *wh-that* can now be put in correlation with the decline of verb second, which started by the middle of the XIV century and culminated in the mid-XV century according to van Kemenade (1993: 236–238). Suppose that in late Middle English the correlative *wh*-elements, like *whan* in (28), developed into true subordinating conjunctions, so that the old correlative structure became obsolete. Then, it was no longer necessary to mark a declarative main clause by verb second topicalization; the assertive force could be assigned by default to a matrix IP. Symmetrically, the "subordination marker" *that* became unnecessary in subordinate clauses.

The reanalysis of the particle *that* as Force° left Top/Fin° empty. Suppose that instead of being eliminated, the empty head survived losing some of its properties: in particular, it lost the ability to attract the verb in main clauses. Suppose, however, that it retained the ability to license a topicalized phrase in its Spec and to support a default declarative reading, due to the specification for a weak [+finite] feature (cf. (20)). This comes very close to the properties that the analysis of § 1 attributes to the head Top/Fin° in the complementizer system of contemporary English.

In conclusion, the preceding discussion can be summarized in three main points:

(1) the phenomenon of V/2 originates in the "correlative" stage of English syntax;

(2) the loss of V/2 is related to the "regression" of the feature [+finite] in the complementizer system (Platzack 1995); this regression is itself related to the decline of the correlative structure, which is substituted by embedding subordination;

(3) the head Top/Fin°, which is responsible for V/2 in main declarative clauses, survives in a residual form in English as the null complementizer optionally alternating with *that* in complement clauses and in zero relatives.

This sketch of the evolution of the English complementizer system is perhaps too ambitious; the proposed hypotheses are admittedly tentative. The architecture of the whole system in Old and Middle English and the exact licensing conditions of the various heads remain to be determined; and more generally, verb second is such a complex phenomenon that its nature and diachronic evolution cannot be dealt with in the narrow scope of this discussion. However, the loss of verb second is a major phenomenon affecting the complementizer system in English: it would be interesting to relate to this general phenomenon the specific properties of the complementizer heads. The preceding discussion was an attempt in this direction.

3. Violations of the doubly filled Comp filter

The main thrust of the preceding argument is that some properties of the highest complementizer head, Force°, can be "reproduced" in a lower head acting as a sort of defective declarative complementizer. It is interesting to examine from this perspective the phenomena of complementizer doubling found in various Germanic and Romance dialects. In this section it will be argued that various dialectal violations of the doubly filled Comp filter similar to the Middle English (30b) are determined by the presence of a complementizer-like head lower than the landing site of *wh*-phrases.

3.1. Quebec French

Some varieties of French, e.g. some dialects spoken in Quebec (Burchert 1993) and popular French (Kayne 1976; Guiraud 1966), exhibit violations of the doubly filled Comp filter in finite embedded questions and relative clauses:

(34) a. *Je me demande qui qui est parti.*
I wonder who *qui* has left
'I wonder who has left.'
b. *La femme à qui que j'ai parlé*
the woman to whom *que* I have spoken
'the woman whom I spoke to'

Burchert (1993: 182–283) argues convincingly that in the interrogative example (34a) the *wh*-phrase is not in the Spec of *que*. In fact, in embedded yes/no questions the interrogative complementizer *si* is overtly realized above *que*:

(35) *Je me demande* [*si* [*que* [$_{IP}$ *Marie est sortie*]]].
I wonder if that Marie has gone out
'I wonder whether Marie has gone out.'
(Burchert 1993: 181)

Therefore, in (34a) the *wh*-pronoun *qui* can be taken to occur in the Spec of the higher interrogative head. Then the problem is to exclude the sequence <*qui si*>. The most simple explanation is that both the *wh*-phrase and *si* are interrogative operators, and therefore, their cooccurrence is ruled out by the Full Interpretation principle (cf. fn. 72 to chapter VI).

If *si* is assimilated to the head Focus/Wh° of the Comp system (5), it follows that in (34)–(35) the purported complementizer *que* is actually a head lower than Focus°.

Note that in (34a) the *que→qui* rule has applied; adopting Rizzi's (1990) analysis, this means that the *wh*-subject has passed through the Spec of *qui* (=*que*$_{[+AGR]}$) before moving on to Spec, FocusP (cf. the discussion in § VI.6.1 and § 7 below):

(36) *Je me demande* [$_{FocP}$ *qui*$_i$ Focus° [$_{XP}$ t$_i$ *qui* [$_{IP}$ t$_i$ *est parti*]]].

This consequence of the analysis is relevant for the syntax of relative clauses. Subject relativization too triggers the *que→qui* rule: this means that the Spec of *qui* is accessible for relative phrases as well as for interrogative ones. If relative phrases moved to Spec,CP above Focus°, then subject relativization could proceed in two steps: first, the relative subject would move to the Spec of *qui*; from there the NP "head" could move to Spec,CP, stranding the relative determiner in the Spec of *qui* (cf. § VI.6). This would predict the grammaticality of a *wh*-relative at least for subject relativization; but this prediction is incorrect:

(37) * *la fille qui qui t'aime bien*
the girl who *qui* likes you very much (Kayne 1976: 275)

In order to exclude (37), it must be assumed that the low complementizer-like head *que* is the highest one in the complementizer system of relative clauses. Therefore, (34a) will have the structure in (38):

(38) $[_{DP}$ *la* $[_{XP}$ $[_{PP}$ *fille* $[_{PP}$ *à qui* t]] $[_{XP}$ *que* $[_{IP}$ *j'ai parlé* t_{PP}]]]]

Note that despite its being synonimous to the Standard French declarative complementizer, this head cannot be classified as declarative, since it appears in interrogative clauses. It may be conceived of as a pure marker of subordination, insensitive to the propositional force of the clause. For unclear reasons, its realization is obligatory: its projection cannot be omitted in interrogative clauses, and it cannot undergo PF deletion in relative clauses.[27]

As suggested by J.Y. Pollock (p.c.), substandard French may be taken to have a limited occurrence of doubly filled Comp violations, attested by the following example, where a relative determiner cooccurs with *que*:

(39) *la fille à qui heureusement que j'ai parlé*
the girl with whom luckily that I spoke
'the girl to whom, luckily, I spoke'

In this structure, the relative PP would be in Spec,ForceP, the adverb in Spec,FocusP, and *que* would be a lower head licensed by Focus°:[28]

(40) [*la* $[_{FP}$ *fille à qui* [Force° $[_{FocP}$ *heureusement* [Foc° $[_{XP}$ *que* IP]]]]]

In case of DP relativization, both Force° and the lower *que* surface giving rise to an instance of complementizer doubling; interestingly, if the subject is relativized, it is the lower *que* that undergoes the *que*→*qui* rule, as in the Quebec French example (36):

(41) $[_{DP}$ *la* $[_{FP}$ *fille*$_i$ [*que* $[_{FocP}$ *heureusement* [Foc° $[_{XP}$ t_i *qui* $[_{IP}$ t_i *est partie*]]]]]]]
the girl *que* luckily *qui* has left
'the girl who, luckily, has left'

This distribution makes sense in Rizzi's (1990; 1995) approach to the *que→qui* rule, whereby the change in the form of the complementizer corresponds to the presence of agreement features coindexed with the subject, which turn the complementizer into a proper governor for the trace in Spec,IP. By relativized minimality, it is the lower head adjacent to IP that must be turned into a proper governor.

3.2. Occitan and Gascon

If *que* is lower than Focus° in the complementizer system, it should also follow dislocated phrases in Spec,TopicP (R. Kayne, personal communication). Although it was impossible for us to verify this prediction for the varieties of French discussed in the preceding subsection, the predicted sequence of elements is attested in Occitan by the structure which Sauzet (1989) calls "prolepse":

(42) a. *Sabi ton paire que vindrà.*
I know your father that will come
'As for your father, I know that he will come.'
b. *Cresi pas los dròlles que li aguesson donat de còcas.*
I don't believe the children that (they) to them gave sweets
'As for the children, I don't believe that they gave them sweets.'

According to Sauzet, Occitan also has a (semantically equivalent) structure where the dislocated phrase follows *que.* However, the possibility for *que* to follow a topic suggests that it is not the highest head Force°; perhaps it is itself an instance of Topic°, hosting a topicalized phrase in its Spec.

In Gascon, a particle *que* is obligatorily realized in main clauses,[29] and the subject obligatorily precedes it, as in (43a). However, in (43b), where there is no lexical subject, a topicalized phrase precedes the particle (Campos 1992: 912–913):

(43) a. *Nosautes qu'aimam la libertat.*
we *que* love freedom
'We love freedom.'
b. *En ivèrn que'ns hè gai d'alucar un huec.*
in winter *que* to us makes happy to light a fire
'in winter it feels good to light a fire'

Campos (1992) analyses the *que* of main clauses as a Propositional head occurring between C° and IP, which expresses the declarative force of a finite clause. In this respect, Gascon *que* differs from the Quebec French particle, which is insensitive to the declarative or interrogative force of the clause.[30] Like the Quebec French particle, however, it instantiates a low head of the Comp system: in fact, in embedded clauses the Propositional particle *e* follows the complementizer *que*, giving rise to the phenomenon of complementizer doubling (44a). The latter is also attestested in other Occitan varieties (44b):[31]

(44) a. *Comprenguètz* qu'*ua lei sola* e *govèrna lo mon ...*
understand that one law alone *e* governs the world
'You understand that only one law governs the world.'
(Campos 1992: 920)

b. *quan credou* que *la mourt* que *tustabe au pourtau*
when (he) thought que death que knocked at the door
'When he thought that death was knocking at the door...'
(Poletto 1994: 19)

3.3. West Flemish

The distribution of complementizers in West Flemish, as described by Haegeman (1992: 46–59), is similar to that of Quebec French:

(45) a. *Kweten nie* [*of* [*dan* [$_{IP}$ *d'joengers geeten een*]]].
I know not whether da_{PL} the children eaten have
'I don't know whether the children have already eaten.'

b. * *Kweten nie* [*wat*$_i$ [*of* [*dan* [$_{IP}$ *d'joengers* t$_i$ *geeten een*]]]].
I know not what whether da_{PL} the children eaten have

c. *Kweten nie* [*wat*$_i$ [∅ [*dan* [$_{IP}$ *d'joengers* t$_i$ *geeten een*]]]].
I know not what da_{PL} the children eaten have
'I don't know what the children have eaten.'

Argument clauses have an obligatory complementizer agreeing in number with the subject (*da* in the singular, *dan* in the plural). Haegeman proposes that this head is obligatory because it assigns the Nominative case to the subject. In yes/no indirect questions like (45a), there is a "complex complementizer" composed of the two particles *of* and *da*. In constituent questions, on the contrary, the particle *of* cannot be realized (45b) and only *da* appears (45c).

Haegeman argues that West Flemish does not comply with the Doubly Filled Comp filter. The complementary distribution of *of* and *wh-*

phrases attested in (45b,c) follows from the assumption that the particle *of* is in Spec,CP, and hence it blocks the movement of a *wh*-phrase to this position; on the contrary, *da* is in C°, and hence it is compatible with a *wh*-phrase in Spec,CP.[32]

The adoption of the Comp system allows an alternative analysis. The interrogative particle *of* can be analysed as an instance of Focus°, whereas *da* is the lowest head of the Comp system, which governs and Case-marks Spec,IP. In (45c), the *wh*-phrase is in the Spec of a null Focus°, and not of *da*. The example (45b) is ruled out by the Full Interpretation principle, in the same way as the sequence <*wh*P, *si*> in Quebec French.[33]

3.4. Bavarian *daß*

In Bavarian, the declarative complementizer *daß* follows both *wh*-phrases, as in (46a), and focalized phrases, as in (46b):

(46) a. I *woaß ned* wer daß *des tõa hod*
I know not who that this done has
'I don't know who has done this.'
b. [*An Mantl daß* da Xaver kaffd hod] hod neamt glaubt
a coat that the Xaver bought has has nobody believed
'That Xaver bought a coat, nobody believed it.'
(Bayer 1984: 212–215)

Therefore, it too seems to instantiate a low functional head of the Comp system.

Bavarian relative clauses raise a number of problems which cannot be adequately dealt with here. First, they show a special complementizer *wo* omophonous to a *wh*-locative adverb:

(47) *der Mõ* der wo *uns g'hoifa hod*
the man who C° us helped has
'the man who helped us'

The origin of this special complementizer and its position in the Comp system are unclear.[34] In particular, it is unclear why Bavarian *daß*, contrary to Quebec French *que*, cannot head relative clauses.

Moreover, as discussed in § VI.2.3.1, in Bavarian the relative determiner can be deleted under specific syntactic conditions. For some speakers, the relative complementizer *wo* is also optional, but as shown in (47), the two elements are not in a complementary distribution as the

PF economy principle (§ VI.8) would predict.[35] Therefore, the omission of *wo* should not be analysed as a process of PF deletion licensed by the external determiner. For the moment, these problems remain open.[36]

3.5. Summary

The data reported in this section show that in many dialects a complementizer-like particle instantiates a low head of the Comp system. In some cases (e.g. in Quebec French and Bavarian) the same particle appears as a complementizer in declarative complements and is also realized below FocusP in indirect questions. For these cases, the null hypothesis is that the particle always instantiates a generic subordination marker, which can also express the declarative force by default (the latter being the unmarked value for propositional force), like Middle English *that.*

In other dialects, e.g. Occitan and Gascon, the low functional head seems to express the propositional force of the clause both in main and in embedded clauses. However, the enunciative particle cannot introduce complement clauses by itself, but it is preceded by a distinct functional head that plays this role, giving rise to the phenomenon of double complementizers. At present, it is unclear whether the function of the enunciative particles may be reconciliated with the universal ordering of features in (14).

This brief diachronic and crosslinguistic survey has enlightened the complex internal structure of the complementizer system. If the proposed characterization is correct, it supports our specific proposal about the status of Top/Fin° as a defective declarative complementizer in Modern English. The following sections will take up again some problems in the syntax of relative clauses which were raised in the preceding chapter, and will provide a solution for them on the basis of the proposed analysis of the complementizer system.

4. English infinitival relatives

Recall that English allows the *wh*-relative in the finite paradigm, but not in the infinitival one, as shown in (25) of chapter VI, repeated here:

(48) a. *I found a topic to work on.*
b. * *I found a topic which to work on.*
c. *I found a topic on which to work.*

This paradigm follows from our hypotheses. In (48a), the relative DP moves to the Spec of a phonetically null infinitival Force°. In (48c) too, the pied piped PP moves to Spec,ForceP, and the NP "head" further raises from the complement position of the relative D° to Spec,PP. In both cases, only one specifier of the complementizer system is involved. On the contrary, the derivation of the *wh*-relative was argued to crucially involve two steps: the relative DP moves to Spec,TopP, and the NP "head" is extracted and moves to Spec,ForceP (§ VI.6.3):

(49) [D° [$_{FP}$ NP [Force° [$_{TopP}$ [$_{DP}$ D$_{REL}$ t]$_i$ [Top° [$_{IP}$ ...t$_i$ ...]]]]]]

Recall from § 1.1 above that the Topic phrase cannot be realised in infinitival clauses, due to the intrinsic [+finite] specification of Top°. The ungrammaticality of (48b) follows without further assumptions: there is no available position below the Force Phrase where the relative determiner can be stranded by the NP "head".[37]

5. The head government requirement

As discussed in § VI.1.1, the optional omission of *that* in complement clauses is usually assumed to be constrained by the ECP. In fact, the complementizer cannot be deleted when it is immediately dominated by a barrier, e.g. in subject clauses, complements of factive and manner of speaking verbs and complements to a noun:

(50) a. [*That John left*] *is a pity.*
b. * [Ø *John left*] *is a pity.*

(51) a. *John whispered* [*that CPs can iterate*].
b. * *John whispered* [Ø *CPs can iterate*].
(Authier 1992: 334)

(52) a. *The proof* [*that this set is recursive*] *is difficult.*
b. * *The proof* [Ø *this set is recursive*] *is difficult.*
(Ross 1967: 80)

As discussed in § VI.1.1, in the traditional "adjunct analysis" of relative clauses, the omission of the complementizer in the zero relative cannot comply with the ECP. In fact, the relative CP is an adjunct, hence a strong island, and its head cannot be properly governed from outside:

(53) * $[_{DP}$ *the* $[_{NP}[_{NP}$ *book*] $[_{CP}$ Op_i [Ø *I read* t_i]]]]

This paradox dissolves in Kayne's analysis: here the relative clause is not an adjunct, but a complement selected by the external determiner. The latter can thus satisfy the head government requirement on the null head:

(54) $[_{DP}$ *the* [*book* [Ø $[_{IP}$ *I read* t]]]]

Note however that the analysis of optional *that* deletion proposed in § VI.5 allows a different approach to (50)–(52) and (54). In fact, the null complementizer was not analysed as a deleted Force°, but as a distinct functional head Top/Fin°. From this perspective, the head government requirement may be related to the syntactic category of the relevant head rather than to its phonetic content.

This alternative conception of the head government requirement is supported by a contrast between the interrogative particles *whether* and *if* parallel to the contrast in (50):[38]

(55) a. [*Whether he'll come*] *is not known.*
b.* [*If he'll come*] *is not known.* (Bresnan 1972: 54)

In § 1.1 it was argued that *whether* is a [+wh] Force°, whereas *if* is the lower head Focus/Wh°. The alternative selection of *whether* or *if* in embedded yes/no questions mirrors the alternative selection of *that* and Topic° in declarative complement clauses. Note that in (55b) Focus/Wh° seems to be subject to the head government requirement, although it has phonetic content.

The comparison of (50) and (55) suggests the equation *whether: if = that*: Ø, leading to the generalization that the lower heads of the Comp system must be head governed, irrespective of their phonetic content.[39]

The rationale for this head government requirement is not entirely clear. The lower heads of the complementizer system can be considered "defective" in that they do not represent the highest possible level of projection of the clausal structure. This somehow recalls the asymmetry between the highest link of movement chains and traces, discussed by Chomsky (1995: 300–304): a trace is invisible for the operation Attract/Move and cannot constitute an independent chain. The analogy can be expressed in Manzini's (1994) Locality Theory: just like a trace must be connected to an antecedent by a wellformed dependency, a defective head cannot be the highest member of a (possibly one-membered) dependency.

6. Embedded topicalization and negative preposing

Another difference in the internal structure of the various relative clause types concerns the possibility of internal topicalization and negative preposing. These syntactic phenomena are possible in *that*-relatives and in *wh*-relatives, but not in zero relatives; cf. (46)–(49) of § VI.3.2, repeated here:

(56) a. *this is the kind of car that* [*for my son*] *I wouldn't even have considered buying.*
b. *a man whom Ø* [*for his brutal insolence and cruelty*] *Robin had long hated*
c. * *this is the kind of car Ø* [*for my son*] *I wouldn't even have considered buying.*

(57) a. ? *I saw a dress that* [*under no circumstances*] *would I have considered buying for my daughter.*
b. *John is the kind of person who Ø* [*under no circumstances*] *would I be willing to talk to.*
c. * *I saw a dress Ø* [*under no circumstances*] *would I have considered buying for my daughter.*

In this respect, the null complementizer of the *wh*-relative patterns with *that* rather than with the equally null complementizer of the zero relative.

If we consider the internal composition of the complementizer system in the three relative clause types, a possible generalization emerges. In the *that*-relative (56a) the Topic Phrase is dominated by the Force Phrase:

(58) *the* $[_{FP}$ *kind of car*$_i$ $[_{FP}$ *that* $[_{TopP}$ *for my son*$_j$ Top° $[_{IP}$ ...]]]]

Similarly, in the *wh*-relative (56b) the Topic Phrase hosting the topicalized PP is dominated by a further TopicP, whose Spec hosts the relative DP, and by ForceP, whose Spec hosts the NP "head":

(59) a. *a* $[_{FP}$ $[_{NP}$ *man*] Force° $[_{TopP}$ $[_{DP}$ *whom* t_{NP}]$_i$ Top° $[_{TopP}$ $[_{PP}$ *for his brutal insolence and cruelty*]$_k$ Top° $[_{IP}$ ...]]]]

On the contrary, in the zero relative (56c) the Topic Phrase that hosts the topicalized PP is not dominated by ForceP, but only by another Topic Phrase providing the landing site for the relative DP:

(60) * $[_{DP}$ *the* $[_{TopP}$ *[kind of car]*$_i$ Top° $[_{TopP}$ *[for my son]* Top° $[_{IP}$ *I wouldn't even have considered buying* t_i t_j]]]]]

A contrast parallel to (58) vs. (60) emerges in complement clauses:

(61) *I didn't know* $[_{FP}$ *that* $[_{TopP}$ *yesterday* $[_{IP}$ *Peter danced*]]].
(62) * *I didn't know* $[_{TopP}$ *yesterday* Ø $[_{IP}$ *Peter danced*]].

The generalization emerging from (58)–(62) is that internal topicalization (and negative preposing) must be licensed by the presence of a c-commanding Force° in the complementizer system.

In § 1.2 above we assumed that the movement of a topicalized constituent is triggered by a feature [topic] which must be matched in an appropriate Spec/head configuration with a head Top° endowed with the same feature, satisfying the F-Criterion:

(63) ...$[_{TopP}$ $XP_{[top]}$ $[_{TopP}$ $Top°_{[top]}$ $[_{IP}$ ...t_{XP}...]]]

Similarly, the preposed negative phrase in the examples (57) is endowed with an operator feature [neg] which must be matched in a Spec/head configuration. The negative phrase moves to Spec,FocusP and the concomitant incorporation of Infl carries the feature [neg] into Focus°, creating the required Spec-head configuration (Rizzi 1991):

(64) ... $[_{FocP}$ $PP_{[neg]}$ $[_{FocP}$ I°+$Focus°_{[neg]}$ $[_{IP}$ ...t_I...t_{PP}...]]]

Note that the features [topic] and [neg] are not lexically selected (contrary to e.g. [+declarative] and [+wh]). However, in an exhaustive study of root transformations, Hooper—Thompson (1973) show that topicalization and negative preposing are sensitive to the propositional force of the clause in which they occur. They argue that these transformations (among others) are fully acceptable only if the clause has an assertive, nonpresupposed force.

Recall that in Rizzi's complementizer system (5) the propositional force is intrinsically specified on the highest head Force°, which constitutes the "interface" with the outer structure. More specifically, the universal ordering of features proposed in (14) implies that the feature [± presupposed] which distinguishes asserted vs. presupposed propositions is realised on Force°. The generalization concerning (58)–(62) can then be restated as follows: the features [topic] and [neg] must be licensed by a c-commanding Force° endowed with [- presupposed] force.

This condition is not satisfied in the zero relatives (56c) and (57c), whose complementizer system contains a Top° acting as a defective complementizer. Recall that by the positive setting of parameter (4), Top° can be assigned the default feature value [+declarative]. However, this feature is not intrinsically specified on the head, and moreover, Top° is not specified for the feature [± presupposed]. Let us assume that this default specification is insufficient to license the features [topic] and [neg]. An alternative generalization on embedded topicalization is proposed by Authier (1992: 334) and by Müller—Sternefeld (1993: 491). These authors observe that some island contexts that disallow *that*-deletion also disallow topicalization.[40] Compare (50)–(52) above with the following examples (from Authier 1992: 334):

(65) a. * [*That this book*$_i$, *Mary read* t_i *thoroughly*] *is true*.[41]
b. * *John regretted* [*that "Gone with the Wind"*$_i$, *we went to see* t_i].
c. * *The fact* [*that Bill*$_i$, *Mary likes* t_i] *makes John very jealous*

Therefore, they propose that the Topic position is only licensed in lexically governed contexts.[42]

Note that if this proposal were correct, it would strenghten the hypothesis of § VI.5 that optional *that*-deletion crucially involves the licensing of a Topic projection. In fact, the ungrammaticality of (50)–(52) could be reduced to the same principle that excludes (65).

However, the possibility of embedded topicalization in *that*- and *wh*-relatives constitutes a counterexample to the lexical government requirement, since the relative clause is not lexically governed either in the adjunct analysis (where the relative FP is a barrier) or in the raising analysis (where FP is governed by a functional head D°).

Furthermore, there is evidence against the assimilation of *that*-deletion and embedded topicalization: in fact, the distribution of the two phenomena diverges in certain cases. For instance, the complement of the verb *realize* allows *that*-deletion but not topicalization; the non-raising complement of the adjective *likely* allows *that*-deletion but not negative preposing:

(66) a. *Harry said* [*Joan realized* [*Ø I hated Max*]].
b.* *Joan realized* [*that Max*$_i$, *I hated* t_i]. (Postal 1972: 221)

(67) a. *It's not likely* [*Ø she speaks Basque*].
b. * *It's likely* [*that seldom did he drive that car*].
(Hooper—Thompson 1973: 483, 479)

Conversely, the complement of the verb *whisper* disallows *that*-deletion (cf. (51)), but it allows internal topicalization (R. Kayne, personal communication):[43]

(68) *John whispered to Mary* [*that that book*$_i$, *he would never have dreamed of reading* t$_i$]

These data show that the lexical government requirement does not correctly characterize the distribution of embedded topicalization and negative preposing: the examples in (66)–(67) suggest that it is not a sufficient condition, and the example (68) suggests that it is not even a necessary condition.

Moreover, the distribution of embedded topicalization and negative preposing in relative clauses (56)–(57) shows that the licensing condition cannot be simply the presence of an *overt* complementizer. Thus, only the analysis based on the flexibility of the complementizer system allows us to draw a generalization from the data and to formulate a licensing condition.

7. Subject relativization

A well known factor conditioning the optional omission of *that* is the A' movement of the subject. In long subject movement, the complementizer must be omitted; this is the so called *that*-trace effect:

(69) a. *the event that I think* [Ø [$_{IP}$ t *took place*]]
b. * *the event that I think* [*that* [$_{IP}$ t *took place*]]

The situation is reversed in local subject relativization; in this case *that* cannot be freely omitted:

(70) a. *the event* [*that* [$_{IP}$ t *took place*]]
b. * *the event* [Ø [$_{IP}$ t *took place*]]

In this respect the null complementizer of the *wh*-relative patterns with *that* rather than with the null head of (70b):

(71) *the event* [*which* [Ø [$_{IP}$ t *took place*]]]

The interaction of subject relativization and complementizer omission is even more complex. Culicover (1993) argues that the *that*-trace effect in long subject movement is suspended if an adverbial topic intervenes between *that* and the trace:

(72) a. *an amendment which*$_i$ *they say* [*that* [*next year* [t$_i$ *will be law*]]]
b. *Which doctor*$_i$ *did you tell me* [*that* [*during an operation* [t$_i$ *had a heart attack*]]]?

Rizzi (1995: 26–30) points out that topics do not have a uniform behaviour in this respect. In (72) the topic intervening between C° and the trace is an adverbial phrase: in this case, the *that*-t effect is avoided. On the other hand, if the intervening topic is an argument, the sentence is ungrammatical:

(73) * *a man who*$_i$ *I think* [*that* [*this book*$_j$ [t$_i$ *knows very well* t$_j$]]]

Pursuing the approach proposed in Rizzi (1990), Rizzi (1995) argues that in case of subject movement the trace in Spec,IP must be properly governed by the immediately c-commanding head in the Comp system. The complementizer heads are unable to properly govern by themselves, but they may do so if they come to be coindexed with the subject trace. Rizzi assumes that the null complementizer in (69a) and *that* in (70a) are endowed with agreement features and thus can get coindexed with the antecedent of the trace when the latter moves through their Spec (cf. § VI.1.3).

As for the anti-adjacency effect in (72), Rizzi (1995) argues that the presence of a Topic Phrase forces the realization of an independent head Finite° below it, which carries the feature [+finite] and constitutes the interface between the complementizer system and the inflectional system (cf. the discussion of (15b) above):

(74) ... [$_{FP}$ *that* [$_{TopP}$ *next year* [Top° [$_{FinP}$ Fin° [$_{IP}$ t ...]]]]]

The head Fin° can be exploited in order to satisfy the head government requirement for the subject trace. The derivation proceeds as follows: the extracted subject moves through Spec,FinP, and by Spec/head agreement Fin° is turned into a proper governor for the trace in Spec,IP. As for the trace in Spec,FinP, it can be properly governed by the same Fin°, which incorporates to the higher Top°. The derivation is synthesized in (75):

(75) a. *that* [$_{TopP}$ *next year* [Top° [$_{FinP}$ t$_i$ [Fin°$_i$ [$_{IP}$ t$_i$ ...
b. *that* [$_{TopP}$ *next year* [Fin°$_i$+Top° [$_{FinP}$ t$_i$ [t$_{FIN}$ [$_{IP}$ t$_i$ ...
(Rizzi's (93))

As for the blocking effect of the argument topic in (73), Rizzi proposes that contrary to adverbial topicalization, argument topicalization cannot be direct raising from the argument position: the topic is generated in Spec,TopP and a null operator in Spec, FinP binds the empty category within IP:

(76) ... *that* $[_{TopP}$ *this book* [Top° $[_{FinP}$ Op_j [Fin° $[_{IP}$ t_i ... t_j]]]]]
(Rizzi's (96))

Consequently, in (73) the Finite Phrase cannot be exploited in order to properly govern the subject trace, and an ECP violation arises.

The representations in (74)–(76) are inconsistent with the hypothesis of § 1.1 that in English the feature [+finite] is syncretically realized on the head Top°. The representations can be maintained if in these examples the feature [+finite] is scattered and projects an independent head below Top°. But as Giorgi—Pianesi (1996) point out, the option of scattering must be constrained by economy principles; it cannot be justified simply by the need of providing an intermediate landing site for movement (or a proper governor for a trace). The scattering of [+finite] in case of subject relativization would be an *ad hoc* stipulation. However, our previous hypotheses allow us to slightly modify Rizzi's insightful proposal, taking a further step toward a unified analysis of these phenomena.

In fact, in our approach it is possible to find a common feature in (69a) and (72): in both cases a Top° is structurally licensed in the Comp system immediately above the subject trace. According to principle (11), in (69a) Top° is licensed under selection by the verb *think*, since it acts as a subordinator for the most deeply embedded clause; in (72), instead, Topic° is licensed by the presence of a [topic] constituent in its Spec. In (69b), on the contrary, no Topic Phrase is structurally licensed, because the verb *think* selects Force° as a subordinator. Thus, it seems that Top° plays a crucial role in the proper government of the trace in Spec,IP.

Suppose that this feature can trigger the abstract incorporation of the immediately lower head AgrS° (cf. Kayne 1989). Since AgrS° is coindexed with the subject by Spec/head agreement, the resulting complex head[44] will turn out to be coindexed with the subject trace, and it will be able to properly govern it.[45]

This analysis can be implemented through the following set of assumptions:

(1) the feature [+finite] can trigger the abstract incorporation of AgrS°;

(2) AgrS° is intrinsically [+V];
(3) Force° is a "nominalizing" head specified as [-V];
(4) Topic° is underspecified for the feature [±V].

The hypotheses (2)–(3) are substantiated by a generalization formulated by Hoekstra (1993), according to whom in the Dutch dialectal area the morphology of number agreement on C° is only found in those dialects where (a) the verbal paradigm has no inflection for person, and (b) the number morphemes are identical in the nominal and in the verbal paradigms. This generalization suggests that the complementizer can only support agreement morphology that is neutral with respect to the categorial features [±N] and [±V]. In English, Agreement-Subject does not have this property. Even assuming with Kayne (1989) that the *-s* morpheme of the present third person singular form expresses the feature of singular number, and not a feature of person, the verbal number morphemes are not identical to those of the nominal inflection:

(77)

	[+V]	[+N]
sing.	s	Ø
plur.	Ø	s

Recall that by the condition of feature compatibility on unification (§ VI.2), two functional heads can be unified only if they bear consistent feature values. The idea is that AgrS° is an intrinsically verb-related head, and hence it cannot incorporate to Force°, but only to Topic°; therefore, in order to properly head-govern a trace in Spec,AgrSP, it is necessary to structurally license a TopP whose head can host AgrS°.

Let us consider how these hypotheses apply to the relevant examples.

In (69a), the complementizer system of the lowest clause contains a Top° licensed by selection. AgrS° can incorporate to Top°, turning it into a proper governor for Spec,AgrSP. The resulting configuration satisfies the ECP:

(78) ...*think* $[_{\text{TopP}}$ AgrS_i+Top° $[_{\text{IP}}$ t_i ...

In (69b), on the contrary, Force° is selected in the lowest complementizer system and Top° is not structurally licensed. By hypothesis (3), AgrS° cannot incorporate to Force°, which is not a proper governor by itself; thus, the trace in Spec,IP fails to be properly governed and violates the ECP:

(79) * ... *think* $[_{\text{ForceP}}$ *that* $[_{\text{IP}}$ t_i AgrS°...

Let us then turn to the topicalization examples in (72) and (73), which exhibit an argument/adjunct asymmetry. It is possible to dispense with the assumption of a null operator, as in (76), and reduce the contrast to an independent asymmetry: namely, the fact that argument topics differ from adverbial topics in that they are endowed with a referential index (Rizzi 1990: 86). In (73), the argument in Spec,TopP bears an index distinct from that of the subject trace; if AgrS° incorporates to Top°, an inconsistent indexing arises: the complex head AgrS+Top° bears the index of the subject, but its Spec bears a distinct index:

(80) * ... *think* [$_{ForceP}$ *that* [$_{TopP}$ *this book*$_j$ [AgrS$_i$+Top° [$_{IP}$ t$_i$ ... [t$_j$ ...

Thus, the incorporation of AgrS° is excluded in this configuration, although Top° is in principle available.

On the other hand, this problem does not arise in (72) because the adverbial topic does not bear any referential index, and no index clash arises:[46]

(81) ... *think* [$_{ForceP}$ *that* [$_{TopP}$ *next year* [AgrS$_i$+Top° [$_{IP}$ t$_i$ ...

In all these examples, the null head Top° plays an indirect but crucial role in licensing the subject trace. The situation is different in cases of local subject relativization (70)–(71). In fact, the Top° introducing the zero relative (70b) cannot avoid an ECP violation. The minimal hypothesis is that in (70b), contrary to (69a), Top° cannot host an incorporated AgrS°.

The difference can be drawn on the basis of the different selectional relations holding in the two cases. Note that in (69a) Top° is selected by a verb, whereas in (70b) it is selected by the external determiner of the relative structure. The different behaviour of Topic° in the two cases can be derived from the following hypothesis:

(5) the external D° of the relative structure is intrinsically [+N], [-V][47] and it is incompatible with a categorial feature [+V] on the head of its complement.

The categorial requirement of the external D° is satisfied in the *that*-relative, headed by a [-V] Force°, and in the zero relative, whose head Top° is underspecified for the categorial feature [±V] (cf. (3)–(4) above). Consider however the effects of the incorporation of AgrS° to Top° in (69b):

(82) * [$_{DP}$ *the* [$_{TopP}$ [*event*]$_i$ [$_{TopP}$ AgrS$_i$+Top° [$_{IP}$ t$_i$ t$_{Agr}$ *took place*]]]]

Since AgrS° is positively specified as [+V], after incorporation the complex head AgrS+Topic° is also endowed with this feature, clashing with the selectional requirements of the external D°. Thus, the incorporation of AgrS° is impossible and the subject trace cannot be properly governed.[48]

Note that this problem does not arise in the *wh*-relative (71): in this structure two layers of the complementizer system are realized, namely ForceP and TopP. Consequently, the incorporation of AgrS° to the lower head Top° does not affect the selectional requirements of the external determiner:

(83) $[_{DP}$ *the* $[_{FP}$ $[_{NP}$ *event*] F° $[_{TopP}$ $[_{DP}$ *which* $t_{NP}]_i$ $AgrS_i$ +Top° $[_{IP}$ t_i t_{Agr} *took place*]]]]

The last case to be considered is the *that*-relative (70a). In this structure Force° acts as a proper governor for the subject trace. By hypotheses (2)–(3), AgrS° cannot have incorporated into it. Therefore it is necessary to assume, following Rizzi (1990: 67–71), that Force° may be endowed with intrinsic agreement features when it is selected by the external D°:

(84) $[_{DP}$ *the* $[_{FP}$ *event*$_i$ $[_{FP}$ *that*$_i$ $[_{IP}$ t_i *took place*]]]]

The hypothesis that Force° may have agreement features is substantiated by the observation that in a configuration like (84) *that* is sensitive to the animacy of the relativized subject; in fact, it is marginal with a [+animate] subject (Kayne 1984: 73):

(85) ? * *I'm looking for somebody that* t_i *can help me with this problem.*

Note that in English the feature [± animate] is typical of pronouns[49] rather than of verbal agreement: this suggests that in (84) and (85) the agreement features on Force° are not the verbal features of AgrS°, but they are nominal features. In other terms, Force° has some of the properties of a determiner. These may be thought of as a residue of its diachronic origin: in fact, in § 2 it was suggested that the Old English particle *ðæt* had an intermediate status between a determiner and a Force°. Let us assume that the residual N-features of Force° can be "activated" by Spec/head agreement only if Force° is selected by a [+N] head, as in (84), but not if it is selected by a verbal head, as in (69b).[50]

In conclusion, the analysis of optional *that* deletion proposed in § VI.5 allowed an assimilation of the two apparently unrelated strategies that eliminate the *that*-trace effect: namely, the omission of *that* in (69a) and the intervening adverbial in (72).

8. Relativization in ancient Tuscan

The analysis of the various relative clause types proposed in § VI.5 establishes a correlation between the zero relative and the *wh*-relative: both structures involve the Spec,TopP position, and hence require a positive setting of parameter (4). More specifically, the analysis predicts that if a language has both the *that*-relative and the zero relative, it also has the *wh*-relative, since it can exploit two Specs of the Comp system for the two-steps derivation of the latter.[51]

It is interesting to note that some diachronic stages of the Tuscan dialect (the ancestor of contemporary Italian) seem to confirm the correlation. In the early XIV century Florentine dialect attested by Dino Compagni's *Cronica*,[52] the complementizer is optionally omitted both in complement clauses and in relative clauses:[53]

(86) a. *La potente e superba famiglia degli Uberti, sentenziò Ø stesse alcuno tempo a' confini...* (*Cronica*, I,3)
the powerful and proud family of the U., he ordered (that) (it) should stay some time at internment
'he ordered that the powerful and proud family Uberti should stay at internment some time'

b. ... *volea da loro trarre denari, opponendo Ø gli aveano vietato il passo...* (*Cronica*, II, 20)
(he) wanted from them to take money, objecting (that) (they) to him had denied the passage
'he wanted to get money from them, objecting to them that they had denied him the right to pass'

c. ...*e convenia loro confessare Ø aveano fatta congiura...*
and it was better for them to confess (that) (they) had made a conspiracy
'and they had better confess that they had made a conspi racy' (*Cronica*, II, 21)

(87) a. *lasciando quella Ø avea tolta e giurata...*
abandoning the one (that) (he) had chosen and sworn
'leaving the one who he had chosen and sworn to marry'

(87) b. *ritorneremo alle cose Ø furono ne' nostri tempi.*
(we) shall go back to the things Ø happened in our times
'we shall go back to the things that happened in our times' (*Cronica*, I, 2)

c. *...e giunsono... a uno luogo Ø si chiama Campaldino [...] in uno luogo che si chiama Campaldino presso Poppi*
and (they) arrived at a place Ø is called Campaldino...in a place that is called Campaldino near Poppi (*Cronica* I, 10)
'and they reached a place called Campaldino near Poppi'

These data suggest that the dialect had a positive setting of parameter (4), like Modern English. The positive setting of the parameter leads to the prediction that the *wh*-relative should be possible as well. The prediction is confimed by the data:

(88) a. *...per raddomandare certe giurisdizioni d'uno castello il quale teneano i Fiorentini...* (*Cronica*, II, 4)
to claim back certain jurisdictions of a village which had the Florentines
'to claim back certain jurisdictions of a village which the Florentines dominated'

b. *.molti poveri infermi uccisono, i quali traeano dagli spedali*
and many poor ill people (they) killed, whom they took out of the hospitals
'and they killed many poor people, whom they took out of the hospitals' (*Cronica* III, X)

By our assumptions, the derivation of the *wh*-relatives in (88) involves Spec,TopP. However, in Romance languages the Topic position is usually connected to an IP-internal resumptive clitic (cf. Cinque 1990: chapter II). The lack of a resumptive clitic in (88) may then seem problematic. But it can be shown that in this stage the Tuscan dialect had English-type topicalization without a resumptive clitic:

(89) a. *...e quelle che chiaramente non vidi, proposi di scrivere* t *secondo udienza...*
and those that (I) didn't see clearly, I decided to write as they were reported to me
'and I decided to write those things that I didn't see personally as they were reported to me' (*Cronica*, I,1)

b. *... la pena pagherò io* t *per te...* (*Cronica* I,2)
the penalty will pay I for you
'I will pay the penalty for you'

(89) c. *Cari e fedeli cittadini, le vostre profferte noi riceviamo* t *volentieri, e cominciare vogliamo a usarle...* (Cronica I,5)
dear and faithful citizens, your offers we accept willingly, and to begin we want to use them
'dear and faithful citizens, we willingly accept your offers, and we want to begin to take advantage of them'

Besides this type type of topicalization, the Florentine dialect also had a resumptive structure which can be identified with Clitic Left Dislocation. Strikingly, the following examples show *wh*-relatives with a resumptive clitic related to the relativized phrase:

(90) a. *però che furon cose notevoli;* le quali *ne' loro principi nullo* le *vide certamente come io.* (*Cronica*, I,1)
for (they) were remarkable things; which in their beginning nobody them_{CL} saw as clearly as I
'for they were remarkable things; which nobody saw in their beginnig as clearly as I did'

b. *...e per le rendite e pe' fedeli volea, l'anno, fiorini III^m*, i quali li *promettesse Vieri de' Cerchi ricchissimo cittadino.*
and for the revenue and for the congregation (he) wanted, each year, three thousand florins, which them_{CL} should promise Vieri de' Cerchi very rich citizen
'and for the revenue and the congregation he wanted three tousand florins per year, which Vieri de' Cerchi, a very rich citizem, should promise' (*Cronica*, I,8)

c. ... *tra voi è nato alcuno sdegno, per gara d'ufici,* li quali, *come voi sappete, i miei compagni e io con saramento v'abiamo promesso d'accomunar*li. (*Cronica*, II, 8)
among you there has arisen some anger because of competition for the offices, which, as you know, my companions and I with an oath have promised to share them_{CL}.
'there has arisen among you a quarrel for the offices, which, as you know, my colleagues and I sworn to share with you'

These data show that the Florentine dialect had two types of *wh*-relative (with and without a resumptive clitic) parallel to the two available topicalization strategies (English-type topicalization and Clitic Left Dislocation).[54] From the perspective of our analysis, this parallelism is not surprising, since the derivation of the *wh*-relative crucially involves the topic position.[55]

The preceding data show that in Italian the head Top° underwent a diachronic change related to the setting of parameter (4). This change may be related to the gradual process of loss of V/2 sketched in (20) above, in a way that remains to be investigated.[56]

9. Some remarks on resumptive relative clauses

All the preceding discussion concerned the syntax of relative determiners appearing in the Comp system of English and Italian relative clauses. However, the relative determiner is not a universal feature of relativization. Another common relativization strategy involves the systematic use of resumptive pronouns, as in the following Hebrew example:

(91) *raʔiti ʔet ha-yeled she-rina hashva ʔalav* (Borer 1984: 220)
saw-I ACC the-boy that Rina thought about-him
'I saw the boy that Rina thought about'

In the traditional approach to relative clauses, resumptive pronouns are thought of as completely distinct from relative pronouns of the Standard English type. In fact, relative pronouns are analysed as operators binding the relative trace from Spec,CP, whereas resumptive pronouns are themselves bound within IP.

The analysis of relative pronouns defended in this book opens a new perspective on this topic. In § III.3, it was argued that relative determiners are not binding operators, but they are non-definite determiners lacking any quantificational force. These determiners head the relative DP, and after raising to the complementizer system, they are stranded by the leftward movement of the NP "head":

(92) a. [$_{DP}$ *the* [$_{FP}$ [$_{NP}$ *boy*] [$_{TopP}$ [$_{DP}$ *who* t_{NP}]$_i$ [$_{IP}$ *I spoke to* t_i]]]]
b. [$_{DP}$ *il* [$_{FP}$[$_{PP}$[$_{NP}$ *ragazzo*] [$_{PP}$ *con* [$_{DP}$ *cui* t_{NP}]]]$_i$ [$_{IP}$ *ho parlato* t_i]]]

In this way, the asymmetry between relative and resumptive pronouns concerning the quantificational force is eliminated. It would be interesting to go one step further, extending the stranding analysis to resumptive pronouns.

Various recent studies assimilate resumptive structures like Clitic Left Dislocation to clitic doubling (Cecchetto 1995: chapter V; Sportiche 1992; Zubizarreta 1994), while other studies analyse doubling clitics as determiner-like heads that are extracted from the constituent that they seem to double. This analysis, anticipated by Kayne (1972: 90) and by E. Torrego in umpublished work, has been recently developed by Belletti (1995), Kayne (1995) and Uriagereka (1994).

Adopting both these hypotheses, a stranding analysis can be proposed at least for resumptive relative clauses featuring clitic pronouns. These clitics originate as the heads of the relative DP and cliticize to an inflectional head; the rest of the relative DP raises to Spec,ForceP in the usual way:

(93) $[_{DP}$ *il* $[_{FP}$ *[ragazzo]*$_i$ $[_{FP}$ *che* $[_{IP}$ *gli ho parlato* $[_{ClP}$ t_{CL} $t_i]]]]]$
the boy that I to him$_{CL}$ have spoken
'the boy that I spoke to'

The main difference between relative determiners and resumptive clitics is then related to the position in which they are stranded: clitics are stranded in an IP-internal position, whereas relative determiners must be licensed in the complementizer system. This approach makes resumptive pronouns compatible with the raising of the relative "head".

Although the analysis in (93) cannot be defended as a general hypothesis, the assimilation of at least some resumptive pronouns to relative determiners is supported by the observation that there are certain pronouns whose status is somewhat ambiguous between the two types (§§ 9.1–9.2). In § 9.3, the analysis is applied to Italian resumptive clitics, whereas § 9.4 discusses a possible extension to the German relative *d*-pronouns.

9.1. Irish resumptive PPs

An exceptional behaviour of resumptive pronouns is found in Irish, as described by McCloskey (1990). Irish has resumptive pronouns in all A' structures and in all positions except for the "local" subject position immediately below the complementizer system that hosts the operator. A' structures with and without resumptive pronouns are distinguished by the form of the complementizer: in resumptive structures the complementizer *a* produces nasalization on the following word, whereas in non-resumptive structures it produces lenition.[57]

As a rule, the resumptive pronoun appears in an IP-internal argument position. However, McCloskey (1990: 226–233) observes that in constituent questions involving a prepositional object, the PP containing the resumptive pronoun may be fronted to the position immediately following the *wh*-pronoun:[58]

(94) *Cé leis a raibh tú ag caint?*
who with-him C° were you talking?
'who were you talking to ?'

Interestingly, McCloskey argues that the fronted PP forms a constituent together with the *wh*-pronoun.[59] In fact, he notes that the nominal modifier *eile* "other" referring to the *wh*-element can follow the resumptive pronoun within the fronted PP:

(95) [*Cé aige eile*] *a bhfuígfeadh siad é* ?
who at-him else C° would-leave they it
'who else would they leave it with ?'

McCloskey proposes that in (94) the PP *leis* is right-adjoined to the *wh*-pronoun *cé* (with the further possibility of right adjoining the modifier *eile* to this complex constituent in (95)). A possible alternative is to adapt to (94) Kayne's (1994) analysis of pied piping relative clauses. The alternative analysis is represented in (96). Suppose that the interrogative pronoun *cé* is a QP generated within the DP complement of the preposition; this QP is extracted from DP and raises to the Spec of the pied piped PP.[60] The resumptive pronoun is a stranded D° cliticized to the preposition:[61]

(96) $[_{PP}\ [_{QP}\ c\acute{e}]\ [_{PP}\ D°{+}P°\ [_{DP}\ t_D\ t_{QP}]]]$

As for (95), let us assume that *eile* originates as a modifier of the *wh*-pronoun and is stranded within the pied piped PP: $[_{PP}[_{QP}\ c\acute{e}]\ [_{PP}\ aige\ [_{DP}\ eile\ t_{QP}]]]$

Note that in the standard approach to resumptive pronouns, a structure like (94) is completely unexpected: it is hard to explain how a resumptive pronoun can form a constituent together with the *wh*-operator that binds it. If instead resumptive pronouns are conceived of as determiners stranded by the A' movement of their complement, this exceptional structure can be accounted for. The peculiarity of Irish resumptive pronouns is that they can either be licensed within IP or be pied piped to the complementizer system.

9.2. Hebrew free relatives

A similar phenomenon is found in Hebrew. In this language headed relative clauses have resumptive pronouns, which are optional in the relativization of subjects[62] and direct objects, but obligatory for the relativization of prepositional complements. The pronouns can appear in the base position corresponding to the relativized phrase, but they can also move to an IP-external position immediately following the complementizer (Borer 1984: 220; Doron 1982; Shlonsky 1992):

(97) a. *raʔitiʔet ha-yeled she-rina ʔohevet ʔoto*
saw I ACC the boy that Rina loves him
b. *raʔiti ʔet ha-yeled (she)-ʔoto rina ʔohevet*
saw I ACC the boy (that) him Rina loves
'I saw the boy Rina loves'

Borer (1984: 241–243) argues that the preposed resumptive pronoun in (97b) cannot be in the Topic position to the right of Comp, but it must be right-adjoined to Comp. The reason is that the preposing of a resumptive pronoun, contrary to the topicalization of a nonresumptive pronoun, licenses the optional deletion of the complementizer. According to Borer, since the resumptive pronoun is in Comp in (97b), the complementizer can be deleted without violating the filter against empty Comps.[63]

This analysis is incompatible with the Linear Correspondence Axiom, since the variant of (97b) with an overt complementizer should have the fronted pronoun right-adjoined to it.

Doron (1982: 10) and Shlonsky (1992: fn.2) propose, on the contrary, that the fronted pronoun occupies a Topic position following the complementizer.[64] This is consistent with the analysis of the complementizer system in (19). The relative clause in (97b) is then analysed as in (98):

(98) *raʔiti ʔet* [$_{DP}$ *ha-* [$_{FP}$[*yeled*]$_i$ [*she-*[$_{TopP}$ *ʔoto* [$_{IP}$ *rina ʔohevet* t$_i$]]]][65]

Note that the structure in (98) closely resembles the structure proposed for the English *wh*-relative (cf. (49) above). It is possible to completely assimilate the two cases by analysing the resumptive pronoun *ʔoto* in (98) as a D° stranded by the movement of the NP "head" to Spec,CP. However, both Doron (1982) and Borer (1984) point out that when the resumptive pronoun is *in situ*, relativization can cross various islands and can violate the Coordinate Structure Constraint. This suggests that the dependency between the NP "head" and the resumptive pronoun is not one of movement after all.[66]

Although the stranding analysis is probably incorrect for resumptive pronouns in headed relative clauses, it is explicitly proposed by Borer (1984: 239) for free relative clauses. The latter are headed by a *wh*-pronoun; resumptive pronouns are only possible in the relativization of prepositional objects (99a,b), but, contrary to headed relatives, not in the relativization of a direct object (99c):

(99) a. *kaniti ʔet ma*$_i$ *she-hexlatet* [$_{PP}$ *ʔalav* t_i]
I bought ACC what that-you-decided on-it
b. *kaniti ʔet ma*$_i$ *she*-[$_{TopP}$ [$_{PP}$ *ʔalav* t_i] [$_{IP}$ *hexlatet* t_{PP}]]
I bought ACC what that-on-it you-decided
'I bought what you decided to buy.'
c. * *kaniti ʔet ma she-raʔit ʔoto*
I bought ACC what that-you-saw it
'I bought what you saw.'

Borer reduces the asymmetry between prepositional objects and direct objects to the different status of the resumptive pronoun in the two cases. In (99a,b) the resumptive pronoun appears cliticized on the preposition: Borer analyses it as a "doubling clitic" generated directly on the preposition, and not as a complement to the latter. Thus, the clitic pronoun is compatible with A' movement of the *wh*-pronoun from the complement position of P°. In (99c), on the contrary, the direct object resumptive pronoun is not a clitic, but it fills the direct object position; therefore, it is incompatible with the generation of the *wh*-pronoun in the same position.

Thus, in Borer's approach resumptive pronouns are ambiguous: in headed relatives they are true pronouns generated in the relativized position; the "head" of the relative clause is generated in its surface position and is coindexed with the resumptive pronoun by a rule of predication. In free relatives, on the contrary, the "head" is a *wh*-pronoun that must raise from an argument position internal to the relative clause; therefore, only doubling clitics are allowed.[67]

The paradigm of Hebrew free relatives in (99) resembles that of Irish interrogative clauses. Here too, the fronting of the resumptive pronoun in (99b) can be analysed as an instance of pied piping by the *wh*-pronoun to a Spec of the Comp system.

These examples of pied piping of a resumptive pronoun by a *wh*-phrase suggest that at least in some cases, the resumptive pronoun forms a constituent with the *wh*-phrase in the base.

9.3. Resumptive clitics in some Romance dialects

In standard English and in the standard Romance languages, the resumptive relativization strategy is not used. However, this strategy is widely attested in the substandard register of these languages and in various related dialects (cf. Cennamo forthcoming; Benincà—Vanelli 1982; Viotto 1994);[68] it is also attested in Castellani's ancient Tuscan texts (cf. also Palermo 1994: 169–180). Some representative examples are reported in (100).

(100) a. *il ragazzo che* gli *ho dato il libro*
the boy that him_{CL} (I) gave the book
'the boy who I gave the book to'
b. *in omu ke nu* ge *se po kunta* (Genova dialect)
man that not on-him you can rely
'a man who you can't rely on'
c. *la persona che i l'hai parlat*ne *a l'é riva* (Turin dialect)
he person that I told you about-her has arrived
'The person I told you about arrived.'
d. *a tosa che te* n'*hoo parlaa* (Milan dialect)
he girl that I about her_{CL} told you
'the girl I told you about'
e. *un peçço di terra che chonperammo... che* v*'à prato e* v*'è la via per meço*
a lot of land that (we) bought... that there_{CL} has grass and there_{CL} is a path in the middle
'a lot of land that we bought... in which there is a meadow and a path in the middle'
(Ancient Tuscan; Castellani, Nuovi Testi, p. 169)

In (100a–e), the relativizated argument is an indirect object. The grammatical function is expressed by the appropriate resumptive clitic, and the preposition that usually introduces the non-clitic indirect object is omitted.

Alongside this relativization strategy, there is also another strategy featuring both a resumptive clitic and a relative determiner (cf. the Ancient Tuscan examples (90)): the preposition introducing a relativized indirect object is not omitted, but it is pied piped to Spec,CP, and it is matched by the appropriate resumptive clitic. The structure is attested in substandard Italian and French. The following French examples are taken from Gapany—Apothéloz (1993) (note that (101b) is a relativization out of a *wh*-island):

(101) a. *Il parait que c'est un endroit* où *on* y *mange mal.*
it seems that it is a place where one there_{CL} eats not well
'It seems that it's a place where one cannot eat well.'

(101) b. *Ce mot,* auquel *Dieu sait quel sense elle* lui *attribuait...*
that word, to which God knows which meaning she to-it$_{CL}$ attributed
'that word, to which God knows what sense she attributed'

The status of the resumptive clitics in (100) and (101) are very different. In (100), the clitic substitutes for the preposition and the relative determiner, whereas in (101) it doubles them. In this respect, the relativization strategy in (101) closely resembles Clitic Left Dislocation: in fact, the dislocation of a PP too disallows the omission of the preposition; the latter is matched by an optional resumptive clitic. Compare (101a) to a parallel Clitic Left Dislocation in Italian:[69]

(102) *Dicono che in quel posto, (ci) si mangia male.*
(they) say that in that place, there$_{CL}$ one eats not well
'They say that in that place, one cannot eat well.'

The different behaviour of resumptive clitics in (100) and (101) may be accounted for by the assumption that they have a different syntactic status in the two relativization strategies. Let us examine a possible implementation of this idea.

In various recent studies, Romance clitics are analysed as "intransitive determiners", namely D° heads that fail to take an NP complement (cf. Belletti 1994; Laetzlinger 1993). Suppose, however, that clitics optionally take an NP complement. This assumption is supported by the observation that in Galician the definite determiner of the direct object cliticizes exactly in the same way as a third person clitic:

(103) a. *Vimo-lo.*
(we) saw-him.
'We saw him.'

b. *Vimo-lo neno.* (Uriagereka 1995: fn. 5)
(we) saw-the child
'We saw the child.'

Then, in the examples (100) the NP "head" can be generated as a complement to the clitic D°, which is overtly marked with an oblique Case. The two elements move independently: the NP "head" raises to Spec,CP, where it checks the phi-features of the external D°; the clitic instead checks its features by adjoining to an inflectional head. This corresponds to the derivation in (93), repeated here:

(104) $[_{DP}$ *il* $[_{CP}[_{NP}$ *ragazzo*$]$ $[_{CP}$ *che* $[_{IP}$ *gli ho parlato* $[_{DP}$ t_{CL} $t_{NP}]]]]]$
the boy that I to him$_{ç}$ have spoken

This analysis does not extend to the relative clauses in (101), where the relativized phrase is larger than the mere NP "head", nor to Clitic Left Dislocation. In these cases, it may be assumed that the constituent is directly generated in its surface A' position, while the clitic is used "intransitively" as a pure pronoun; the two categories are not related by movement, but only by coindexing. For instance, the relative clause (101b) will be analysed as in (105):

(105) $[_{DP}$ *un* $[_{CP}[_{PP}$ *mot* $[$*au* $[_{DP}$ *quel*$]]]_{i}$ $[_{IP}$ *Dieu sait quel sense elle lui*$_{i}$ *attribuait* $t_{CL}]]]$

Note that this approach requires the stipulation that the relative phrase can be generated in Spec,CP, in the same way as the dislocated phrase is generated in the Topic position in Cinque's (1990: chapter II) analysis of Clitic Left Dislocation.

In this approach, the two relativization strategies are reduced to the syntactic ambiguity of clitic pronouns: these can be used "transitively" as stranded determiners of the NP "head", as in (104), or "intransitively" as well-behaved resumptive pronouns, as in (105).

However, this approach implies that the type of doubling attested in (101) and (102), corresponding to the intransitive use of the clitic, is only possible when the doubled constituent is generated in a position outside IP, and the clitic fills the IP-internal position. But there are at least two structures where a doubled constituent sits in the IP-internal position.

The first case is the clitic doubling of Romanian (Dobrovie Sorin 1994: chapter VI) and Spanish (Suñer 1988; Zubizarreta 1994):

(106) a. *Fiecare profesor îi va examina pe zece elevi*
every teacher them$_{CL}$ will examine P° ten students
'Every teacher will examine ten students.'
(Dobrovie Sorin 1994: 230)

b. *Lo vimos a Juan.*
(we) him$_{CL}$ saw P° Juan
'We saw Juan.'

The authors argue that in (106) the doubled object is in the base A position. This hypothesis is also forced by the Linear Correspondence

Axiom, since there is no right-hand adjunct position available for the doubled object (cf. Kayne 1994: 78–83).[70] Therefore, in these structures the doubling clitic is not a theta-marked pronoun.

The second case is French complex inversion, where a subject clitic doubles a preverbal lexical subject. Kayne (1972: 82–83) has shown that this structure cannot be assimilated to left dislocation: in fact, the two constructions contrast with respect to the requirement of specificity for the doubled phrase:

(107) a. * *Quelqu'un, depuis quand m'attend-il ?*
somebody, since when me_{CL} waits-he_{CL}?
b. *Depuis quand quelqu'un m'attend-il* ?
since when somebody me_{CL} waits-he_{CL}?
'since when is somebody waiting for me?'

This contrast suggests that the subject in (107b) is not dislocated, but it must be in the preverbal subject position; then the latter cannot be filled by an independent DP headed by the subject clitic.

The point is that in clitic doubling and in French complex inversion neither of the analyses sketched above is adequate. The clitic cannot be analysed as the D° head of the doubled constituent, since the latter is larger than NP. The clitic cannot be analysed as a pronoun generated in the theta-position either, since that position is filled by the doubled constituent.

A unifying analysis of clitics that can accommodate all the structures in (100)–(107) has been recently proposed by Belletti (1995) and Kayne (1995). In this analysis, the clitic and the doubled argument are generated as a single constituent; more specifically, the clitic constitutes the highest functional head of the constituent, and its maximal projection envelops the argument.[71] A clitic doubling structure like (106b) will have the following complex constituent as a direct object in the base:

(108)
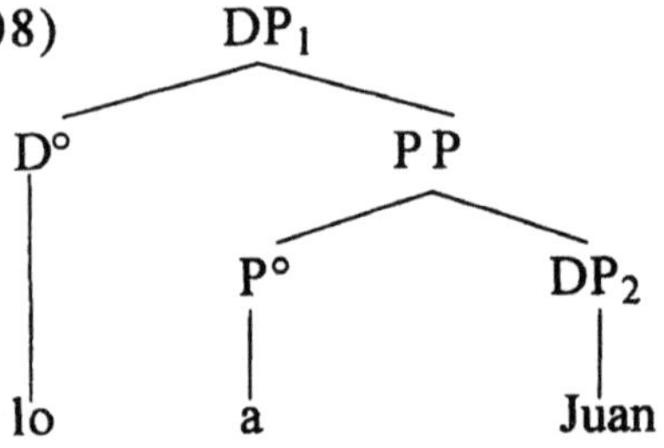

(Belletti 1995)

According to Belletti, the clitic is a D° endowed with a strong Case feature, which must be eliminated before Spellout. For this reason, the D° must adjoin to the head Agreement-Object, which is responsible for the checking of the Accusative case. In doubling structures like (108), the clitic D° selects a Case-marker (the dummy preposition) which marks the embedded argument DP_2. This dummy Case-marker is required in order for DP_2 to satisfy the Case filter, because after the incorporation of the clitic Agr-O is empty and it is unable to license a DP in its Spec at LF.[72]

Let us assume that doubling structures like the relative clauses in (101) and Clitic Left Dislocation in (102) involve the generation of a complex constituent like (108). For instance, the derivation of (101b) involves a complex constituent headed by the clitic: $[_{DP}$ *lui* $[_{PP}$ *a* $[_{DP2}$ *lequel* $[_{NP}$ *mot*$]]]]$; the highest D° is cliticized and its complement moves to Spec,CP:

(109) $[_{DP}$ *un* $[_{CP}[_{PP}$ *mot* $[_{PP}$ *auquel* $t_{NP}]]$ $[_{IP}$ *Dieu sait* [*quel sens*$_i$ *elle lui attribuait* t_i $[_{DP}$ t_{CL} $t_{PP}]]]]]]$

This analysis resembles the one in (104) in that it takes the clitic to be the original head of the relative phrase; however, the analysis in (109) can account for the presence of both the resumptive clitic and the relative determiner.

With respect to the analysis in (109), the problematic examples are (100), where the clitic seems to double a bare NP "head". In this case, it appears that the clitic D° should be allowed to directly select NP, omitting the preposition and the relative determiner, as proposed in (104).

Note, however, that alongside the relativization strategy in (100), where the deleted preposition can be "recovered" by means of the clitic, there is also another strategy lacking resumptive clitics where the deletion of the preposition remains unrecoverable (Gapany—Apothéloz 1993: 127; Cennamo forthcoming):[73]

(110) a. *il ristorante che abbiamo mangiato ieri* (substandard Italian)
the restaurant that we ate yesterday'
'the restaurant in which we ate yesterday'

b. *l'homme que tu as parlé récemment* (substandard French)
the man that you spoke recently
'the man of/with whom you have spoken recently'

(110) c. *el gran che se fa el pan bianch* (Milan dialect)
the wheat that one makes white bread
'the wheat with which one makes white bread'

d. *lo mio leto grande ch'eo dormo* (Ancient Venetian)
the my bed big that I sleep
'my big bed in which I sleep'

e. *la lana que vòli far lo drap* (Occitan; Wheeler 1988:275)
the wool that I want to make the cloth
'the wool with which I want to make the cloth'

It is unclear how this "unrecoverable deletion" of the preposition can be licensed; however, note that it is limited to relative structures, both with and without a resumptive clitic. This suggests that the external determiner of the relative structure may play a role in licensing it. Suppose that the relative phrase in (100) and (110) is generated as a full PP, with the preposition and the relative determiner deleted in the PF representation. The only difference would be that in (100) the relativized PP is enveloped in a clitic DP. Then, (110) would be to (100) what the standard pied piping relatives are to the doubling relatives in (101).

An important question remains unanswered for the moment: namely, under what syntactic conditions a complex constituent like (108) is generated in the place of a simple argument. It has been frequently observed that clitic doubling correlates with the specificity of the doubled argument (cf. Suñer 1988). In various recent proposals, specificity is taken to be an intrinsic property of the clitic (cf. Cecchetto 1995: chapter 5; Corver—Delfitto 1994; Sportiche 1992; Uriagereka 1994: 87–91). In the analysis in (108), the specificity requirement on doubled arguments can be analysed as a selectional requirement imposed by the clitic D° on its complement. But this hypothesis seems to be too strong, since it cannot account for the lack of a specificity effect in French complex inversion (cf. (107b)). For the moment, this problem remains open.

9.4. German relative determiners

In the preceding subsections it was proposed that a resumptive pronoun can be analysed as the highest head of the constituent that it "doubles", stranded by the leftward movement of the latter.

It is interesting to reconsider from this perspective the status of the German relative pronoun *der-die-das*. This pronouns does not only appear in relative clauses, but it is also used as a resumptive pronoun in left dislocation:

(111) a. *der Mann, der in seinem Büro arbeitet*
the man who in his office works
'the man who works in his office'
b. *den Hans, den habe ich gestern gesehen.*
the Hans, him have I yesterday seen
'As for Hans, I saw him yesterday.'
(Cardinaletti 1983: 113)

A left dislocation structure like (111b) apparently violates the V/2 constraint: in fact, if the resumptive pronoun is topicalized to Spec,CP, the dislocated DP must occur in a CP-external position (recall that double adjunction to CP is ruled out by the Linear Correspondence Axiom). Assuming the complementizer system, it is possible to take the dislocated DP to be in the Spec of a phonetically null functional head above the head that the verb has raised to: [$_{XP}$ *den Hans* [$_{XP}$ X° [$_{TopP}$ *den* [$_{TopP}$ *habe* [$_{IP}$ *ich* ...]]]]]. But this would not account for the ungrammaticality of (112), where the phrase in Spec,TopP is not the resumptive pronoun:[74]

(112) * *Den Hans, gestern habe ich den gesehen.*
the Hans, yesterday have I him seen
(Cardinaletti 1983: 113)

Kayne (1994: fn. 15 to ch. 8) proposes a solution to this problem based on an analysis of German *d*-pronouns parallel to that proposed in (108) for Romance resumptive clitics: the dislocated phrase and the resumptive pronoun form a single constituent that is raised to the highest Spec of the sentence. The peculiarity of the resumptive *d*-pronouns is that it can host its DP complement in Spec:

(113) [$_{XP}$[*den Hans*$_i$ [*den* [e]$_i$]]$_j$ [$_{XP}$ *habe ich gestern* t$_j$ *gesehen*]]

In this analysis, V/2 topicalization uniformly targets the highest functional projection in the complementizer system of the clause. The ungrammaticality of (112) follows from the fact that the dislocated DP does not form a constituent with the topicalized adverb.

Let us then consider *d*-pronouns in their relative function. Superficially, their behaviour resembles that of English and Italian relative determiners: they appear in front of the relative clause and linearly follow the relative "head". Thus, German relativization can be analysed as a specific type of A' movement similar to English and Italian relativization. The morphological definiteness of the relative determiners raises no problem if they are analysed as expletive determiners close to the final stage of Greenberg's cycle (cf. § III.5.3).

However, it is also possible to assimilate (111a) to (111b), and to conceive of *d*-pronouns as resumptive pronouns even in relative clauses. From this point of view, German relative clauses would be closer to the resumptive relatives in (100) than to the English *wh*-relatives. *D*-pronouns would not be "relative" in any sense, in the same way as the resumptive clitics in (100) are not distinguished from those of Clitic Left Dislocation. In relative clauses, the *d*-pronoun simply happens to form a constituent with the "head".

A consequence of this view is that German relativization can be assimilated to any type of A' movement that *d*-pronouns can independently undergo. In other terms, the relative phrase does not raise to an A' position that is specific to relativization, but to one that is independently available to the *d*-pronoun. The only constraint would be that the target position must be prominent enough for the NP "head" to be properly connected to the external D° via the Spec of the *d*-pronoun.

From this perspective, German relativization may be analysed as an instance of scrambling. In fact, German has a scrambling position above the preverbal subject (cf. Franks—Lee—Rambow 1992), which constitutes an available landing site for the relative phrase headed by the *d*-pronoun. This scrambling position must then be the most prominent Spec of the relative clause, immediately below the external D°.

The following subsections discuss some evidence in favour of this hypothesis.

9.4.1.

A first argument in support of the scrambing analysis comes from the phenomenon of "internal *wh* movement" (van Riemsdijk 1985). In German heavy pied piping, the *d*-pronoun cannot remain embedded within the pied piped constituent, but it must be fronted to the beginning of it. Van Riemsdijk (1985: 168) arrives at the generalization that the NP "head" can be separated from the *d*-pronoun at most by a preposition:

(114) a. *Mohammed Ali ist ein Mann, [mit dem es aufnehmen zu wollen], reiner Wahnsinn wäre.*
Mohammed Ali is a man with whom EXPL to take up to want sheer madness would be
'Mohammed Ali is a man, to challenge whom would be sheer madness'

(114) b. * *Mohammed Ali ist ein Mann, es mit dem aufnehmen zu wollen, reiner Wahnsinn wäre.*
Mohammed Ali is a man, EXPL with whom to take up to want sheer madness would be (van Riemsdijk 1985: 167)

c. *der Mann mit dem nach Frankreich zu fahren ich vorhatte*
the man with whom to France to drive I planned
'the man with whom I planned to drive to France'

d. *Forscher auf die stolz zu sein ich jedem empfehle*
researchers of whom proud to be I everybody$_{DAT}$ recommend
'researchers of whom I recommend that everyone should be proud' (Webelhuth 1985: 219)

In this respect, the German *d*-pronoun contrasts with English and Italian relative determiners, which are stranded within the pied piped phrase. Compare (114a–b) to the parallel Italian examples in (115):

(115) a. ?* *Mohammed Ali è un uomo col quale voler litigare è pura pazzia.*
Mohammed Ali is a man with whom to want to argue is sheer madness

b. *Mohammed Ali è un uomo voler litigare col quale è pura pazzia.*
Mohammed Ali is a man to want to argue with whom is sheer madness
'Mohammed Ali is a man to want to argue with whom would be sheer madness.'

In (114a,c), the PP [*Mann* [*mit dem*]] appears in front of the infinitival clause. Van Riemsdijk (1985) proposes that the infinitival clause is pied piped to the Spec of the relative CP, and the PP contining the relative pronoun moves to Spec,CP of the infinitival clause. This second movement he calls "internal *wh*-movement".

Webelhuth (1985: 223–227) points out a problem with this analysis: it is not obvious that German infinitival clauses have a Spec,CP available for internal *wh*-movement, since German lacks infinitival questions and infinitival relatives.

Webelhuth also criticizes a proposal by H. Haider, according to whom in (114a,c) the relative pronoun is extracted from the infinitival clause and moves to the Spec of the relative CP, whereas the infinitival phrase is in a scrambling position. Webelhuth notes that this analysis incorrectly predicts the possibility for lexical material of the relative clause to intervene between the relative pronoun in Spec,CP and the scrambled infinitival clause. The prediction is falsified by the example (116), where the adverb *wahrscheinlich* cannot have scope over the whole relative clause, but only over the infinitival clause; this means that it must be analysed as an internal constituent of the latter:

(116) *das Buch [das wahrscheinlich zu lesen] niemand versuchen wird*
the book which probably to read nobody try will
'the book that, probably, nobody will try to read'

Webelhuth concludes, in agreement with van Riemsdijk, that the relative pronoun must be in a position internal to the infinitival clause. Since by hypothesis the infinitival clause lacks a Comp position, he proposes that the relative pronoun is adjoined to the infinitival IP. In other terms, internal *wh*-movement of the relative pronoun within the pied piped phrase is an instance of scrambling. This immediately explains why internal *wh*-movement is impossible in English, Italian or French, which lack scrambling (cf. (115)).[75]

But if the relative pronoun can be licensed in a scrambling position in (114a), it is possible to argue that it does so generally. The heavy pied piped constituent may be itself scrambled to the most prominent position of the relative clause. In this way, all the relativization process is accomplished through scrambling.[76]

9.4.2.

It has been occasionally pointed out that German relativization is constrained more strictly than *wh*-movement of an interrogative pronoun, at least in the Northern dialects (cf. for instance Müller—Sternefeld 1993: fn. 19; van Riemsdijk 1985: fn. 6). The following example illustrates the ungrammaticality of relativization out of a *daß* clause:

(117) * *Ich habe meinen Regenschirm, den ich weiß, daß mein Onkel Dagobert gestohlen hat, gestern im Marschall Fields wieder gefunden.* (Loetscher 1972: 50)
I have my umbrella, which I think that my uncle Dagobert stolen has, yesterday in Marschall Fields refound

This example contrasts with the grammaticality of *wh*-movement and topicalization in (118), but interestingly, it mirrors the impossibility of long distance scrambling in (119):

(118) a. *Was*$_i$ *sagt sie [daß der Fritz* t_i *mag]* ?
what said she that the Fritz liked
'What did she say that Fritz liked?'
b. *Pudding*$_i$ *sagt sie [würde der Fritz* t_i *mogen]*
pudding said she would the Fritz like
'Pudding, she said that Fritz would like.'

(119) a. * *weil sie Pudding*$_i$ *sagt [daß der Fritz* t_i *mag]*
since she pudding said that the Fritz liked
b. * *weil Pudding*$_i$ *niemand sagt [würde der Fritz* t_i *mogen]*
since pudding nobody said would the Fritz like
(from Müller—Sternefeld 1994: 332)

Although judgements on the extraction from a *daß*-clause are in general delicate and subject to dialectal variation, there seems to be a contrast between relativization and scrambling on the one hand, and *wh*-movement on the other, with respect to the extraction from a finite clause.

On the other hand, it is possible to scramble out of certain infinitival clauses, like ECM complements, raising infinitivals, and certain subject control infinitivals like the complement to *versuchen* 'try' and *hoffen* 'hope' (Grewendorf—Sabel 1994: 264–266):. It seems that relativization is also possible from at least some of these infinitival complements: cf. van Riemsdijk (1985: 169–171).[77]

Although the parallelism between scrambling and relativization in this respect should be carefully verified, the few data that can be collected from the literature suggest that relativization patterns with scrambling rather than with *wh*-movement with respect to locality constraints, contrary to what the standard analysis of relativization would predict.

9.5. Concluding remarks

The preceding discussion is simply tentative; it is not meant as a positive proposal of an analysis for resumptive relatives, but rather, as an indication of an interesting consequence of the approach to relative pronouns elaborated in the raising analysis. The consequence is that the traditional clearcut distinction between operator relative pronouns and bound resumptive pronouns is called into question. It was suggested that this consequence is not so negative as it may seem at first, because in some languages resumptive pronouns exhibit certain properties that one would rather expect of relative pronouns, e.g. the possibility of moving to the Comp system.

On the other hand, note that certain *wh*-relative determiners exhibit properties typical of resumptive pronouns: for instance, in the following English example, reported by Kayne (1994: n. 16 to ch. VIII), the relative determiner *which* is stranded in an IP-internal position:

(120) ?? *This is the book that I would like very much for* which *to be sent to me.*

The common syntactic feature of relative determiners and of at least some resumptive pronouns would be that both are nominal functional heads stranded by the leftward A' movement of their complement. This unconventional perspective on resumptive pronouns also raises interesting questions w.r.t. languages like German, where relative determiners are morphologically indistinguishable from resumptive pronouns.

An outstanding problem is why relative clauses featuring a resumptive pronoun require the realization of the Force Phrase level in the Comp system (cf. Kayne 1984: ; Rizzi 1990: fn. 29 to chapter II).[78] If resumptive relativization is an instance of topicalization, as suggested by Haberland—van der Auwera (1990), then the impossibility of zero resumptive relatives could be related to the impossibility of topicalization within zero relatives (cf. § 6): resumptive pronouns must be related to a [topic] constituent, but the feature [topic] must in turn be licensed by a c-commanding Force°. Once again, this hypothesis cannot be developed without a deeper understanding of the syntactic relation linking resumptive pronouns to the Topic position.

10. Summary

This chapter discussed various issues in the syntax of the complementizer system of English and Italian. In the preceding chapter, the possibility of "optional *that* deletion" and of the *wh*-relative in English was reduced to the availability of a low head in the complementizer system acting as a defective declarative complementizer. The discussion in this chapter justified the identification of this head with Top/Fin°: this head can be structurally licensed independently of topicalization because it contains the intrinsic specification [+finite]. This specification can be thought of as a "residue" of the gradual process of loss of verb second in English (cf. Platzack 1995). In Modern Italian, Top° has lost this residual property and it can only be structurally licensed in topicalization structures; but in earlier stages of the language, Top° probably had the same status that it has in Modern English.

Various distinct phenomena were reduced to the properties of the head Top/Fin° in English. The intrinsic specification [+finite] makes it unavailable in infinitival clauses, accounting for the impossibility of the infinitival *wh*-relative. The "default" declarative force of Top/Fin° is insufficient to license the features [topic] and [focus], accounting for the ungrammaticality of topicalization and negative preposing in zero relatives and zero complement clauses. The head government requirement on Top/Fin° was argued not to depend on the lack of phonetic content, but rather on the syntactic defectiveness of this head. Finally, it was proposed that Top/Fin° plays a crucial role in subject relativization; and more generally, the Topic position may be involved in the resumptive relativization strategy of various other languages.

Chapter VIII
Stacking, coordination, and extraposition

0. Introduction

The preceding chapters were mainly devoted to the internal structure of headed relative clause. This brief chapter discusses some aspects of their "external" syntax, namely stacking, coordination, and extraposition. An approach to stacking and extraposition form the perspective of the raising analysis has already been proposed by Kayne (1994: chapters 8 and 9). However, the analysis of the various relative clause types elaborated here in chapters 6 and 7 differs from Kayne's (in particular, with respect to the lack of an overt relative determiner in *that-* and zero relatives); therefore, it is necessary to verify to what extent his approach to stacking and extraposition is consistent with our hypotheses.

1. Stacking

Kayne (1994: 92) discusses the following example, featuring the stacking of two restrictive relatives:

(1) *I just read the book that's about your ancestors that your son gave me last year.*

Stacking requires the recursion of the raising relative structure: the string [*book that's about your ancestors*] consisting of the "head" and the leftmost relative clause must be analysed as a constituent raised to the Spec,ForceP of the rightmost relative clause from the complement position of *gave*.

(2) $[_{DP}$ *the* $[_{FP}$ [*book that's about your ancestors*]$_i$ $[_{FP}$ *that your son gave me* t_i]]]

In this structure, the external D° selects the rightmost relative FP; but it is unclear whether the inner FP is selected by a null external D°. Kayne seems to assume that it is not.[1] But if we adopt principle (10) of chapter II, the phrase originating in the complement position of *gave* cannot be a FP; it must be a DP headed by a null determiner. Moreover, according to the discussion in §§ VI.2.1–2.2, the relative "head" *book* must be introduced by an underlying relative D° subject to PF deletion. Then, the structure of (1) must be something like (3):

(3)

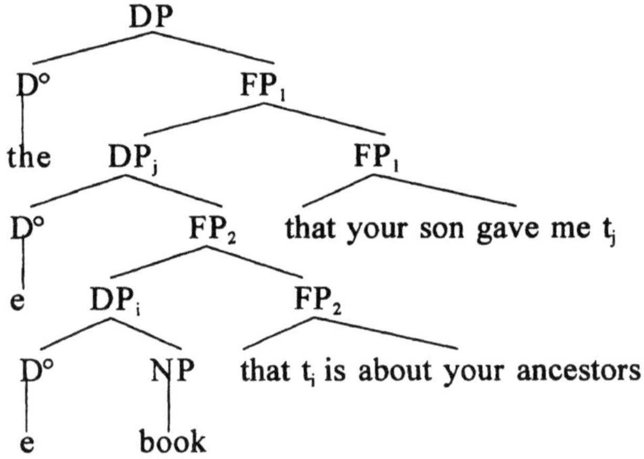

This structure contains two phonetically null determiners. The licensing of these empty Ds can be reduced to the mechanism of PF deletion proposed in § VI.2: the lower relative determiner of DP_i incorporates to the higher determiner of DP_j, which selects the inner FP_2, and the resulting complex head incorporates to the highest external determiner *the*. Both steps are triggered by the principle of economy of representation ((95) of chapter VI). By the first step, the minimal domain of $D°_j$ is extended to include the NP complement to $D°_i$, and by the incorporation of $D°_j$, its minimal domain is included in that of the highest D°.In this way, all the determiners may check their features against the [+interpretable] features of the NP "head". [2]

Consider then the stacking of two *wh*-relatives:

(4) *the book which John wrote which Bill read*

The derivation is slightly more complex. The base structure is (5):

(5) $[_{DP}$ *the* $[_{FP2}[_{IP2}$ *Bill read* $[_{DP2}$ *which* $[_{FP1}[_{IP1}$ *John wrote* $[_{DP1}$ *which book*]]]]]]]

Once again, the inner relative clause is included in the DP "head" of the outer one. The derivation proceeds as follows: the lowest relative DP_1 raises to the complementizer system of the inner relative clause, where the NP $[_{NP}$ *book*] strands its relative determiner *which* in Spec, TopP (cf. § VI.6):

(6) [*the* [$_{FP2}$[$_{IP2}$ *Bill read* [$_{DP2}$ *which* [$_{FP1}$ *book* [$_{TopP1}$ [$_{DP1}$ *which* t]$_i$ [$_{IP1}$ *John wrote* t$_i$]]]]]]

Then the relative DP$_2$, including the inner relative CP$_1$, raises to the complementizer system of the outer relative CP$_2$:

(7) [$_{DP}$ *the* [$_{FP2}$ [$_{TopP}$ [$_{DP2}$ *which* [$_{FP1}$ *book which John wrote* t]]$_j$ [$_{IP2}$ *Bill read* t$_j$]]]]

As a final step, the relative FP$_1$ moves to the left of the relative determiner *which* selecting it. The resulting representation is (8):[3]

(8)

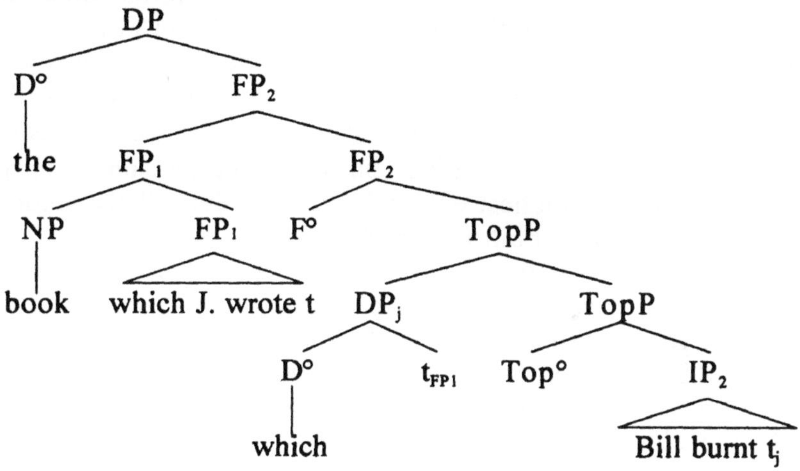

Note that in (8) the NP [$_{NP}$ *book*] originating within FP$_1$ has reached a position that falls in the minimal domain of the highest external determiner, because it is only dominated by one segment of FP$_1$ and one segment of FP$_2$ (cf. § II.5.2); thus, the required checking relation between the external D° and NP is ensured.[4]

One aspect of stacking that does not follow directly from this proposal is the impossibility of having a rightmost zero relative introduced by a Top° (cf. the impossibility of omitting the second *that* in (1)). In fact, in (3) the rightmost relative clause is selected by the highest D°, so that the head government requirement on Top° (§ VII.5) is satisfied. It is therefore necessary to adopt for Top° the constraint against a complex specifier proposed by Kayne (1994: 92, 98) for null complementizers.[5]

A further problem is the stacking of appositive relatives. An appositive relative can be stacked to the right of a restrictive relative, but not to the left of it:[6]

(9) a. *The man that came to dinner, who was drunk, fainted.*
b. * *The man, who was drunk, that came to dinner fainted.*
(Jackendoff 1977:171)

Moreover, the stacking of two appositive relatives is impossible:

(10) * *John, who goes to MIT, who likes maths, will get a job.*

Kayne (1994: 114) proposes the following structure for (9b):

(11) D° [$_{FP1}$[$_{FP2}$ *man who* [$_{FP2}$ F° IP]] [$_{FP1}$ F° IP]]

Recall that the appositive interpretation is obtained by moving the IP constituent out of the c-command domain of the external determiner in LF. This movement is possible for the rightmost IP in (11).[7] However, the movement of the leftmost IP would constitute an illicit extraction from a left branch, since FP_2 is embedded within the phrase in the specifier of FP_1. Thus, the left branch constraint excludes an appositive interpretation for all stacked relatives but the rightmost one, i.e. the one that is selected by the highest external D°, as in (9a). This accounts for the ungrammaticality of both (9b) and (10).

This analysis can be transposed to the revised stacking structures (3)/(8). However, it is interesting to note that these structures also allow a slightly different explanation. In fact, each stacked relative is here selected by its own external D°. Let us assume with Kayne that the appositive IP/TopP moves to Spec,DP.[8] By relativized minimality, it will move to the closest position of this type, namely to the Spec of the closest c-commanding external D°. This implies that in (7), the IP of the inner relative FP_1 will move in LF to the Spec of the internal DP_2; but in this position, it is still c-commanded by the highest external determiner. By minimality, only the IP of the rightmost FP_2 can move to the Spec of the highest D°, receiving a truly appositive interpretation.[9]

2. Coordination

A second problem is the coordination of two relative clauses, as in (12) and (13):

(12) *the picture* [*which Bill liked*] *and* [*which Mary hated*]
(13) *the picture* [*that Bill liked*] *and* [*that Mary hated*]

In the raising analysis, these examples would involve an across-the-board extraction of the NP/DP "head" from the coordinated relative clauses. But in Kayne's (1994) approach, the coordinated strings are not even constituents: in fact, in (12) the relative determiner *which* must have the NP "head" in its Spec (cf. § VI.6.2), and in (13) the F° *that* must have the relative NP in its Spec.

Note that the problem concerning (12) does not arise under the analysis of the *wh*-relative proposed here. In fact, (12) may involve the coordination of two TopPs with across-the-board extraction of the NP "head" from Spec, TopP:

(14) $[_{DP}$ *the* $[_{FP}[_{NP}$ *picture*$]$ F° $[_{\&P}[_{TopP}[_{DP}$ *which* t$]_i$ *Bill liked* $t_i]$ $[_{\&P}$ *and* $[_{TopP}[_{DP}$ *which* t$]_j$ *Mary hated* $t_j]]]]]$

But the problem holds for the coordination of *that*-relatives in (13).[10]

A solution comes from the null operator analysis of across-the-board gaps proposed by Munn (1992). Munn argues that coordination is syntactically an asymmetric relation mediated by a "Boolean" functional head. The across-the-board gap in the second conjunct is bound by a null operator which lands in the Spec of the Boolean head; the whole Boolean phrase is right-adjoined to the first conjunct:

(15)
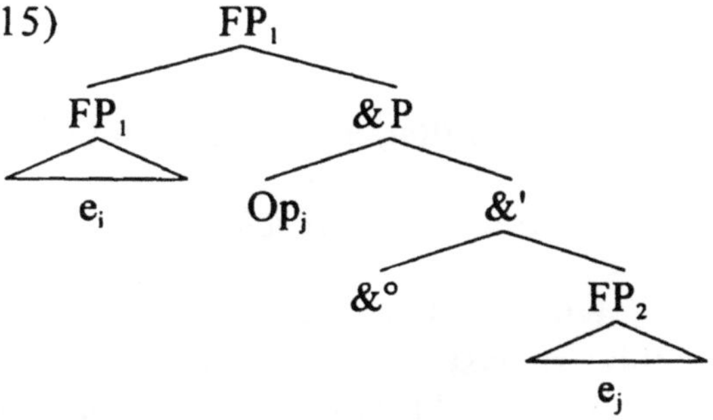

Though this proposal is not directly compatible with the Antisymmetry theory, it can be slightly revised in the following way (Kayne 1994: 59):

(16)
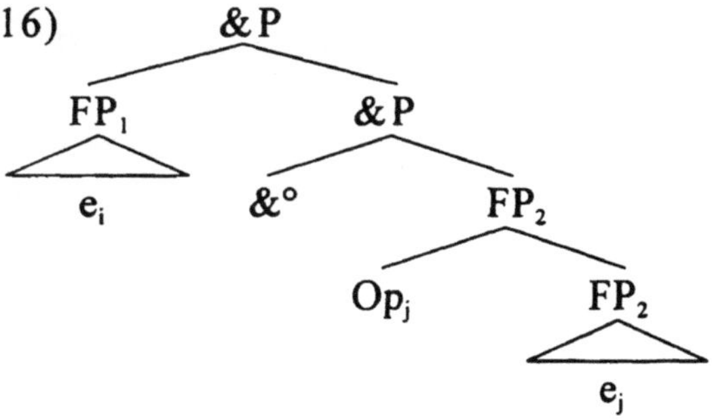

The null operator lands in the Spec of the second conjunct FP_2, and the first conjunct FP_1 is in the Spec of the Boolean head.

With this analysis of across-the-board gaps, we can assume that (13) involves the cordination of two FPs complement to the external determiner: the Spec of the first conjunct contains the relative DP, while the Spec of the second conjunct contains the null operator:[11]

(17) $[_{DP}$ *the* $[_{\&P}[_{FP1}$ $[_{DP}$ D° *picture*$]_i$ $[_{FP1}$ *that Bill liked* $t_i]]$ $[_{\&P}$ *and* $[_{FP2}$ Op_j $[_{FP2}$ *that* [*Mary hated* $t_j]]]]]]$

Finally, recall that in the analysis proposed in chapters VI and VII, *that*-relatives and *wh*-relatives are Force Phrases, whereas zero relatives belong to a different syntactic category Topic Phrase. This may account for the deviance of the coordination of a zero relative and a *that*-relative, as already suggested by Weisler (1980: 630–631):

(18) ? *The book John read and that Bill wrote is boring.*[12]

3. Extraposition[13]

A third problem is relative clause extraposition, exemplified in (19):

(19) [*A man*] *came into the bar* [*who we knew in school*].

The standard analysis takes the extraposed relative clause to be right-adjoined to VP or IP; but this analysis is inconsistent with the Antisymmetry theory, which excludes right adjunction on principled grounds. The raising approach opens a new perspective on the phenomenon: this can be thought of as the leftward movement of the "head" which strands the relative clause in the base position (Kayne 1994: chapter 9).[14]

(20) $[_{QP}$ *A man*$]_j$ *came into the bar* [[e]$_j$ *who we knew in school*].

Note that in (20) the extracted QP must have originated as a complement to the relative determiner; therefore, it is necessary to generate a complex DP like (21):

(21) $[_{DP}$ wh $[_{QP}$ *a* $[_{NP}$ *man*$]]]$

This structure is never overtly realized; the indefinite article cannot follow another determiner.[15]

This obligatory omission of Q° recalls our discussion of the the phenomenon of split topicalization in § VI.2.3.2. It is possible to extend to English the proposal made in that section for German: namely, the indefinite determiner is the spellout of a Q° bearing the feature [singular]; when this Q° is governed by a D°, as in (21), the principle of economy of representation forces its PF deletion by incorporation; in (20), on the contrary, incorporation is impossible, and Q° must be spelled out as an independent morpheme.[16]

Another problem concerns the clause-final position in which the relative clause is stranded. On the one hand, the extraposed relative follows all the verbal complements and modifiers; in (20), for instance, it follows the locative PP. On the other hand, the relative clause cannot be stranded in the intermediate position of a chain. In this respect, relative clause extraposition sharply differs from another instance of stranding, namely quantifier floating (cf. Sportiche 1988).[17]

(22) a. * [*A man*] *was* [[e] *who has no relatives*]$_j$ *injured* t_j *in the accident.*
b. [*The men*] *were* [*all* [e]]$_j$ *injured* t_j *in the accident.*
(Kayne 1994: fn. 4 to ch. 9)

Starting from the second problem, a possible approach to (22) emerges if we slightly revise the structure in (20). By principle (10) of ch. II, an argument must be a DP, and hence in (20) the stranded relative must be introduced by a null external determiner which allows the extraction of the QP "head". More accurately, the representation is (23):

(23) [$_{QP}$ *A man*]$_i$ *came into the bar* [$_{DP}$ e [$_{FP}$ t_i *who we knew in school*]]

From this perspective, the restriction to the clause-final position may be related to the licensing requirements of the empty determiner.

This hypothesis is supported by the parallelism between relative clause extraposition and another "splitting" construction which is usually analysed as the leftward movement of a DP subconstituent, leaving an empty category in front of the stranded DP:

(24) *Combien a-t- il conduit* [[e] *de voitures*] ?
how much did he drive of cars
'how many cars did he drive?'

Besides extracting *combien,* it is also possible to front the whole *wh*-phrase; in this case the past participle agrees with the fronted phrase:

(25) [*Combien de voitures*]$_i$ *a-t-il conduites* t$_i$?
how much of cars has he driven-AGR

Assuming Kayne's (1989) analysis of participial agreement, the *wh*-phrase in (25) must have passed through the Spec of the participial Agr. Note now that the remnant DP cannot be stranded in this intermediate position by the extraction of *combien* alone:

(26) * *Combien a-t-il* [[e] *de voitures*]$_i$ *conduites* t$_i$?[18]
how much has he of cars driven (Rizzi 1990: fn. 13 to ch. 1)

In this respect, *combien* extraction patterns with relative clause extraposition (cf. (22)) and against quantifier floating:

(27) a. *Il a consulté* [*tous* [*les livres*]].
he has consulted all the books
'He consulted all the books.'
b. *Il les*$_j$ *a* [*tous* [e]$_j$] *consultés* t
he them-CL has all consulted-AGR
'He consulted all of them.'

Note that (24) and, under our assumptions, (23) differ from (22b)/(27b) in that they have an empty category in front of the stranded DP.[19] The contrast may be reduced to the hypothesis that this empty category can only be licensed in a base position. Though this constraint recalls the old notion of theta-government, its status in the present syntactic theory is unclear. Whatever the ultimate explanation may be, what is relevant to our argumentation is that with respect to the unavailability of intermediate chain positions, relative clause extraposition can be assimilated to a structure like (24), for which a stranding analysis is relatively straightforward.

As for the relative ordering of the extraposed relative and of verbal complements and modifiers, Kayne (1994: 120–122) suggests that the relative clause is stranded in a non-Case marked position which is below the normal Case marked DP positions; this low position is crossed over by verb raising and by the scrambling of various PP complements or adverbials. (If it can be identified with the base VP-internal position, Kayne's hypothesis would converge with the conclusion independently reached in the preceding paragraph.)

The reason for the rearrangement of constituents can be found in the theory of prosodically driven reordering of constituents recently proposed by Zubizarreta (1994). Zubizarreta argues that the scrambling of VP-internal constituents can be triggered by the need of obtaining a prosodically wellformed structure; the target positions of this scrambling she dubs "P-positions".

The hypothesis that the clause-final position of the extraposed relative is obtained by prosodic rearrangement is supported by the observation that at least in Italian, an extraposed relative can be followed by a PP if the latter is sufficiently "heavy":

(28) ? [$_{QP}$ *Che libro*] *hai trovato* [[$_{QP}$ e] *che ti serviva per l'esame*] [$_{PP}$ *in quella famosa libreria di Firenze*] ?
which book did you find that you needed for the exam in that famous bookshop in Florence?
'Which book that you needed for the exam did you find in that famous bookshop in Florence?'

Thus, though various aspects of the problem remain open for further research, the constraints on the surface position of the extraposed relative do not seem to undermine the stranding hypothesis.

A further problem is that the stranding of the relative clause seems to involve an instance of improper movement:

(29) [$_{QP}$ *A man*]$_j$ *walked in* [$_{FP}$ t_j *that we knew in school*]].

In Kayne's (1994) analysis, the "head" of the *that*-relative is a QP, and in (29) this QP directly raises from Spec,FP of the relative clause (an A' position) to Spec,IP of the matrix clause (an A position).

However, in chapter VI it was argued that even in the *that*-relative the QP "head" is introduced by a null relative determiner. Thus, more accurately, the structure is (30):

(30) [$_{QP}$ *A man*]$_j$ *walked in* [$_{DP}$ D° [$_{FP}$[D° t_{QP}] *that we knew in school*]]

The QP "head" is extracted from the complement position of the relative determiner, which is plausibly L-related,[20] and hence an A position. Thus, we have an instance of proper A movement.

Consider then the extraposed *wh*-relative (23), repeated here:

(23) [$_{QP}$ *A man*]$_j$ *came into the bar* [$_{DP}$ e [$_{FP}$ t_j *who we knew in school*]]

In this structure, the overt relative D° is stranded in a Spec position below Spec,FP (§ VI.6.3). Its QP complement moves to Spec,FP in order to establish a checking relation with the external D°; then, from Spec,FP it moves on to Spec,IP of the matrix clause:

(31) QP ... $[_{DP}$ D° $[_{FP}$ t_{QP} Force° $[_{TopP}$ $[_{DP}$ *who* $t_{QP}]$ Top° IP]]]

At first sight, the second step seems to be an improper movement. Hovever, given the definition of L-relatedness in Chomsky (195: 196),[21] in (31) Spec,FP is actually L-related, since the QP is in a local relation to the phi-features of the external determiner (cf. § II.6). With this assumption, no improper movement arises in (31).[22]

The most puzzling problem for a stranding analysis of extraposition is the "interpretive nesting requirement", recently discussed by Koster (1995), Nakajima (1992), Rochemont—Culicover (1990: 53–60), Wiltschko (1995).[23] This requirement bars intersecting dependencies between the base position and the right-adjoined position of the extraposed clause, as in (32a), and only allows nested dependencies as in (32b):

(32) a. * $[_{IP}$ DP_{SUBJ} ... $[_{VP}$ $DP_{OBJ}]$ ... $CP_{SUBJ}]$ ...$CP_{OBJ}]]]]$
b. $[_{IP}$ DP_{SUBJ} ... $[_{VP}$ $DP_{OBJ}]$ ... $CP_{OBJ}]$ ...$CP_{SUBJ}]]]]$

This accounts for the surface order of extraposed relatives: a relative extraposed from the subject must be to the right of a relative extraposed from an object:

(33) a. *Someone*$_i$ *brought beans*$_k$ *to the party* $[_{CPk}$ *cooked in oil im ported from Italy*] $[_{CPi}$ *who we didn't know*].
(Nakajima 1992)

b. [*Die man*]$_i$ *heeft* [*de vrouw*]$_j$ *ontmoet* [*die alles wist*]$_j$ [*die haar bewonderde*]$_i$
the man has the woman met who all knew who her admired
'The man who admired her met the woman who knew everything.'
c. * [*Die man*]$_i$ *heeft* [*de vrouw*]$_j$ *ontmoet* [*die haar bewonderde*]$_i$ [*die alles wist*]$_j$
the man has the woman met who her admired who all knew
(Koster 1995)

The problem for the stranding analysis is that in (33a–b) the object relative clause would have to be stranded in a position higher that that of the subject relative; but since the base position of the object is lower than that of the subject, it follows that the object relative has been stranded in a derived position. This is inconsistent with our approach to the ungrammaticality of (22a) above.

An alternative to the stranding analysis is proposed by Koster (1995), who analyses optional extraposition as an instance of "specifying coordination" akin to the coordination in (34b):

(34) a. *Zij heeft Jan en Peter gezien*
she has John and Peter seen
b. *Sij heeft Jan gezien en Peter*
she has John seen, and Peter
'She saw John and Peter.'

According to Koster, (34b) involves the coordination of a full VP to an elliptical VP in which all the material except the direct object undergoes deletion under parallelism:

(35) *Zij heeft* [[$_{VP}$ *Jan gezien*] [*en* [$_{VP}$ *Peter*]]]

The following example, with the second conjunct coordinated to the subject, must instead involve the coordination of two IPs:

(36) [[$_{IP}$ *Jan ging weg*] [*en* [$_{IP}$ *Peter*]]].
John went away and Peter
'Jan went away, and Peter too.'

In case of extraposition, the structure is similar but the head is an asyndetic specifying coordination (represented by a colon). This head mediates the relation between the NP "head" and the relative clause in (37a); in case of extraposition from an object (37b) it coordinates two VPs and in case of extraposition from a subject, it relates two IPs (37c).[24] Extraposition is thus conceived as an instance of ellipsis under coordination:

(37) a. *Hij heeft* [$_{DP}$ *de* [[$_{NP}$ *vrouw*] [: [$_{CP}$ *die alles wist*]]]] *gezien.*
he has the woman who everything knew seen
b. *Hij heeft* [[$_{VP}$ *de* [*vrouw*]] *gezien*] [: [$_{VP}$ [*die alles wist*]]]].
he has the woman seen who everything knew
'He saw the woman who knew everything.'

(37) c. [[$_{IP}$ [*De vrouw*] *heeft hem gezien*] [: [$_{IP}$ [*die alles wist*]]]].
The woman has him seen who everything knew
'The woman who knew everything saw him'

The coordination analysis has various advantages: it immediately derives the right roof constraint (namely, the fact that extraposition never crosses a clause boundary) and the impossibility of extraposition in the "middlefield", as in (22a) (since the elliptical conjunct containing the extraposed element is always to the right of the first conjunct); but above all, it reduces the nesting requirement to the different levels of coordination involved. In fact, "extraposition from an object" involves VP coordination, whereas "extraposition from a subject" involves IP coordination: thus the VP coordination is included in the first conjunct of the IP coordination.[25]

The analysis of extraposition as ellipsis under coordination can perhaps be combined with the raising analysis of relatives. From this perspective, the second VP conjunct in (37b) would contain a raising relative structure whose "head" is deleted along with the rest of the VP. Note however that this requires the deletion of a subconstituent in the Spec of the relative FP. For the moment, the feasibility of such an approach remains an open problem.

4. Summary

The preceding discussion is admittedly sketchy. The phenomena of coordination and extraposition are not limited to relative clauses; a comprehensive approach to them goes far beyond the narrow scope of this discussion. However, the data reviewed here seem to be consistent with the raising analysis of relatives (with the notable exception of the "nesting requirement").

Another phenomenon which appears problematic for the raising analysis is "heavy" pied piping. This phenomenon cannot be dealt with here for reasons of space; the reader may find the relevant discussion in Bianchi (1995: chapter VI), where a non-raising and a raising analysis are compared.

This chapter concludes our discussion of the major aspects of the syntax of headed relative clauses.

Chapter IX
Concluding remarks

This study explored the consequences of a very general theory within a restricted domain, that is, the syntax of headed relative clauses. As was repeatedly pointed out, various aspects of the analysis of this syntactic domain depend on more general consequences of the Antisymmetry theory, some of which have not been assessed yet (cf. for instance the locality problem discussed in chapter II). With respect to these, the conclusions reached here must be considered provisional, and open for further investigation.

In stating our hypotheses, we have always tried to carefully and explicitly state all the assumptions that they depended upon. Some of these assumptions may look problematic: for instance, the analysis of the agreeement relation between the external determiner and the relative "head" in Spec,CP assumes head government as a "checking" configuration, which is not allowed in Chomsky's (1993; 1995) theory. Moreover, some empirical problems were left open: in particular, the nesting requirement for extraposition (chapter VIII), and the proper analysis of heavy pied piping (see Bianchi 1995: chapter VI).

From a more general perspective, one problem that could not be addressed is the semantic interpretation of the proposed relative structures. In chapter III it was suggested that the relative DP is a non-definite phrase, which in restrictive relatives gets interpreted within the relative CP under reconstruction (perhaps along the lines of Reinhart's (1987) proposal for internally headed relatives). But the precise compositional interpretation of the raising structure remains undetermined, as well as the status of the appositive relative which moves out of the c-domain of the external determiner in LF (chapter V).

Alongside these open problems, however, it is also fair to point out that the raising analysis allows for a unified approach to various relative structures: these include headed relatives, internally headed relatives, correlatives, degree relatives with adjectival "heads", and perhaps also resumptive relatives. Moreover, it allows an original approach to various traditional problems, such as the syntactic difference between appositive and restrictive relatives, the deletion phenomena affecting the Comp of relative clauses, and such puzzling phenomena as Case attraction and "inverse attraction".

It is also worth pointing out that, apart from its relevance to the Antisymmetry theory, the raising analysis bears on some aspects of current research in the Principles and Parameters framework. The hypothesis of the external determiner fully brings out the potential semantic significance of the DP hypothesis (Abney 1987; Longobardi 1994, Stowell 1989), allowing the generalization that a determiner selects its restrictive term. This generalization admits of one exception: the appositive relative, though selected by the determiner, moves out of its c-command domain in LF. This LF process, and the consequent asymmetry between restrictive and appositive relatives with respect to the reconstruction of the relative "head", were shown to support Chomsky's (1993) view of reconstruction as a phenomenon intimately related to the quantificational structure of the sentence.

The discussion of the complementizer system of relative clauses confirmed the high specificity of the landing site of various types of A' movement; this issue is related to a general trend towards the elimination of "multi-purpose" syntactic positions (like, typically, the adjunct position). The approach to PF deletion of the relative determiner and of the complementizer rests on the hypothesis that functional heads are "seen" by the syntax as sets of features which can undergo operations in the course of the derivation. Furthermore, this approach leads to a specific conception of economy, whereby economy principles do not only select among the possible steps of a derivation (Chomsky 1995: 227-29), but they may also trigger specific syntactic processes.

As Kayne (1994: 132) emphasizes, the Linear Correspondence Axiom, by limiting the class of possible X-bar structures, also restricts *a priori* the possible analyses of many phenomena of natural language. If the raising analysis of relative clauses ultimately proves superior to the adjunct analysis, it will provide empirical support for the restrictiveness of the Antisymmetry theory.

Notes

Notes to Chapter I

1. On the development of X-bar theory from phrase structure grammars see Chomsky (1981: 31–32, 48–55; 1986a: 56–64, 80–84); Chomsky—Lasnik (1991: § 3.2).
2. The Kleene star in the right-hand side of rules (1)–(2) indicates the free iteration of a symbol any number of time (including zero).
3. As Gazdar et al. (1985: 44) emphasize, rewriting rules whose right-hand side is an unordered set of elements are not, strictly speaking, context free rules; in the GPSG framework, they are called immediate dominance rules.
4. The directionality parameters provide an explanation for the implicational universals of word order formulated by Greenberg (1963) (cf. Rizzi 1987). If the value of the Head-Complement parameter is consistent across categories, this may explain the correlation between the verb-objects order and the use of prepositions, and between the objects-verb order and the use of postposition (Greenberg's Universal 4).
   However, some languages seem not to be consistent with respect to the value of the directionality parameters. For instance, German and Latin are essentially prepositional, but they have a base object-verb order. It is possible to assume that the parameters' values are fixed with respect to specific categories; but the choice of the value for every category remains essentially arbitrary. Another logically possible analysis is to take one order as the basic one, reflecting the parameter value of the language, and to derive the inconsistent sequences by movement. Thus, for instance, German could be underlyingly a head-initial language, and the objects-verb order could be derived by obligatory leftward movement of the verbal objects (cf. Zwart 1994 on Dutch). This possibility cannot be excluded on theoretical grounds, and the choice of either analysis must be empirically justified against the other one.
5. Contrary to specifiers and complements, adjuncts seem not to be subject to a directionality parameter. In Italian, for instance, sentential adverbs are usually considered to be freely generated left or right-adjoined to IP:

   (1) [$_{IP}$ *Probabilmente* [$_{IP}$ *Gianni non verrà*]].
   Probably Gianni not will-come.
   'Probably, Gianni will not come'.

   (2) [$_{IP}$ [$_{IP}$ *Gianni non verrà*], *probabilmente*].
   Gianni not will-come probably.
   'Gianni will not come, probably'.

   It follows that the basic order of constituents is not completely and uniquely determined by directionality parameters.

On the other hand, Kayne (1991: 649) argues that adjunction to the head is always left-adjunction, hence it is not subject to parametric variation. Baker's (1988) Mirror Principle apparently presupposes that adjunction to a head is directionally consistent in any language.

6. Where the terminal symbols are read in the order in which they are reached.
7. Although movement of intermediate projections has been occasionally proposed, e.g. by van Riemsdijk (1989) in his analysis of the German "Split topicalization". Opinions also diverge on the possibility for the intermediate projection to be the target of adjunction.
8. Successive cyclic adjunction is exploited by Chomsky (1986b) in his approach to locality. On movement of adjoined heads see Roberts (1991).
9. The postulation of a fixed number of projection levels for every category is what Kornai—Pullum (1990) define the uniformity property. It is possible to define X-bar theories that do not meet this requirement; cf. for instance Fukui—Speas (1986) and Chomsky (1995: 245; 336).
10. The condition that the relevant node be branching is not assumed by all the authors.
11. Kayne's definitions are the following:
    Let a path P in a tree T be a sequence of nodes ($A_0 \ldots A_i, A_{i+1} \ldots A_n$) such that
    $\forall$ i,j $0 \leq i,j \leq n$ $A_i = A_j \rightarrow i = j$;
    $\forall$ i $0 \leq i < n$, A_i immediately dominates A_{i+1} or A_{i+1} immediately dominates A_i;
    $\forall$i $0 \leq i < n$
    if A_i immediately dominates A_{i+1} then A_i immediately dominates no node in T other than A_{i+1}, with the permissible exception of A_{i-1};
    if A_i is immediately dominated by A_{i+1}, then A_i is immediately dominated by no node in T other than A_{i+1}.
12. This fact is emphasized by Kayne (1984: 133).
13. Hoekstra (1991) derives binary branching for complements from the Uniqueness of Licensing Principle, which requires that the relation of selection between a head and a complement be biunique: thus, he argues that the binary branching constraint must not be formulated in X-bar theory. But his approach does not derive binary branching for specifiers. As mentioned in § 1.1.2, Hoekstra assumes that a head can enter into different agreement relations with various categories adjoined to its maximal projection, thus licensing multiple specifiers: in French complex inversion, for instance, the verb is incorporated to C, and it licenses the preverbal subject by Case assignment, and the fronted *wh* phrase by agreement for the feature [+wh] (cf. Rizzi—Roberts 1989):

    (1) [$_{CP}$ *Quel livre* [$_{CP}$ *Jean* [$_{CP}$ *a-t-il lu*]]]?
    which book Jean has-he read
    'Which book did Jean read?'

    In this representation the two adjoined phrases are not sisters, but they are separated by a segment. In this case, apparently, binary branching must be stipulated, or in any event, it does not derive from the Uniqueness of Licensing Principle.

14. Manzini (1992: 112) postulates that "each maximal projection is associated with at most one A' escape hatch (whether it is a Spec position or an adjoined one)". This uniqueness suggests that at least in the A' system the Spec/adjunct distinction is spurious.
15. The impossibility of multiple sister specifiers (as opposed to complements) is stipulated in the X-bar rules of Stowell (1981), and it appears to be presupposed by many authors.
16. The alternative analysis is rightward adjunction of the thematic adjective to N' or NP, but this type of movement is in general optional, and it predicts that the adjective should also follow the noun's complement.
17. The different position of the noun with respect to the color adjective in (14b)–(14c) is due to the absence of noun raising in English.
18. The constrast between (16f) and (16h) shows that there is a head position between the quantifier *tottu* and the adverb *bene* which the participle obligatorily incorporates to.
19. Cinque also proposes a label for every functional head in the participial structure and argues that each modifier is semantically related to the head whose specifier it fills. The nature of these heads, however, is not immediately relevant to the present argument.
20. The adjunct status of the topicalized constituent is associated to the hypothesis that the topic moves to its surface position, rather than being base-generated there and binding a null operator (as proposed by Chomsky 1977). The problem of base generation versus movement is not directly relevant to the present argument.
21. The same argument holds for English (cf. Lasnik—Saito 1992: 96), but not for Italian, which has expletive *pro* in Spec, IP and extracts the subject from the postverbal position.

    According to Lasnik—Saito (1992: 85–88), the subject trace in (1) below cannot be properly governed because adjunction of the topic to IP creates an extra node which qualifies as a barrier. The barrierhood induced by the topic is confirmed by the marginality of (2), which has the status of a weak island effect.

    (1) * *Who*$_i$ *do you think* [*that* [$_{IP}$ *this book*$_j$ [$_{IP}$ t_i *likes* t_j]]]?
    (2) ?? *the man* [$_{CP}$ *to whom*$_i$ [$_{IP}$ *that book*$_j$ [$_{IP}$ *I gave* t_j t_i]]].

    But this analysis does not carry over to French, where left-dislocated elements do not induce a weak island effect, as shown by the grammaticality of (3)–(5) (from Déprez 1990: 57–59):

    (3) *ces vieilles histoires qu'à ses enfants, Marie racontait souvent.*
    'those old stories that to her children Marie often told'
    (4) *un secret qu'à cet homme, Marie peut confier*
    'a secret that to that man Marie can confide'

(5) *Je me demande ce que pour cet homme, cette femme donnerait.*
I wonder that that for that man that woman would give
'I wonder what that woman would give for that man'.

22. Fanselow (1990) argues that scrambling has the properties of A movement, but analyses it as adjunction to IP. This is inconsistent with the common view that adjunct positions are intrinsically A'.
23. Independently of these structures, the possibility of adjunction was crucially exploited in Chomsky's (1986b) locality theory, where adjunction to a barrier was a way to neutralize barrierhood in A' movement. However, this aspect of Chomsky's theory has been extensively criticized by Cinque (1990: chapter I). Manzini (1994) also proposes a locality theory that does not have recourse to successive cyclic movement, and *a fortiori* it eliminates successive cyclic adjunction (cf. § II.5).
24. Kayne (1984: 152) assumes that the rule of *of* insertion can only apply when the noun governs NP.
25. Müller (1992) presents an alternative analysis of the German and English double object construction, based on a VP-shell structure with a head position empty in the base:

    (1) $[_{IP}$ subject I° $[_{VP1}$ α $[_{V}$ Ø $]$ $[_{VP2}$ β $[_{V'}$ δ $[_{V'}$ γ V$]]]]]$.

    In this structure, γ is the (recursive) base position of adjunct PPs; δ is the base position of the indirect object, and β, i.e. the Spec of VP2 is the base position of the direct object (as in Larson's analysis). Müller proposes that an indirect object can be Case-licensed in two ways: if it remains in its base position, a preposition must be inserted; however, it can also raise to α, the Spec of the higher VP-shell, where the structural Dative case is assigned without the insertion of any preposition. This yields in the IO-DO sequence the same structural relations as Larson's structure. However, the goal argument does not move to the position of the direct object, so that argument demotion is not required. In the English double object construction, Dative case is no longer morphologically distinct from the Accusative, but the derivation is essentially the same; the leftward movement of V to I obscures the fact that the argument closer to the verb is in fact structurally higher than the rightmost one.
26. Kayne (1984) did not take into account the structural asymmetry between the complements and proposed a headless small clause constituent $[_{S}$ NP PP$]$ in which the two complements are in fact sisters. However, this type of constituent is incompatible with current versions of X-bar theory, where the clausal constituent S is analysed as the projection of a head (see Chomsky 1986b: 3–4).
27. Recall that movement to a complement position is excluded in the Principle and Parameters framework.
28. Unless the indirect object in (47) is taken to overtly move to a position c-commanding the VP.

29. An exception is principle C effects with extraposed relative clauses (though not with clausal complements):

    (1) *I sent her$_i$ many gifts last year* [*that Mary$_i$ did not like*].
    (2) * *It bothered her$_i$* [*that Rosa$_i$ had failed*].

    Haider concludes that "there must be a non-structural reason which immunizes the relative clause against principle C violations". A contrast in reconstruction between relative clauses and clausal complements is discussed by Lebeaux (1991) (cf. § IV.8); however, his proposal does not account for the contrast in (1)–(2), where no recostruction is involved.
30. In case of extraposition from a DP, a rule of predication relates the extraposed phrase to the relevant DP. A different analysis is proposed by Kayne (1994: chapter 9); cf. § VIII.3.
31. The difference between VO and OV languages is determined by a directionality parameter on licensing: if a head licenses constituents to its left, it can remain in its base position (with an empty complement position); if it licenses to the right, it has to raise to a position c-commanding, and hence preceding, all the constituents to be licensed.

    (1) [WP [YP [ZP X°]]].
    (2) [X° [WP [t_X [YP [t_X ZP]]]]].

    In fact, the hierarchical relations between objects is the same in OV and VO languages: asymmetric c-command always correlates with linear precedence:

    (3) *that they will send everybody$_i$ a package to his$_i$ home address.*
    (4) *dass sie jedem$_i$ ein paket an seine$_i$ Privatadresse schicken werden.*
    that they to-everyone a pack to his private address send will.
    'that they will send everyone a pack to his private address'.

    This means that the German sentence (4) cannot be analysed as a case of regressive branching where the rightmost element c-commands the leftmost one.
32. However, the Branching Constraint does not *per se* determine other essential restrictions on X-bar theory, e.g. the binary branching requirement, which must be independently stipulated.
33. Although directional relations between nonterminal symbols have been proposed: e.g. Kayne's (1984: 168) canonical government, Koopman's (1984) directional Case/Theta assignment, and Haider's (1993) directional licensing.
34. Recall that the images are sets of ordered pairs <x,y> of terminal symbols.
35. A node α c-commands a node β iff neither α dominates β nor vice versa, and the first node dominating α dominates β as well.
36. This restricts the possible projections to the ones generated by rule (1); on specifiers and adjuncts will be discussed below.
37. X excludes Y iff no segment of X dominates Y (Chomsky 1986: 9).

38. R, S and T do not c-command either M or Q, since they are dominated by the whole category P which does not dominate M and Q.
If P is immediately c-commanded by a head B, the specifier M of P and B c-command each other, but B asymmetrically c-commands the head Q of the specifier (and its possible complement), hence linear order obtains.
39. This gives a principled solution to the problematic phenomenon of "specifier binding" discussed by Reinhart (1987); cf. Kayne (1994: 23–24).
40. Note that contrary to Chomsky (1986: 4), it is no longer necessary to consider two types of movement, substitution and adjunction. Movement is always to an adjoined position.
41. Another aspect of X-bar theory that derives from the LCA is the "structure preservation hypothesis": an adjoined category must match the projection level of the host category. (Cf. the stipulation above that in rule (3) α and β must be both heads or both maximal projections). Adjunction of a nonhead to a head, as in (1), yields a violation of antisymmetry:

(1)

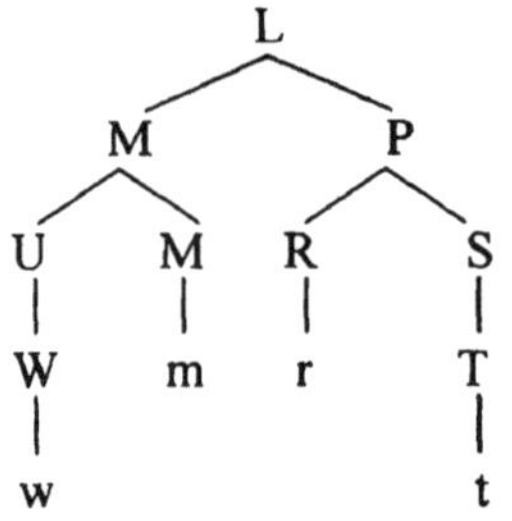

Here P asymmetrically c-commands W within the adjoined nonhead U; on the other hand, U asymmetrically c-commands all the categories dominated by P. Thus the terminal *w* cannot be ordered with respect to the terminals *r,t* dominated by P.
Consider then the adjunction of a head M to a nonhead P in (2):

(2)

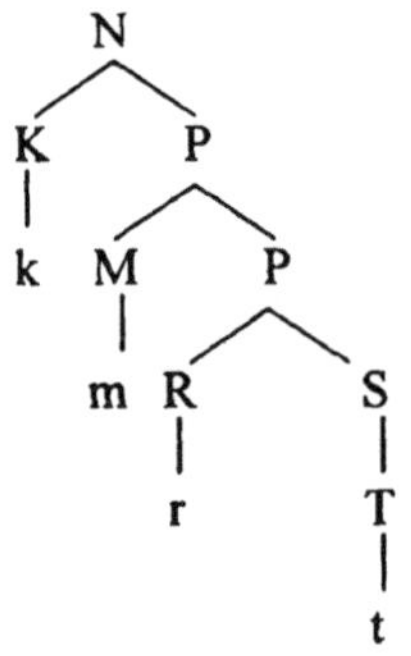

The adjoined head M asymmetrically c-commands all the constituents dominated by P; however, a problem arises with respect to the immediately c-commanding head K, which takes P as its complement. In fact, K c-

commands M, but M also c-commands K, since it is not dominated by the whole category P, but only by a segment of it, which is irrelevant for c-command as defined in (56). The same violation of antisymmetry arises if the subtree with root P is the specifier of a category which is the complement to a higher head.
This means that the adjunction of a head to a nonhead is only allowed if the nonhead is a root node, or the specifier (of the specifier...) of the root node. Kayne (1994: 31–32) excludes this possibility by stipulating that the highest element of a head chain must have a specifier. However, Kayne (1994: 36–37) postulates an abstract node that asymmetrically c-commands every other node in the tree; this abstract node is explicitly exempted from the condition of having a specifier.

42. And also the possible complement of Z.
43. Cf. fn. 41 above.
44. For a critical discussion of the latter, see Gaertner (1995).

Notes to Chapter II

1. Cf. for instance Safir (1986), Browning (1987) and Fabb (1990); see also § V.1.4. These differences are irrelevant to the present discussion.
2. The term "head" in enclosed between inverted commas in order to avoid confusion with the X-bar theoretic notion of head, namely the preterminal projecting element.
3. This is why the pronoun can be freely deleted in Comp without violating the recoverability condition. The deletion of the relative pronoun gives rise to relative clauses introduced by the declarative complementizer: cf. Chomsky—Lasnik (1977: 434–435; 446) on English, Kayne (1976) on French and Cinque (1982) on Italian. This aspect of the syntax of relative clauses will be examined in chapter VI.
Note that the mechanism of deletion is restricted to apply to recoverable elements in Comp, and it is blocked if the fronted phrase contains unrecoverable material, e.g. a pied piped preposition. With the development of the theory of empty categories, the deletion of the *wh*-pronoun is replaced by the generation of a phonetically empty null operator, which is subject to a strong binding requirement (Chomsky 1986a: 85; Chomsky 1991: 439–440): lacking any intrinsic semantic or grammatical specification, it is essentially interpreted as a free variable; but since free variables are excluded by the Full Interpretation principle, the null operator must have its reference fixed by an antecedent.
4. The term 'phi–features' refers to the grammatical features of gender, number and person. Cf. Chomsky (1981: 330).
5. Cf. Chomsky (1986b: 34–35).
6. A further consequence of the adjunct analysis is that relative clauses, not being selected, are invisible for the Projection Principle. It has been argued by some authors that this allows them to be inserted not in D-structure, but

in a later stage of the derivation (Safir 1986; Lebeaux 1991; Chomsky 1993: 36–37). This hypothesis will be discussed in § IV.8.

7. The earliest version of this hypothesis is due to P. Schachter according to Stockwell—Schachter—Partee (1973: 435–440). Cf. also § V.1.4.

8. Actually, in Jackendoff's (1977) three-level X-bar theory, appositive modifiers are sisters of the determiner under N''', whereas restrictive modifiers are daughters of N'' (Jackendoff 1977: 61–64; 169–173). Thus although there is a structural asymmetry between the two types of relative clause, it cannot be formulated in terms of the c-command domain of the determiner.

9. The complement of D° may be another functional projection rather than simply NP (cf. § I.2.1). This is irrelevant for the present argument.

10. In the sense of Higginbotham (1985: 561–62).

11. Although the syntax-semantic mapping is not so straightforward: cf. Stowell (1989) and Zamparelli (1995).

12. In the sense of Barwise—Cooper (1981: 182–184).

13. That is, the subtree rooted in the head's sister.

14. Or else, the relative clause is included within the determiner's complement, with a functional head F° intervenes between D° and CP:

    (1) $[_{DP}$ D° $[_{FP}$ F° CP]]

    In the lack of specific evidence, (16) constitutes the null hypothesis; but see also § III.2.3.

15. Similarly, in Siloni's (1991) analysis of Hebrew and French participial relatives D° acts as a sort of complementizer and selects a participial VP.

16. However, the external determiner of the relative structure (16) differs from the clausal determiner of (18)–(19) in that semantically it must be related to a nominal projection denoting a set of individuals, rather than to a clausal projection denoting a proposition: cf. § III.2.3.

17. Jackendoff (1977: 177–182) objects that the ability to license a determiner is not limited to relative clauses, but it is shared by other restrictive modifiers:

    (1) *the Paris of my youth*
    (2) *the three books of John's on the table*
    (3) *the usual four of the boys*

    Actually, Kayne's approach to relative clauses extends to other restrictive modifiers, so that the parallelism between (1)–(3) and (21) can be captured: see Kayne (1994: 97–105) and § 3.1 below.

18. Kayne's analysis differs from Schachter's and Vergnaud's analysis in that it does not involve the generation of an empty NP position outside the relative clause as the landing site of the "head".
    If a functional head intervenes between D° and CP (cf. fn. 14), its specifier can be the landing site of the "head":

    (1) $[_{DP}$ *the* $[_{FP}$ *book* $[_{FP}$ F° $[_{CP}$ *that I read* t]]]]

Since there is no clear evidence for the postulation of an abstract functional head, the null hypothesis (23) can be retained. But see again § III.2.3.

19. Longobardi (1994: 622) points out that in certain varieties of Italian proper names are introduced by a definite determiner, which in his analysis is an expletive. The definite determiner in (21a), however, is not expletive, since it binds a variable ranging over a set of abstract objects (stages of Paris?).

20. Kayne (1994: 97–05) extends this approach to reduced relatives, both participial and adjectival, taking them to be introduced by a null C°:

 (1) *the* $[_{CP}$ *book*$_i$ [C° $[_{IP}$ t_i *sent to me*]]]

 Prenominal reduced relatives are derived by raising to Spec,CP the complement of the inflectional head rather than the subject of predication:

 (2) *the* $[_{CP}$ [*recently sent*]$_i$ [C° $[_{IP}$ *book* [I° t_i]]]]

 This accounts for the fact that reduced relatives too are able to license a definite determiner in contexts like (21): cf. fn. 17 above.
 This analysis can also account for the following Bulgarian structure, where the enclitic definite determiner is attached to the participial head of a reduced relative (Giusti—Vulchanova 1994):

 (3) *kupeni-te vcera knigi*
 bought-the yesterday books
 'the books bought yesterday'

 This may be derived if the participial constituent raises to Spec,CP, as in (2), and its head then incorporates to the external D°:

 (4) $[_{DP}$ *kupeni*$_X$-*te* $[_{CP}$ $[_{XP}$ t_X *vcera*] $[_{IP}$ *knigi* $[_{IP}$ I° t_{XP}]]]]

 The assumption that the Spec of the relative CP is transparent for extraction is independently required: see below § 5.

21. An outstanding problem is the fact that the indefiniteness of the relative trace is not general. Both Carlson (1977: 524–527) and Heim (1987: 36–39) argue that the trace is indefinite only in *amount relative clauses*, whose "head" denotes a quantity rather than an individual or a set of individuals. For recent discussion see Grosu—Landman (1996). The problem is discussed in detail in § V.4.2.
 The presentational structure featuring a motion verb:

 (1) *There walked into the room a strange man.*

 is irrelevant for the present discussion, because it does not impose a requirement of indefiniteness on the indefinite DP, but rather a requirement of "heaviness" (Rochemont—Culicover 1990: 28–31).

22. The grammaticality of (26b) also implies that the raised "head" is not intrinsically definite. This will be argued in more detail in § III.3.1.
23. Cf. also § III.3.
24. According to Giusti, the numeral in (30b) is an adjectival category. Following Kayne (1994: chapter VIII) and Zamparelli (1994), it is instead possible to analyse the numeral as a head lower than D° (called Q° and Num° respectively by the two authors).
25. Cf. the following paradigm:

(1) ? *Non ho mai letto suoi libri.*
(I) have never read his books
'I've never read any book of his.'

(2) *Non ho mai letto* [$_{DP}$ *i* [*suoi* [$_{NumP}$ *due libri*]]].
(I) have never read the his two books
'I've never read his two books.'

(3) ? *Non ho mai letto* [$_{DP}$ *due* [*suoi libri*]].
(I) have never read two his books
'I've never read two books of his.'

(4) * *Non ho mai letto* [$_{DP}$ Ø [*suoi* [$_{NumP}$ *due libri*]]].
(I) have never read his two books
'I've never read two books of his.'

26. Sportiche (1988: 436) gives an example of floating quantification in an apparently restrictive relative:

(1) [*Les livres que j'ai lus presque tous* t] *sont de...*
the books that I have read almost all are of...
'The books I've read almost all belong to...'

27. On the possibility of floating quantifiers in appositive relatives cf. § V.4.3. The distributive quantifier *tutti/e* should not be confused with the non-distributive *tutto (intero)* 'whole': the latter can be stranded in a restrictive relative by the leftward movement of the "head":

(1) *la bistecca che ha mangiato tutta (intera) in un solo boccone*
the steak that (he) has eaten all (whole) in a single mouthful
'the steak that he ate in a single mouthful'

Contrary to the distributive Q°, this can take an indefinite DP complement:

(2) *Ho mangiato tutta una bistecca.*
(I) have eaten all a steak
'I ate a whole steak.'

Moreover, the floating quantifier *tutti* can quantify over an indefinite DP if a modifier forces a "collective" reading of the indefinite DP, as in (3); the corresponding relative clause is possible as well, as expected (4):

(3) *Ho letto cinque libri tutti in un solo giorno.*
(I) have read five books all in a single day
'I read five books in a single day.'

(4) ? *i libri che ho letto tutti in un solo giorno*
the books that (I) have read all in a single day
'the books that I read in a single day'

Also the floating quantifier *ciascuno* may quantify over an indefinite DP, and it is allowed within restrictive clauses:

(5) *Ho consegnato dieci libri ciascuno a uno studente diverso.*
(I) have delivered ten books each one to a student different
'I delivered ten books each to a different student.'

(6) *i libri che ho consegnato ciascuno a uno studente diverso*
the books that (I) have delivered each one to a different student
'the books that I delivered each to a different student'

28. Let us assume that the demonstrative pronoun (*quello/ce*) raises to Spec,CP and then incorporates to a phonetically null external D°.
With respect to the selection of CP, the bare quantifier *tout* contrasts with the bare negative quantifier *rien*:

(1) *Il ne dira rien qui puisse l'intéresser.* (Kayne 1975: 17)
he not will- say nothing that could him interest
'He won't say anything that may interest him.'

Recall that in Barwise—Cooper's (1981) terms, *rien* is weak and *tout* is positive strong. Bianchi (1992) argues that this semantic distinction corresponds to a different syntactic categorization, namely D° versus Q°. But this leaves unexplained the status of English *all*, a positive strong quantifier which, nevertheless, can select a relative clause:

(2) *All I want is you.*

29. Some examples of this structure are also found in Dante's literary Tuscan:

(1) *...e coronarmi allor di quelle foglie che la materia e tu mi farai degno* (Paradiso I, 26-27)
and to crown myself of those leaves that the subject and you me will-make worth
'and to crown myself with those leaves of which the subject ject and you will make me worth'

(2) ... *tant'era pien di sonno a quel punto che la verace via abbandonai* (Inferno I, 11-12)
so (I) was full of sleep at that point that the right way I abandoned
'so sleepy I was at that moment at which I lost the right way'

(3) *... di quel che udire e che parlar vi piace noi udiremo e parleremo a vui* (Inferno V)
about that that to hear and to speak to- you pleases, we shall-hear and speak to you
'we shall hear and speak to you about the subject about which you wish to hear and speak'

Other examples from Catalan and in XVII century French are quoted by Hirschbühler—Rivero (1981: 601–602).

30. A similar phenomenon is found in English free relatives:

(1) *I will live in whatever town you live.*

Larson (1987) proposes an analysis based on antecedent-contained deletion: the free relative raises in LF, stranding the preposition within the matrix VP (2). The residual PP is then copied in the relative clause (3):

(2) $[_{IP}$[*whatever town you live* t$]_i$ $[_{IP}$ *I will* $[_{VP}$ *live* $[_{PP}$ *in* t_i]]]]
(3) $[_{IP}$[*whatever town you live* $[_{PP}$ *in* t_i]] $[_{IP}$ *I will* $[_{VP}$ *live* $[_{PP}$ *in* t_i]]]]

However, Grosu (1996) argues convincingly that the visible preposition is internal to the free relative, and the free relative is adjoined to a null PP antecedent, a proposal similar to Cinque's, though restricted to free relatives.

31. Cf. also Chomsky (1981: 89–90).

32. This argument is criticized by Brody (1995: 135–137). Brody argues that in the idiomatic interpretation of *take pictures*, the idiomatic DP must contain a PRO subject coindexed with the subject of the idiomatic verb; this PRO obviously binds the anaphor (cf. the discussion in § IV.3). Therefore, the impossibility of the binding option in (42b) is not due to reconstruction, but to the intrisic control properties of the idiomatic expression.
The presence of a controlled PRO in the picture-DP is confirmed by the following principle B effect:

(1) * *John wondered* [*how many* [PRO_i *pictures of* us_i]] we_i *expected to take* t.

33. In the typology of features proposed by Chomsky (1995), the phi-features of the DP subject (apart from number) are intrinsic, being specified in the lexical entry of the noun head, whereas all the phi-features of the adjective are optional, namely, they are added to the lexical entry when the latter is included in a numeration of lexical elements. Furthermore, the phi-features of the DP are [+interpretable], being relevant for the LF interpretation, but those of the adjective are [-interpretable], and hence they must be checked in order for the derivation to converge.

34. Moreover, Spec,CP must be transparent in order to allow the noun of the "head" to incorporate to the clitic external D° in Rumanian. Cf. Kayne (1994: fn. 12 to chapter 8).
35. The transparency of Spec,CP is independently required for Kayne's (1994: chapter 9) analysis of relative clause extraposition, in which the relative clause is not moved to the right, but it is stranded in the base position by the leftward movement of the relative "head", which is then extracted from the relative CP.
36. A L-marks B iff A is a lexical category that theta-governs B;
A theta-governs B iff A is a zero level category that theta-marks B and A,B are sisters.
(Chomsky 1986b: 15)
37. Assuming instead Cinque's (1990: 24–30) condition of 0-subjacency for chain links, it is necessary to postulate that the relative "head" in Spec,CP is transparent, possibly adopting Lasnik—Saito's (1992: 100–105) proposal that A' moved phrases never qualify as barriers.
38. This will not affect the derivation of (46), where the first step to Spec,CP involves the whole relative "head", and the PP is only involved in the second step.
39. Actually, Manzini (1994) elaborates further this locality condition in order to take into account branching dependencies, like those of parasitic gaps. This aspect of her proposal is not immediately relevant to the present discussion.
40. For the time being, the "head" in Spec,CP has been provisionally labelled NP. The syntactic category of the "head" is discussed in detail in chapters III and VI.
41. Manzini analyses the dependency between these heads as related to the temporal interpretation of the clause: I° binds the event variable of V°, and C° serves as an anchoring point for the interpretation of the Tense. The exact motivation for the dependency does not affect the argument.
42. Williamson (1987) mentions, without developing the point, that in Lakhota questions the *wh*-phrase remains *in situ;* LF movement both in questions and in relative clauses freely violates Subjacency.
43. Parametric variation concerning the level of application of movement also plays a crucial role in Chomsky's (1993; 1995) system. According to Chomsky, the parametric variation is ultimately linked to the strength or weakness of the feature to be checked by movement. A feature checking approach to internally vs. externally headed relative clauses will not be developed here; it is problematic for languages like Quechua, which exhibit both internally and externally headed relative clauses (see (60) below).
Recently, Cole—Hermon (1994) have argued that the LF raising analysis is appropriate for internally headed relative clauses in Quechua, but not for *wh*-phrases *in situ.* For the latter, they assume the raising of a null operator in overt syntax, following Aoun—Li (1993). However, they exclude in a footnote the LF pied piping approach to *wh*-phrases *in situ* recently proposed by Moritz—Valois (1994), without actually discussing it.

44. Cf. also Ross (1967: 215–217).
45. According to Rivero (1980: 437–445), in the contemporary language the structure requires the insertion of the preposition *de* in front of the adjective:

    (1) *No puedes imaginar lo que dicen que son de inteligentes.*
    you cannot imagine the that (they) say that (they) are of intelligent
    'You cannot imagine how intelligent they say that they are.'

    Rivero analyses this structure as A' movement of a phonetically null degree adverb from the specifier of the adjectival phrase, parallel to the following interrogative example:

    (2) *¿ Como dices que es* [t *de inteligente*] *?*
    how (you) say that (you) are of intelligent?
    'how intelligent do you say you are?'

46. A raising analysis is also proposed by Rivero (1980: 450). It is also straightforwardly consistent with the analysis of degree relatives proposed by Grosu—Landman (1996), which requires that the "adjectival" head get interpreted within the relative CP.
47. The status of the comparative complements in (73) is beyond the scope of this discussion.
48. In Italian and French restrictive clauses with an adverbial "head" are possible with antecedent-contained deletion of the VP, as in the following examples:

    (1) *Il fit demi-tour et se sauva le plus vite qu'il put.*
    he made half-turn and himself saved the most fast that he could
    'He made half a turn and ran away the fastest that he could.'
    (Sandfeld 1965: 163)

    (2) *Se ne andò il più velocemente che poté.*
    (he) went-away the most fast that (he) could
    'He left the fastest that he could.'

49. For unclear reasons, Italian disallows the equivalent of (75).
    P. M. Bertinetto (personal communication) observes another occurrence of a definite article with a comparative in the English "correlative" structure (1). The Italian equivalent in (2) lacks the neuter article:

    (1) *The more you study, the more you learn.*
    (2) *Più studi, più impari.*
    more (you) study, more (you) learn

50. This hypothesis is supported by the observation that in a degree relative with a nominal "head", the neuter article actually agrees with the "head".

This follows from Chomsky's (1995) assumption that the phi-features of the noun are [+interpretable]:

(1) *No puedes imaginar los niños que se presentaron.*
you cannot imagine the children that presented themselves
'You cannot imagine how many children presented themselves.'
(Rivero 1980: 440)

51. R. Kayne (personal communication) points out that in the following example the head *troppi* semantically modifies the determiner *pochi* but agrees with the noun head for phi-features:

(1) *Ho letto troppi pochi libri.*
(I) have read too- MPL few- MPL books
'I read too few books.'

This agreement pattern is excluded when the adjectival phrase is in predicative, rather than attributive, position:

(2) *Questi libri sono troppo pochi.*
these books are too few- MPL
'These books are too few.'

(3) * *Questi libri sono troppi pochi.*
these books are too-MPL few-MPL

Suppose that the phi-features on *troppi* in (1) must be checked by the D° *pochi.* It is possible to assume that in (1) the phi-features of D° are [+interpretable], hence available for checking; in (2)–(3), instead, *pochi* is not a determiner but an adjective carrying [-interpretable] phi-features; these features are checked by the subject of predication and then eliminated, so that they are not available to check the features on *troppi* in (3).

52. Note that besides the order in (76b), in Italian and French the superlative adjective can also follow the noun phrase that it modifies:

(1) *la città più bella che io abbia mai visto*
the town most nice that I have-SUBJ ever seen

(2) l*a ville la plus belle que j'aie jamais vue*
the town the most nice that I have-SUBJ ever seen
'the nicest town I've ever seen'

In order to extend to (1)–(2) the analysis in (77), the NP "head" and the adjective must be assumed to be generated in a small clause configuration; after the small clause has raised to Spec,CP, the NP further moves to the Spec of superlative Deg° (or of the lower definite determiner in French):

(3) $[_{DP}$ *la* $[_{DegP}$ $[_{NP}$ *città*$]$ $[_{DegP}$ *più* $[_{CP}$ $[_{AgrP}$ t_{NP} *bella*$]$ $[_{CP}$ *che* ...$]]]]]$

Alternatively, one should give up the hypothesis that the superlative morpheme selects the relative clause, contrary to the discussion in § 2.1, and generate the whole superlative adjectival phrase within the raised "head". For the moment, the problem of (1)–(2) remains open.

Notes to Chapter III

1. This analysis partially resembles Cinque's proposal (§ II.3.6) in that the overt determiner preceding the relative "head" is an internal constituent of it. Hence, like Cinque's proposal, it is problematic with respect to the data discussed in §§ II.3.1- 3.5
2. The author does not explicitly state which syntactic relation links the antecedent and the relative TP. It is presumably adjunction, as in the standard analysis.
3. The data concerning idiom chunks are also confused. The idiomatic expression *make fun (of s.o.),* which exhibits a contrast parallel to (40) of chapter II, seems to be incompatible with the relative pronoun *which* in (3):

   (1) *You cannot imagine the fun that they made of me.*
   (2) * *They made of me fun that offended me.*
   (3) * *You cannot imagine the fun which they made of me.*

   At least for some speakers relative *which* retains one property of its interrogative counterpart, the incompatibility with mass nouns. This could independently rule out (3), where the idiom chunk has an abstract mass interpretation. However, Fabb (1990: 71) and Wilder (1991: 125) give some examples where a mass idiom chunk allows restrictive relativization by means of *which:*

   (1) *The headway which the students made last week was phenomenal*
   (2) *The advantage which they took of me last week...*
   (3) *The heed which they paid to us was insufficient.*

   It is unclear whether these idiom chunks exhibit the contrast in (40) of chapter II. All these factors must be carefully verified before any conclusion can be drawn from these data.
4. It is also possible to assume that the whole relative DP must move out of the base argument position, as in (9b), in order to create a predication variable that turns the IP category into a predicate (an open formula). The predicative IP is c-commanded by the external determiner, and hence it is included in the determiner's restrictive term.
   However, this hypothesis is problematic in two respects. First, the gap created by relativization does not qualify as a suitable predication variable in case of pied piping, e.g. in (11b), where it is of category PP (cf. the

discussion in Koster 1987: 81–85). Second, it will be argued in chapter IV that in restrictive relative clauses the relative DP is obligatorily reconstructed in the base position at LF (cf. also Kayne 1994: 87): this obligatory reconstruction would destroy the predication configuration at LF.

5. Alternatively, it can be assumed that the NP "head" must check some strong feature against a D° category, but the relative D° is unable to check it; this defectiveness of the relative D° forces the NP "head" to move out of its c-command domain in order to check its features against the external D°. Note that this would be inconsistent with Chomsky's (1995: 232) hypothesis that strong features only belong to non-substantive categories (A. Belletti, personal communication). This possible alternative does not affect the core of the problem: the movement of the NP "head" is triggered after the external D° has been merged with the CP structure.
6. In the simple DP structure $[_{DP}$ the $[_{NP}$ book]], the checking of the phi and categorial features of D° may exploit the option of covert feature attraction (Chomsky 1995: 261–276). This minimalist solution is instead unavailable in (9c) and (11c): there, one would have to assume that the NP category is overtly pied piped by the relevant attracted features to a position from which these features can be attracted by the external D°. Plainly, this would be a reformulation of the non-minimalist notion of head government.
7. Apart from the elimination of the Agr heads proposed in § 4.10.
8. Which also dominates an Agr_D. This detail is omitted in the representations in (17).
9. Furthermore, this proposal implies that the derivation of this type of relative clause is more economical than those in (9)/(11) (or (17)), in that it only involves one step, the raising of the NP "head".
10. Law (1991) points out that whereas interrogative *who* is intrinsically singular, relative *who* is underspecified for number, being compatible with a singular or plural "head".
    From a diachronic perspective, it must be noted that *which* was not restricted to [+human] antecedents until the XVII century. In Middle English headed relative clauses also used the forms *whose* and *whom*, specialized for [+human] antecedents, but for unclear reasons the nominative form *who* becam common only in the late XV century (Traugott 1972: 154; Lightfoot 1979: 320–321; 334).
11. Shlonsky (1991) shows that in Hebrew the universal quantifier *kol* is inflected for gender and number in floating quantification: he proposes that Q° agrees with its complement DP when the latter moves through Spec,QP:

    (1) [*Kol* [*ha-yeladim*]] *?ohavim le-saxek.*
    'All the children like to play.'
    (2) [*Ha-yeladim*]$_i$ *yašnu* $[_{QP}$ t_i $[_{QP}$ *kulam* t_i]]
    the children slept $all_{3m.sg.}$
    'The children all slept.'

12. This ambiguity is captured by the DP hypothesis, in which both determiners and pronouns are of category D°. A possible hypothesis is that determiners are "transitive" D°s, taking an NP complement, whereas pronouns are "intransitive" D°s. Otherwise, it is possible to assume that in the pronominal use the D° licenses a *pro* NP complement.
13. Although this analysis of the relative DP was originally proposed by Reinhart in connection with internally headed relative clauses (cf. § II.7) it can be straightforwardly extended to externally headed relative clauses in Kayne's analysis.
14. The Old English ancestor, the pronoun *hwa-hwæt*, was used both as an interrogative and as an indefinite pronoun (Campbell 1987: 292–294).
15. In Reinhart's analysis, the indefinite expresses a set variable, rather than an individual variable, and all the predicates contained in the sentence are predicated of this set variable. For instance, the relative clause in (1) would be interpreted as in (2), denoting a clausal singleton set, and the set variable would be bound by the external determiner as in (3).

    (1) $[_{DP}$ *the* $[_{CP}[_{DP}$ *boy* $[_{DP}$ *who* t]] $[_{IP}$ *I met* t]]]
    (2) X= {x | *boy*(x) & *I met*(x)} & |X| = 1
    (3) the X [X= {x| *boy*(x) & *I met*(x)} & |X| = 1]

    The idea that an indefinite denotes a set variable can be dispensed with by assuming a lattice-theoretical approach to noun denotations, as in Link (1983) or Ojeda (1990: 368–378). What is important here is that all the categories thematically related to, or predicated of, the indefinite DP be interpreted as conjoined predicates of the variable introduced by the indefinite DP.
16. This LF process is unavailable for the frozen determiner of the idiomatic expressions in § II.3.3, under the assumption that this determiner lacks any semantic content and the idiomatic NP has no variable position (cf. § II.3.3 and Bianchi 1993: 375–377).
17. Doron (1982) points out that in Hebrew relative clauses the presence of a resumptive pronoun imposes a *de re* reading of the relative phrase when the latter originates within an intensional context. This suggests that resumptive pronouns force a specific interpretation of the "head". On resumptive relatives see § VII.9 .
18. Note in passing that the grammaticality of (22b) and the ungrammaticality of (22d) confirm the hypothesis that the definiteness of the external determiner does not affect the indefiniteness of the relative DP (§ II.3.2).
19. Because of this semantic defectiveness, Horvath (1986: 36–44) suggests that the morpheme *a* is the residue of a complementizer incorporated to the relative pronoun, but without much motivation, as observed by Kenesei (1994: 298–299).
20. As discussed by Szamosi (1976) and Ingria (1990), when a main verb selects a tensed clausal complement it is in the definite conjugation; in this case, the extraction of an indefinite object from the clausal complement leads

to ungrammaticality. The problem is obviated if the main verb is in a form in which the definite and indefinite conjugations collapse.
The so called indefinite conjugation actually occurs with pronominal direct objects of first and second person, which are semantically definite. It may be argued that the category of definiteness that triggers the definite conjugation is a purely formal, morphological one; but note that the relative determiners in (22) are formally definite, yet they do not trigger the definite conjugation. The evidence quoted in footnote 22 also suggests that the two conjugations are sensitive to a semantic property of the direct object (specificity). The problem of first and second person pronouns remains open.

21 On the contrary, Marcantonio (1985: 291) states that *amely* is indefinite like *ami* and *aki*, but she does not give an example with *amely* triggering the indefinite conjugation.

22. Marcantonio (1985) proposes that the definite conjugation developed from a marking of the topicality of the direct object: in earlier stages of Hungarian the definite conjugation appeared not only with formally definite direct objects, but also with "pragmatically definite" objects, that is, either specifics or generics. In those stages relative pronouns could trigger both conjugations; Marcantonio assumes that the possibility of the definite conjugation was related to the intrinsic topicality of relative pronouns, but she does not take into account an important factor, the restrictive vs. appositive interpretation.
She also reports that in the colloquial language the two conjugations sometimes appear to be sensitive to a semantic category of definiteness, rather than a formal one. In the following examples, the definite conjugation on the main verb gives a specific interpretation of the object free relative (1), whereas the indefinite conjugation gives a generic interpretation:

(1) *Azt csinálom, amit akarok.*
this$_{ACC}$ (I) do$_{DEF}$ what$_{ACC}$ (I) want.
'I'm doing exactly what I want to do.'

(2) *Azt csinálok, amit akarok*
this$_{ACC}$ (I) do$_{INDEF}$ what$_{ACC}$ (I) want.
'In general, I do what I want to do.'

In the following pair, the direct object is a proper name:

(3) *J. Attilát olvasom.*
J. Attila$_{ACC}$ (I)read$_{DEF}$

(4) *J. Attilát olvasok.* (Marcantonio 1985: 294)
J. Attila$_{ACC}$ (I)read$_{INDEF}$

(3) means 'I am reading something specific written by J. Attila', whereas (4) means 'I am reading something not specified from J. Attila's works'.
These data support the hypothesis that the two conjugations are sensitive to the feature [specific], rather than [definite].

23. Albanian thus differs from languages like Modern Greek, Arabic and Hebrew, which have only a definite form of the adjectival article, morphologically identical to the definite determiner.
24. The only exception is the dative feminine singular, which has the definite form *së* rather than the expected *të*. The opposition of definiteness is neutralized for all other dative forms, since indefinite *të* has replaced the original *së* in the definite paradigm for all forms except the feminine singular. The full paradigms are the following (Morgan 1972: 66):

Adjectival articles

	NOM		ACC		DAT	
	def	indef	def	indef	def	indef
MSG	*i*	*i*	*e*	*të*	*të*	*të*
FSG	*e*	*e*	*e*	*të*	*së*	*të*
MPL	*e*	*të*	*e*	*të*	*të*	*të*
FPL	*e*	*të*	*e*	*të*	*të*	*të*

Relative pronouns

	NOM	ACC	DAT
MSG	*i cili*	*të cilin*	*të cilit*
FSG	*e cila*	*të cilën*	*së cilës*
MPL	*të cilët*	*të cilët*	*të cilëvet*
FPL	*të cilat*	*të cilat*	*të cilavet*

25. This is because adjectival agreement must be locally licensed under m-command by the noun, and it cannot be licensed by the noun's trace if the latter raises; the adjective ia taken to occur in a NP-internal position (Santelmann 1993: 164–165). Kester (1993: 147–151) proposes a different analysis for the idea that in (26a) the prenominal adjective prevents the noun from raising to D°.
26. The dummy preposition can also be realized:

(1) *el señor a quien María entrevistó ayer* (Toribio 1992: 292)
the gentleman *a* whom Maria interviewed yesterday

In (1) the relative determiner cannot be omitted, as is the case in pied piping in general: cf. § VI.2..
Toribio (1992), from whom the examples are drawn, does not discuss any semantic difference between (30) of the text and (1) with respect to existentiality. However, in the Porteño Spanish variety spoken in Buenos Aires the relative determiner *(a) quien* is incompatible with clitic doubling; since clitic doubling is only possible with [+specific] direct objects, this

incompatiblity suggests that the relative DP is not specific (Suñer 1988: 418–419).

27. The term "coreferent" is inaccurate: at least for Hindi correlatives, Srivastav Dayal (1991; 1995) and Grosu—Landman (1996) propose that the relation involves quantification (cf. § 4.1.1)
28. On the morphological shape of these elements see below § 5.1.
29. Note that the NP "head" and the relative D° bear two morphologically distinct Cases. This problem will be discussed in § 4.4.
30. A third type of relative clause is derived from the embedded type by extraposition:

    (1) *vo larki lambi hai jo khari hai*
    that girl tall is who standing is
    'That girl who is standing is tall'

    This type is identical to the embedded relative with respect to all the syntactic properties discussed by Srivastav.
31. Apart from relative clause extraposition, mentioned in the preceding note.
32. Actually, Srivastav (1991: 661–662) proposes a type shifting operation: the dependent CP denotes the unique individual that satisfies the predicates expressed by the NP "head" and by IP, which contains a trace; type shifting turns the dependent CP into a generalized quantifier, the set of all the sets containing this unique individual. The details of Srivastav's semantic analysis are not directly relevant for the present discussion.
    Similar to the correlative structure are the following (left-dislocated?) free relatives, which explicitly involve universal quantification (cf. Srivastav 1991: 680–782):

    (1) [*Qualunque cosa tu voglia*], *te la darò.*
    whichever thing you want, (I) to - you it will give
    (2) [*Whatever you want*], *I will give it to you.*
33. Cf. Reinhart (1987) on the pair reading in donkey anaphora.
34. In both structures, the DP headed by the relative determiner raises to Spec,CP, though for different reasons. The raising of the "head" in (39) is obligatorily overt, being triggered by the strong features of the external D° (§ 2.1), whereas it can be either overt or covert in the correlative structure.
35. It must be pointed out that Haudry's proposal concerns not only relative clauses, but also many other types of subordinate clauses, like temporal, causal and concessive adjuncts. The diachronic development proposed by Haudry (1973) for Latin also straightforwardly extends to Old English (cf. §§ 5.1 below and VII.2).
36. The hypothesis that the Case morphology corresponds to the functional head D° is defended by Giusti (1993). Cf. also § 4.4 below for a more detailed discussion of the Case morphology of the relative determiner.
37. Recently, Pittner (1995: 218–221) has proposed a different diachronic evolution from the correlative structure to the headed relative clause in Old

and Middle High German. The evolution starts from an "inverse" correlative dyptic, with a cataphoric correlative pronoun in the main clause pointing forward to a following asyndetic relative clause:

(1) [NP_i correl [pro_i ...

In this stage, the correlative pronoun obviously bears the Case required by the matrix clause. At the next stage the correlative pronoun is reanalysed as a relative pronoun in the Comp of the relative clause, though it still bears the Case required by the matrix:

(2) NP_i [(cor)rel_i ...]

Finally, the reanalysed pronoun takes the Case assigned within the relative clause and becomes a well-behaved relative pronoun:

(3) NP_i [rel_i ...

The reanalysis of the correlative pronoun as a relative pronoun in Spec,CP is motivated by a condition of C-visibility holding in New High German, whereby the Comp projection must be filled by a lexical element endowed with a [+C] feature. In stage (3), this function is accomplished by the relative pronoun and accordingly, the old invariable relative particles are eliminated. When fulfilling the C-visibility condition, the relative pronoun can not be attracted to the Case required by the matrix clause, because it must be clearly identified as belonging to the relative CP. This accounts for the lack of "Case attraction" of the relative pronoun in New High German.
Though the correlation between the C-visibility requirement and the elimination of Case attraction is interesting, the proposed development appears problematic. First, note that it requires a radical reanalysis of the structure, symmetric to that in (45): here a constituent of the matrix clause (the correlative pronoun) must be reanalysed as belonging to the relative CP. Second, the stage (2) of the derivation is problematic, as Pittner herself acknowledges: the reanalysed pronoun in Spec,CP must still be assigned the matrix case "by the same rules" which determined Case marking in stage (1). This Case attraction of the relative pronoun is much less problematic from the perspective of the raising analysis: cf. below § 4.4.

38. Kiparsky (1995: 154–155) also argues that inverse attraction is an instance of the correlative structure, with optional omission of the correlative pronoun in the main clause; he suggests that the inversion of the relative determiner and the "head" may have arisen from the movement of the latter to a topic position, stranding the relative determiner in a lower focus position.
39. Cf. Giusti (1993) for the claim that demonstratives originate in a specifier position below D°.
40. This hypothesis is discussed in detail in § VI.2.3.
41. Or whose minimal domain it is included in (cf. (51) of chapter II).
This hypothesis is supported by the observation that in languages like e.g. Russian, the Case assigned to a DP can be realized on the determiner only,

while NP realizes the Genitive Case assigned by the determiner itself under government (cf. Babby 1987, Franks 1994):

(1) *poslednie* *pjat'* *butylok*
last$_{NOM}$ five$_{NOM}$ bottles$_{GEN}$
'the last five bottles' (Russian; Babby 1987: 92)

42. Or, alternatively, the relative D° agrees with the NP "head" in its Spec, which has received the Case feature from the external D°.
43. Though not in an IP-internal position, as discussed in § 2.2.
44. Cf. Harbert (1982; 1983) for a discussion of the attraction phenomena from the perspective of the adjunct analysis.
45. In the Old English example the relative and correlative elements are derived from the same pronominal stem, but the dependent clause is marked by doubling the relative element *ða* and by the lack of Verb Second (cf. Stockwell—Minkova 1991: 369–373, and § VII.2).
46. Cf. Hock (1988) and Pittner (1995) for somewhat different proposals which, however, maintain the idea that the headed relative derived from an original correlative structure.
47. In many cases the adjectival article only appears with modifiers embedded in a definite DP:e.g. Ancient and Modern Greek, Arabic, Hebrew, Rumanian. In the first three languages the definite adjectival article is omophonous to the definite article.
In Old Slavic languages the definite adjectival article, a descendant of the Indo-European pronominal stem **yo-*, was incorporated to the adjective giving rise to the "long" adjectival declension; a similar evolution took place in Old Germanic languages (Haudry 1981).
48. However, Shlonsky (1992) analyses this morpheme as an inflected complementizer rather than a relative pronoun.
49. "Il ne peut plus être question ... de voir un développement secondaire dans l'emploi du pronom comme déterminant du nom ou d'adjectif. Bien plutôt c'est là, à l'origine même, sa fonction propre, dont l'emploi comme "pronom relatif" n'est qu'une extension à la phrase verbale. Dans les deux cas le rôle du pronom est le même, celui d'un déterminant, qu'il soit déterminant d'un terme nominal où d'une phrase complète." (Benveniste 1966: 221).
50. The relevance of the referential force of the * *yo-* morpheme is also discussed by Longobardi (1980), in a criticism of Benveniste's paper.
51. Note that Haudry's (1973) proposal does not account for the non-correlative use of the pronouns with adnominal modifiers discussed by Benveniste and Brunel. The indefinite theme **kwo-* also had this function in Latin:

(1) *Salvete, Athenae, quae nutrices Graeciae!*
Hail, Athens, which foster-mother of Greece!
'Hail, Athens, foster-mother of Greece!'
(Benveniste 1996: 217–220)

This is unexpected, given the hypothesis that the theme was indefinite and not anaphoric. Therefore, Haudry (1973: 180) is forced to conclude that in the adnominal use, the theme * *kwo-* was not original, but it replaced an original anaphoric pronoun.

52. Another case would be *el* in Spanish:

(1) *el tren con el que vas viajar*
the train with the that you are going to travel
'The train by which you are going to travel'

assuming with Donati (1994) and Brucart (1992) that *que* is a declarative complementizer and not a part of a complex relative pronoun *el que*.

53. Kunstmann (1991: 662) quotes the following late Latin example, where the relative determiner precedes the NP "head":

(1) *Qualis locus... solet a Graecis Cherronesos vocari*
which place... is used by Greeks Cherronesos be called
the place which is called Cherronesos by Greeks

According to Palermo (1994: 169–192), the form *il quale* originated in the literary language, then it was adopted in officialese and finally in the colloquial language. However, in the colloquial language the original form *quale* survived for a long period; for instance, it is still attested in some XVII century non-literary texts reported by Petrucci (1982).

54. R. Kayne (personal communication) suggests another possible implication of the hypothesis that in the type *il quale* the definite article is expletive: the non-existence of * *the which* in Modern English may follow from the hypothesis, independently proposed by Longobardi (1994: 653–654) and Vergnaud—Zubizarreta (1992: 634–638), that in English the definite determiner cannot function as an expletive with proper names, in generic DPs, and in constructions expressing inalienable possession (cf. *John raised his/*the hand* versus *Gianni ha alzato la mano*).

Notes to Chapter IV

1. Vergnaud (1974: 227) points out the following contrast in reconstruction:

(1) * *The few attacks against Nixon$_i$ that he$_i$ deplored in his last press conference were actually vicious.*

(2) *The girl that Bill$_i$ had been dating that he$_i$ finally married was Italian.*

The restrictive relative in (1) shows the expected Principle C effect caused by the reconstruction of the "head". But in (2), the stacked relatives do not show a Principle C effect. This is not predicted if the "head" modified by the innermost relative clause is raised from the complement position of the outermost relative clause (cf. chapter VIII):

(3) $[_{DP}$ *the* $[_{CP}[$*girl that Bill*$_i$ *had been dating*$]_j$ $[_{CP}$ *that he*$_i$ *finally married* $t_j]]]$...

If this is the structure of (2), the innermost relative clause should be reconstructed along with the "head", and as a result, the R-expression *Bill* should be c-commanded by the coindexed pronoun *he*. But the expected Principle C effect is not observed.
An alternative possibility is to take the rightmost relative clause to be stranded within the leftmost relative clause by a process of extraposition (cf. Kayne 1994: 117–126 and § VIII.3):

(4) $[_{DP}$ *the* $[_{CP}$ *girl*$_j$ $[_{CP}$ *that Bill*$_i$ *had been dating* $[_{DP}$ D° $[_{CP}$ t_j $[_{CP}$ *that he*$_i$ *finally married* $t_j]]]]]$ ...

In this analysis, the leftmost relative clause cannot be reconstructed within the rightmost one, and no Principle C violation is predicted. The analysis in (4) also predicts that, like extraposed (i.e. stranded) relative clauses, the rightmost relative in a stacking structure disallows complementizer deletion (see Kayne 1994: 92 for a different analysis).

2. The observation that the relative operator is not a restricted quantifier is also made by Lasnik—Stowell (1991: 706–707) in the context of the adjunct analysis.

3. Belletti—Rizzi (1988) demonstrated that the psych verb *piacere* takes the Theme argument as a D-structure complement and the Experiencer argument as an indirect object. Therefore, in (5a) the Theme "head" is reconstructed in a position c-commanded by the dative argument.

4. In fact, Cardinaletti (1990: 46–50) points out that in a null subject language there is an asymmetry between lexical and null pronouns: strong pronouns are only licensed in contexts that exclude deficient pronouns (cf. Cardinaletti—Starke's principle Minimize Structure). Among these contexts are focalization, coordination and modification. If this reasoning is correct, then the strong pronouns in (8) should be licensed by some sort of focalization or emphasis.
The hypothesis that in null subject languages strong subject pronouns are emphatic is also put forward by Rigau (1988: 504). According to Rigau, these pronouns do not occupy Spec,IP, but rather an A' position, binding a *pro* in Spec,IP. This explains why strong pronouns do not allow a resumptive use and cannot act as quantifier-bound variables.

5. English speakers' judgements about Principle C effects under reconstruction of the relative "head" are not systematic. This may be due to the fact that in English the subject pronoun is always overt and allows an optional emphatic reading.

6. E.g.:

   (1) * *il tema di ogni studente*$_i$ *che lui*$_i$ *ha consegnato ieri...*
   the essay by every student that he gave in yesterday

7. Of course, emphatic pronoun can also behave as normal pronouns subject to Principle B: cf. the following example:

   (1) *Gianni*$_i$ *ha detto* [*che non hanno invitato neanche lui*$_i$].
   Gianni said that (they) didn't invite even him
   'Gianni said that they didn't invite him either.'

8. Reinhart (1983: 164–170) replaces Principle C with the rule of Intrasentential Coreference:
NP α cannot corefer with NP β if replacing α with γ, γ a variable A-bound by α, yields an undistinguishable interpretation.
For instance, (10a) is excluded because the R-expression *Gianni* within the direct object can be replaced by a possessive pronoun, which gets bound by the emphatic subject pronoun:

   (1) *Non comprerebbe neanche lui*$_i$ [*il suo*$_i$ *quadro*].
   not would buy even he his picture
   'Not even he would buy his picture.'

But the same holds of the LF representation of (10b), in which the relativized object is by hypothesis reconstructed in its base position. Therefore, if Reinhart's principle applies to the output of LF, it cannot draw a distinction between (10a) and (10b).
Actually, it is not obvious that in (10a) binding and accidental coreference give two undistinguishable interpretations. Consider the predicates corresponding to the binding option in (2) and the one corresponding to accidental coreference in (3):

   (2) λx (x *would buy x's picture*)
   (3) λx (x *would buy Gianni's picture*)

Focalization on the subject (*neanche lui* 'not even he') gives the presupposition that the negation of the predicate expressed by the VP applies to other people in addition to the subject. But then the two predicates in (2)–(3) give different presuppositions, and hence the two options do not give an equivalent interpretation. In this case Reinhart's rule would not block accidental coreference, and (10a) is predicted to allow the coreferential interpretation.

9. The notion of syntactic predicate is defined as follows (Reinhart—Reuland 1993: 678):

The syntactic predicate formed of a head P is P, all its syntactic arguments, and an external argument of P (subject).

10. Another type of logophoric anaphor, discussed by Reinhart—Reuland (1993: 672–673), is the so called focus anaphor:

    (1) *Bismark$_i$'s impulsiveness has, as so often, rebounded against himself$_i$.*

    Here the anaphor appears in the argument position of a syntactic predicate, but in LF it undergoes Focus movement to an A' position, whereby it loses the ability to reflexive-mark the predicate and it is interpreted logophorically. This logophoric use too disallows true anaphors in Italian:

    (2) *L'impulsività di Bismark$_i$ si è rivolta, come spesso accade, contro lui stesso$_i$/ * se stesso$_i$.*

11. Reinhart—Reuland (1993: 682–683) note that in English the logophoric use of an anaphor should be impossible if the NP has an overt subject. This is because an NP with an overt subject qualifies as a syntactic predicate, and therefore an anaphor contained in it must be used reflexively. But the grammaticality judgement is uncertain:

    (1) */? *Lucie$_i$ liked [your picture of herself$_i$].*

    In Italian, however, true anaphors and logophoric elements are morphologically distinct, and the judgement is quite clear: the logophor is marginally acceptable, but the anaphor is ruled out (as any formulation of Principle A predicts):

    (2) *Lucia$_i$ apprezza molto [il tuo ritratto di *se stessa$_i$/ ?lei stessa$_i$].*

    The marginal acceptability of the logophor in (2) is not predicted by Reinhart—Reuland's approach, since the NP qualifies as a syntactic predicate.

12. Reinhart—Reuland (1991: 316) acknowledge the fact that the logophoric use of anaphors is not universal as their analysis seems to imply: "... the logophoric use of anaphors [...] is possible as long as no binding rule is violated and an antecedent can be found. Nevertheless, whether a language allows its anaphors to be used logophorically is still subject to variation [...] we are not able, at present, to explain these variations".

13. Belletti—Rizzi (1988: 312–324) argue that both long-distance and short-distance anaphors must be c-commanded by their antecedent. Therefore, besides *self*-anaphors, also the anaphors *sé* and *proprio*, which have both a

short-distance and a long-distance use (Giorgi 1984), will appear in the reconstruction examples below.

14. This point is extensively discussed by Giorgi—Longobardi (1990: chapter IV).
15. The same point is made by Giorgi (1984: 325).
16. Of course (22), (23c) and (24c) are acceptable in the reading in which the anaphor is bound by the matrix subject.
17. Consider also the following contrasts:

(1) *L'[immagine di se stesso*$_i$*] che Gianni*$_i$ *cerca di trasmettere* t *agli altri è molto lontana dalla sua realtà interiore.*
the image of himself that Gianni tries to convey to other people is very different from his inward reality

(2) ?* *Un'immagine positiva di se stesso*$_i$*, che Gianni*$_i$ *cerca di trasmettere* t *agli altri, è necessaria al suo equilibrio interiore.*
a positive image of himself, that Gianni tries to convey to other people, is necessary to his serenity

(3) *L'unica [fotografia di se stesso*$_i$*] di cui Gianni*$_i$ *ha autorizzato la pubblicazione* t *è di grande valore.*
the only picture of himself of which Gianni authorized the publication is of great value

(4) * *L'unica fotografia di se stesso*$_i$*, di cui Gianni*$_i$ *ha finalmente autorizzato la pubblicazione* t, *è di grande valore.*
the only (existing) picture of himself, of which Gianni finally authorized the publication, is of great value

Although the nouns *immagine* and *fotografia* may in principle license an implicit possessor, anaphor binding is allowed only in the restrictive relatives (1) and (3), and not in the appositives (2) and (4). This suggests that the possibility of binding the anaphor does not depend on the implicit possessor, which should be equally present or absent in both structures; it requires the reconstruction of the "head".

18. See Kayne (1994: 112–113) for a different conclusion.
19. Even if we assume that certain quantified phrases may enlarge their scope by moving in LF, this possibility is irrelevant for the following examples, since the quantified phrase is always embedded in a strong island, the relative CP, and it cannot move to a position that c-commands Spec,CP. (See Longobardi 1991 for the claim that Subjacency holds in LF).
20. The issue of relativization of predicate nominals was discussed in § II.4.2.
21. A further problem with the English examples in (31) is that the relative pronoun *which*, obligatory in appositive clauses, seems to be incompatible with an amount "head" (cf. fn. 3 to chapter III; this restriction is explicitly stated by Carlson 1977: 526). But many idiom chunks have abstract mass nouns as heads: see for instance (31a,b,d). It is unclear to what extent this

factor may independently rule out the relativization of these idioms in appositive clauses.

22. Cf. Postal (1994) for a proposal along these lines.

23. This dependency cannot be completely assimilated to the null resumptive pronoun strategy proposed by Cinque (1990: chapter III), because appositive relatives exhibit only a subset of the diagnostic properties identified by Cinque for resumptive *pro*. The A' structures studied by Cinque are parasitic gaps, complement object deletion constructions, and apparent extractions from islands. Appositive relatives share with these the following properties: the antecedent of the gap cannot be nonreferential, nor idiomatic; it cannot be reconstructed in the gap position; the dependency does not show Weak Crossover effects. However, appositive clauses do not share two other properties of these structures: the fact that the gap is necessarily of category DP, and the impossibility of a successive cyclic derivation.

24. According to Chomsky (1993: 28), adjunction differs from substitution in that it can be countercyclic. Therefore, the adjunction of the relative clause to the DP in Spec,CP in (37b) is licit. However, countercyclic adjunction seems to be incompatible with Chomsky's (1995: 248) stricter hypothesis that the operation Merge always applies at the root.

25. The hypothesis that relative clauses are invisible for the Projection Principle is also adopted by Brody (1995: chapter 5).

26. Also Watanabe (1995) reports the following pair of examples, which he attributes to H. Lasnik:

    (1) [*Which piece of evidence* [*that John*$_i$ *was asleep*]] *was he*$_i$ *willing to discuss* t?
    (2) [*Which piece of evidence* [*that John*$_i$ *discovered*]] *was he*$_i$ *willing to discuss* t?

    The coreference between *John* and the matrix subject pronoun is possible both in (1), featuring a noun complement clause, and in (2), featuring a relative clause.

    Moreover, Watanabe points out that the countercyclic insertion of the relative clause in Spec,CP, required by Lebeaux's analysis, is inconsistent with the "preference system" of Chomsky (1993), which requires the minimization of the restriction in the operator position (i.e., it requires reconstruction of the restrictive term whenever possible). The option of cyclically inserting the relative clause in the base position of the *wh*-phrase should be preferred, giving rise to reconstruction effects.

Notes to Chapter V

1. Emonds accepts Ross's (1967: 239–241) and Jackendoff's (1977: 198–199) argument that the underlying conjunction cannot be *and*, because in that

case an appositive clause embedded in an imperative main clause (1) should be derived from an ungrammatical coordination (2):

(1) *Go to Cincinnati, which is on the Ohio River.*
(2) * *Go to Cincinnati, and it is on the Ohio River.*

Emonds proposes that (1) is instead derived from an underlying structure containing a phonetically null conjunction:

(3) *Go to Cincinnati; it is on the Ohio River.*

2. This is different from the deep structure conjunction analysis of restrictive relatives discussed by Stockwell—Schachter—Partee (1973: 440–441): in the latter, the relative clause is not coordinated to the matrix clause, but rather it expresses a predicate that is conjoined to the predicate expressed by the NP "head" and to the predicate of the matrix clause.
3. This recalls the proposal by Lebeaux discussed in § IV.8. The problem of countercyclic merging of the relative clause with the "head" arises in Safir's analysis as well. Moreover, given the independence of the PF and LF branches, it is unclear how the appositive clause attached in LF' can be inserted in the clause containing its antecedent in the phonetic representation.
4. See however Safir (1986: 673) for some counterexamples.
5. A true quantifier phrase is composed of a quantifier Q and a nominal term T defining a range R that Q quantifies over, such that R is a possibly nonsingleton set (Lasnik—Stowell 1991: 704).
6. One problem for Lasnik—Stowell's approach is that in appositive relatives a pied piped phrase, contrary to the "head", gives rise to a weak crossover effect (Kayne 1984: fn. 25 to chapter 8; Safir 1986: 669):

(1) ?* *John*, [*a sister of whom*]$_i$ *her*$_i$ *child loved* t_i, ...

In Safir's analysis, this follows from the fact that the coindexing between the genitive pronoun and the trace cannot be deferred to LF', and hence it violates the PCOB in LF. Lasnik—Stowell's approach does not predict the ungrammaticality of (1), since the pied piped DP is not a quantifier phrase. For a criticism of Lasnik—Stowell's proposal cf. also Postal (1993).
7. The sentence may be excluded by Reinhart's rule of intra-sentential coreference (cf. fn. 8 to chapter IV), if the latter applies at the LF' level.
8. This aspect of the adjunct analysis is reviewed in § VI.1.
9. Cf. Browning's (1987) analysis mentioned in § II.3.2. More recently, Stroik (1994: 46) has argued that restrictive relatives are adjoined to a functional projection NumP lying between D° and NP.
10. On the stacking order see Kayne (1994: 113–114) and § VIII.1.
Another property that sharply distinguishes appositives from restrictive relatives is the impossibility of stacking:

(1) *People who go to MIT who like math will get a job.*
(2) ** John, who goes to MIT, who likes math, will get a job.*

This problem too will be discussed in § VIII.1.

11. In English, appositives also differ from restrictives with respect to the distribution of relative pronouns. This aspect of their syntax will be discussed in detail in chapter VI.
12. Actually, Milsark (1977) shows that the existential construction excludes not only definite DPs, but also indefinite specific ones. In other terms, it excludes strong DPs in the sense of Barwise—Cooper (1981): see § 4 for discussion.
13. Actually, an AP can also be the "head" of a "degree" relative (cf. § II.7.2).
14. Another possibility is to take the DP "head" to be left-adjoined to the relative CP, allowing the latter to act as a nominal argument (cf. the analysis of restrictive relatives by Åfarli 1994 reviewed in § III.1). This possibility will not be taken into account here.
15. X° could be taken to exceptionally "inherit" the denotational properties of its specifier; but once again, this would reduce the abstract head to a notational variant of rightward adjunction.
16. This is partially similar to Koster's (1995) analysis of extraposed relative clauses discussed in § VIII.3,
17. The relative suffix on the verb is likely to be an incorporated C°; it also appear as a complementizer in indirect questions (de Rijk 1972: 116).
18. The hypothesis that the free relative is coordinated to the antecedent seems to be confirmed by the fact that the case marking on the antecedent must be repeated on the free relative.
19. The distribution of asyndetic vs. explicit coordination is unclear. Non-referential categories, like adverbials and adjectives, allow pure asyndetic coordination (1–2); on the other hand, referential categories require explicit coordination (3)–(4), apart from multiple coordination, as in (5):

(1) *Gianni ha agito consapevolmente, in piena libertà.*
Gianni acted consciously, at liberty
(2) *Una lunga, piacevole serata*
a long, agreeable evening
(3) ** Marco, Lucia sono partiti.*
Marco, Lucia have left
(4) *Marco e Lucia sono partiti.*
Marco and Lucia have left
(5) *Marco, Lucia e Valentina sono partiti.*
Marco, Lucia and Valentina have left

20. In the same way as in the raising structure the external determiner checks its features against the NP "head" in Spec,CP.
21. In Safir's analysis, this LF movement is required for the relative pronoun to be R-bound by the "head".

22. The relevance of these examples was pointed out by P. Acquaviva. Cf. also Cinque (1988: 473).
23. In certain Italian dialects, appositive relatives differ from restrictive ones in that they can contain a resumptive clitic pronoun (Cennamo forthcoming). But resumptive clitics cannot be assimilated to the relative operator in Spec,CP postulated in (18). The issue of resumptive clitics will be briefly discussed in § VII.9.
24. Besides these empirical problems, there is also a theoretical problem: free relatives are probably introduced by an external determiner in the same way as restrictive relatives (see Kayne 1994: fn. 13 to chapter 8). But no evidence justifies the postulation of an abstract external determiner introducing the appositive relative in the structure (18).
25. Note that leftward movement of IP destroys the chain (DP_j, t_j), so that the reconstruction algorithm discussed in § IV.2 does not apply.
26. However, as Kayne acknowledges, the intonation break does not precede the raised constituent IP, but rather the whole CP (including the complementizer or relative determiner).
27. Following Longobardi (1994), the proper name is assumed to originate in the N position of the relative "head" and to incorporate to the external D°, where it acquires referential uniqueness. Cf. § II.3.1.
28. Kayne (1994: 115) offers a somewhat different explanation, which does not presuppose the impossibility of reconstructing the relative "head". He proposes that an idiom chunk must c-command its trace in LF, and this relation is destroyed in the appositive structure because the relative IP, containing the trace of the idiom chunk, raises to Spec,DP in LF. The c-command condition is independently confirmed by the ungrammaticality of (2):

    (1) *Advantage is* [t *likely* [t *to be taken* t *of us*]].
    (2) * [*How* t *likely* [t *to be taken* t *of us*]] *is advantage* ?

29. Moro (1994: 61–67) proposes a symmetric splitting of the determiner and the restrictive term in the existential construction. In his analysis, the NP moves to a subject-like position and the weak determiner is stranded in the postcopular position and acts as a predicate. As in Heim's analysis, this splitting is only allowed if the determiner of the postcopular DP is weak.
30. Cf. also Frampton (1991: 39–44), who argues that only an A' chain involving an individual variable can cross a weak island. The correlation between the splitting of the determiner and its restrictive term and semantic weakness is formalized in Dobrovie Sorin (1994: 242–250).
31. This analysis of (14) resembles in part Postal's (1994) proposal that the A' dependency of appositive relatives involves a null resumptive pronoun: both analyses predict the impossibility of partial reconstruction of an indefinite category in the postcopular position.
32. Heim (1987) and, more recently, Grosu—Landman (1996) argue that both restrictive and appositive relatives have an individual variable in the trace

position in LF. The definiteness effect is only avoided in degree relatives (Carlson's (1977) amount relatives), where the gap position contains a degree variable rather than an individual variable. For Grosu—Landman, the "head" reconstructed in the gap position is a sortal predicate, defining the sort of the plural individual whose degree is relativized.

If this is the correct generalization about the definiteness effect, it can be reconciled with the raising analysis of restrictives in two ways. The first possibility is to assume that in restrictive relatives, as well as in appositives, the "head" is not reconstructed, leaving an individual variable in LF; this, however, would contrast with the empirical evidence for reconstruction presented in chapter IV (at least in Chomsky's (1993) approach to reconstruction). Alternatively, it can be assumed that in English restrictive relatives the relative DP is non-definite but [+specific], and it violates a "specificity constraint" of the existential structure under reconstruction. This would require a reconsideration of the empirical evidence reviewed in § III.3, sorting out true restrictive relatives from degree relatives. From this perspective, the English relative determiners ***who*** and ***which*** would be more similar to the Hungarian *amely* rather than *amit*. (In the next chapter it will be argued that *who/which* is overtly or covertly present in all types of restrictive relatives.)

33. The sentence is acceptable if the *wh*-phrase ranges over a presupposed set of sets of books (e.g. the books of linguistics, the books of logic, etc.).

34. It is unclear why the relative morphemes allow such an anaphoric use. Note that this use is restricted to specific morphemes: in Italian, for instance, only *il quale* but not *cui* allows it (Cinque 1978).

35. E.g. when the appositive relative is embedded in a PP:

(1) *Ne ho parlato* $[_{PP}$ *con* $[_{DP}$ *Gianni, che è il mio migliore amico*]].
(I) of-it have spoken with Gianni; that is my best friend
'I talked about it with Gianni, who is my best friend.'

36. In this respect, this focus construction contrasts minimally with the pseudo-cleft, as shown by the following minimal pair (Toribio 1992: 301):

(1) *Juan compró fue un libro.*
Juan bought was a book

(2) *Lo que Juan compró fue un libro.*
that that Juan bought was a book
'What Juan bought was a book.'

37. Toribio also points out that a subject cannot be focussed in this structure (1), whereas it can be pseudo-clefted (2):

(1) * *Ganó el premio fue el estudiante cubano.*
won the prize was the Cuban student

(2) *El que ganó el premio fue el estudiante cubano.*
'The one that won the prize was the Cuban student.'

In the analysis proposed in the text, the ungrammaticality of (1) could be reduced to the ECP if it could be demonstrated that subject extraction takes place from Spec,IP. In fact, raising of IP would leave the trace in Spec,IP ungoverned. Toribio (1992: fn. 4) assumes that subject movement to Spec,CP must move through Spec,IP because the latter is a potential A' position.

Notes to Chapter VI

1. In a study of French relative clauses, Kayne (1976) first argued that the invariable particle *que* is not a relative pronoun, but the complementizer of finite subordinate clauses. His arguments also hold for English *that* and Italian *che* (Cinque 1978). The invariable particle differs from relative pronouns in at least two respects: it cannot pied pipe a preposition, as shown in (1)–(2), and it cannot appear in infinitival relatives, as shown in (3)–(4):

   (1) *l'argomento* [*del quale* / *di cui* / * *di che*] *mi voleva parlare*

   the subject about which he wanted to speak to me

   (2) *the way* [*in which* / * *in that*] *he was acting*

   (3) *Cerco un libro* [*da* / * *che*] *leggere.*

   (4) *I'm looking for a book* [*to read* / * *that to read*]

   The impossibility of pied piping follows from the hypothesis that *that/che* is not generated as a complement to the preposition, but it is generated in the head C°. The ungrammaticality of *that/che* in infinitival relatives follows straightforwardly if the particle is analysed as the complementizer of finite clause.
2. This holds for the relativization of any bare DP, with the only exception concerning subject relativization (cf. Cinque 1981; Bianchi 1995: ch. VI).
3. Cf. Haegeman (1992: 54).
4. Complementizer deletion is possible only in subjunctive complement clauses under specific conditions: cf. Poletto (1994), Giorgi—Pianesi (1996: chapter 5), and fn. 56 below.

5. Actually, Cinque's analysis departs from Kayne's in assuming that the rule of Relative NP Deletion is not subject to cross-linguistic parametrization, but it can only apply to a structure where the relative pronoun is c-commanded by the NP "head" (cf. the next footnote).
6. Other authors argue that in appositive relatives the pronoun is unrecoverable because it is not properly related to its antecedent at the point of the derivation where deletion applies. As discussed in § V.1.3, Safir (1986: fn. 11) proposes that the relative pronoun is R-bound by the antecedent at S-structure (i.e. before Spellout) in restrictive relatives, but only at a later stage of the derivation in appositives.
   Cinque (1982: 258–268; 278–279) instead proposes that all *wh*-relatives have a "parenthetical" structure NP,...S'..., in which the relative pronoun is not c-commanded by the antecedent. He assumes that the rule of Relative NP Deletion applies up to recoverability; in the 'parenthetic' structure, the relative pronoun is not c-commanded by the "head", and hence it cannot be deleted. The drawback of this proposal is that in a restrictive relative like (1b) the parenthetical structure must be allowed to receive a restrictive interpretation; this is inconsistent with the DP hypothesis, according to which the restrictive term of the determiner is in the c-domain of the latter. Moreover, it must be stipulated that in Italian the parenthetical structure does not allow the restrictive interpretation (cf. (2b)).
7. It is unclear how the ECP effects in LF can be formulated in the analysis of "feature attraction" of LF movement proposed by Chomsky (1995: 297–312).
8. Cf. Sportiche's (1992) reinterpretation of the doubly filled Comp filter.
9. Alternatively, Grimshaw (1975) and Allen (1977; 1980) argue that in Old and Middle English the *wh-* and pied piping relatives are derived by *wh-*movement, while the *that*-relative is derived through unbounded deletion of the relative noun phrase in the gap position.
10. In normal adult grammars, Recoverability is ranked higher than any other principle.
11. That is, Telegraph and Left Edge(C) respectively.
12. The abbreviation <*wh that*> refers to the sequence of a relative pronoun and an invariable clausal particle. In Old English the relative pronouns of headed relatives were morphologically demonstrative rather than *wh*.
13. A possible answer to this problem is that in these languages complementizer deletion is excluded by some principle ranked higher than the previously discussed ones, say the ECP. But this solution is unavailable, since both Old and Middle English allowed *wh*-relative clauses introduced only by a relative pronoun and lacking a complementizer. On the other hand, if one stipulates that Left Edge(C) was not effective in these stages of the language, then the *wh*-relative would be the only optimal candidate, incorrectly excluding the *that* and <*wh that*> relatives, since contrary to these, it does not involve a violation of Telegraph.
14. Toribio (1992: 292) shows that the relativization of an animate direct object allows both the *that*-relative and the *wh*-relative:

(1) *el señor que María entrevistó ayer*
the gentleman that Maria interviewed yesterday
(2) *el señor a quien María entrevistó ayer*
the gentleman *a* whom Maria interviewed yesterday
'the gentleman whom Maria interviewed yesterday'

Actually, the second example must be analysed as a pied piping relative, because the animate direct object is introduced by the preposition *a*. On the deletion of *a* in (2) cf. § III.3.1.

15. The relative determiners of the two languages differ in at least two respects:

(1) in English, the relative determiners *who* and *which* are not inflected for phi-features, but they are distinguished by the value of the feature [± animate];
(2) the relative determiners *who* and *which* are identical to interrogative morphemes.

In contemporary Italian, on the contrary, *cui* and *il quale* are specialized for the relative function; neither of them expresses the feature [± animate], but *il quale* is inflected for phi-features, *cui* is not.
Neither of these properties can be directly related to the availability of Spec,DP for the NP "head", if one takes into account the behaviour of French relative pronouns. In French, *lequel* is also used as an interrogative pronoun, but it has the same distribution as Italian *il quale*; similarly, the relative pronoun *qui* closely resembles English *who* in being interrogative and [+animate]; nevertheless, it cannot appear in a restrictive *wh*-relative:

(1) * *Le garçon qui Marie prefère s'appelle Georges.*
the boy who Marie likes is called Georges
(Kayne 1976: 260)

16. One possible answer is that overt relative determiners induce a specific interpretation of the relative "head" (Bundrick 1989), and this is incompatible with the intensional interpretation of infinitival relatives, whose "head" is obligatorily indefinite and non-specific (cf. Cinque 1988: 452–453; Stowell 1982: 562). But although this hypothesis may explain the ungrammaticality of the *wh*-relative (25b), it incorrectly rules out the pied piping example (25c), which also involves an overt relative determiner.

17. Kayne (1994: n. 18 to chapter VIII) assumes that the NP "head" can bind an empty category within Spec,CP only if the latter is governed by a *wh*-determiner.
An alternative approach to the contrast in (26) can be based on a feature percolation requirement. Suppose that a phrase can be pied piped only if it inherits a functional feature from one of its internal constituents (cf. Ortiz de Urbina 1990 on interrogatives). In (26a), the relative determiner is overtly specified as [+*wh*] (or [rel]) and this feature can percolate to the dominating

node PP. In (26b), instead, the lexical category NP cannot be specified for any functional feature, and therefore it cannot license the pied piping of a larger constituent. Note that this implies that in a configuration like (24a) no Wh/Relative Criterion need be satisfied (cf. § III.2.4).

18. The generalization holds for relativization, question movement and topicalization. Postal's relativization examples are the following:

    (1) *Max, who I believe* [*Mary thinks* [*Joan talked* [$_{PP}$ *to* t]]]...
    (2) *Max,* [$_{PP}$ *to whom*] *I believe* [*Mary thinks* [*Joan talked* t_{PP}]]
    (3) * *Max, who I believe* [[$_{PP}$ *to* t] *Mary thinks* [*Joan talked* t_{PP}]]
    (4) * *Max, who I believe* [*Mary thinks* [[$_{PP}$ *to* t] *Joan talked* t_{PP}]]

    Postal (1972) regarded these data as decisive evidence against the successive cyclicity of unbounded movement.

19. The problem of how to relate the NP "head" to the external determiner in (29) will be discussed in § 2.3.
20. The distribution of the declarative complementizer is examined in sections 5–8 below.
21. Cf. the operation of "fusion" proposed by Halle—Marantz (1993: 116).
22. This proposal is close to Pesetsky's (1995) analysis of zero morphemes as affixes.
23. A similar view of the incorporation of functional heads seems to be presupposed in Kayne's (1993) analysis of the possessive verb (*have*), corresponding to the spellout of the copula (*be*) plus an incorporated D°/P°.
24. Moreover, the Case feature on the relative DP is [-interpretable] in the sense of Chomsky (1995), and it has already been checked off and erased when the relative DP reaches Spec,CP (§ III.4.4).
25. The particle *wo* is a relative complementizer (cf. § VII.3.4).
26. Cf. Kayne's (1994: 85–86) analysis of the genitive structure *three books of John's*, summarized in § II.3.1; cf. also Ritter (1991), Szabolsci (1994) and Zamparelli (1995: 118–119).
27. On the obligatoriness of deletion see below § 8.
28. By hypothesis (§ III.2), the relative D° is not involved in the checking relation between the external D° and the NP "head"; by Chomsky's (1995) last resort principle, its incorporation cannot be triggered by the need to mediate a checking relation between two distinct categories.
29. This constraint has been studied from a diachronic viewpoint by Bever—Langendoen (1971) and Erdmann (1980). The latter author mentions some contemporary dialects allowing this structure in limited contexts. This issue will be discussed in § VII.7.
30. According to Müller—Sternefeld (1993: 492), some speakers allow the deletion of the complementizer in this context. Unfortunately, it is impossible for the moment to investigate the properties of C° in the dialect of such speakers.
31. With respect to embedded topicalization and negative preposing, pied piping relatives (with a null complementizer) pattern with *wh-* relatives, as shown by the following examples (Culicover 1996: 456):

(1) *Terry is the person to whom [only books like these] would I give.*
(2) *Terry is the person for whom [on my vacation], [not even a postage stamp] did I remember to buy.*

32. Cf. § I.2. 1. This hypothesis has also been argued for on independent grounds by Müller—Sternefeld (1993: 485).
33. This example is originally due to J. Grimshaw.
34. The order topic-focus is also found in Bulgarian (Rudin 1990–1991), Hungarian (Puskàs 1993) and Welsh (Tallerman 1993). A FocusPhrase immediately above AgrP in Welsh is also argued for by Roberts (1995). Kiparsky (1995: 153–155) credits K. Hale with the hypothesis that the clausal structure of ancient Indo-European languages had a Topic and a Focus position in the same order, while lacking the CP level.
35. Italian focalization differs from English negative preposing in that it does not trigger the incorporation of Infl to Focus°. In other Romance varieties, e.g. European Portuguese, focalization instead triggers subject inversion:

(1) *Creio que ISSO disse ele.* (Manzini 1993: 12)
I believe that this said he
"I believe that he said THIS.'

36. Chomsky (1977: 92–94) considers topicalization ungrammatical both in interrogative and in relative clauses, in any relative order of the constituents:

(1) * *To whom the books did John give away?*
(2) * *This book, to whom should we give?*
(3) * *The boy to whom the books John gave away*

Watanabe (1993: fn. 8) instead quotes (2) as simply marginal.
These uncertain judgements may be related to the island effect induced by the lower A' moved phrase for the higher one. Different analyses of these island effects have been proposed by Lasnik—Saito (1992: chapter III), Müller—Sternefeld (1993: 481–498), Rochemont (1989). See below § 5.2.
As Chomsky (1977) emphasizes, in English topicalization must be distinguished from left dislocation, featuring a resumptive pronoun. He finds the latter marginally possible in relative clauses:

(4) *I want to find a corporation to which, as far as my new invention is concerned, I can offer it with a feeling of security that it will be exploited for the good of mankind.*

On the contrary, Baltin (1982: 21) quotes the following less elaborated example as ungrammatical:

(5) * *He's a man to whom (as for) liberty, we could neve grant it.*

These contrasting judgements are probably due to some interfering factor. Since these constructions are related to the informational structure of the sentence, an example could be pragmatically infelicitous without being syntactically illformed. However, Lasnik—Saito (1992: 77) state that in English, left dislocation is only possible in root clauses. In this, it would resemble the Romance structure which Cinque (1977) calls 'hanging topic'. Given this possible interfering factor, the following discussion will leave aside English left dislocation.

37. The pattern in (53) is also observed in Catalan (Vallduví 1992: 324) and the pattern in (53c–d) in Spanish (Campos—Zampini 1990: 59–63).

38. The Wh Criterion (Rizzi 1991):

    (a) A wh-operator must be in a Spec/head configuration with a [+wh] X°.
    (b) A [+wh] X° must be in a Spec/head configuration with a wh-operator.

39. For independent reasons, Infl hosts only auxiliaries in English, whereas in Italian it also hosts lexical verbs (Pollock 1989, Belletti 1990). Moreover, according to Rizzi (1991), the inverted subject is in Spec,IP in English, whereas in Italian it is in the base VP-internal position. These differences are irrelevant to the present discussion.

40. Rizzi identifies the landing site of interrogative and focussed phrases. In fact, interrogative phrases are intrinsically focussed, at least in direct questions. However, it may be necessary to distinguish two functional projections for the two types of phrases: in Rumanian, for example, topics precede the interrogative particle *oare* but focussed phrases follow it (Motapanyane 1994). This further refinement of the Comp area will be disregarded in the following discussion.

41. Note that this hypothesis is contradicted by the phenomenon of "postposition of the relative pronoun", discussed in § III.4.5.

42. It remains to be determined whether the verb selects the feature [+wh] on C° or on Focus°. The former hypothesis would imply the possibility of the sequence <*wh*-phrase, topic> in English embedded questions.
Hiberno-English allows subject-auxiliary inversion in embedded questions (McCloskey 1991: 294–298):

    (1) *Did he tell you how did he do it?*
    (2) *Ask your father does he want his dinner.*

This inversion is incompatible with the presence of the interrogative particle *if*:

    (3) * *Ask your father if does he want his dinner.*

The peculiarity of Hiberno-English may be explained by assuming that the feature [+wh] can be specified on Infl in embedded questions, as in Italian, and hence Infl moves to Focus°. If the feature [+wh] is instead selected directly on Focus°, the latter is spelled out as the particle *if*; in this case Infl does not bear the feature [+wh], and therefore it does not move.

43. Note that Rizzi's hypothesis (c) does not extend to Italian yes/no embedded questions, where the feature [+wh] is directly realized on Focus°, spelled out as *se* 'if', nor to indirect questions in the subjunctive mood, where there is no subject inversion, indicating that Infl° has not raised to Focus°. In these cases, a TopicP can (marginally) appear above FocusP:

(1) ? *Non so a Gianni se gli hanno dato quel libro.*
I don't know to G. if (they) to-him gave that book
'I don't know whether they gave that book to Gianni.'

(2) ? *Non so di Gianni che cosa Maria pensi veramente.*
I don't know about G. what M. thinks_{SUBJ} really
'I don't know what Maria really thinks about Gianni.'

It appears TopicP blocks the selectional relation between the governing verb and the head Focus° in English (55), but not in Italian. This difference could be related to the parameter concerning TopicP to be discussed in § 5.

44. Cf. for instance the following example:

(1) *Credo* [$_{CP}$ *che* [$_{TopP}$ *a Gianni*$_j$ [$_{FocP}$ *QUESTO*$_i$ [$_{TopP}$ *domani* [$_{IP}$ *gli dovremmo dire* t_i t_j]]]]].
I think that to G THIS tomorrow (we) to-him should tell
'I think that to G, tomorrow, we ought to tell THIS.'

45. The lower Topic position seems to be unavailable in direct questions (cf. (54f)), for unclear reasons.

46. The possibility of multiple topics is also discussed by Culicover (1996: 453) who, however, criticizes the postulation of different types of A' position

47. According to Rizzi, however, this projection is only realized when other projections intervene between Force° and IP: cf. § VII.1.2.

48. Hoekstra (1992b) and Vikner (1991) demonstrate that the sequence of complementizers cannot be analysed as a single complex complementizer. Alber (1994) argues that the particles *wia* and *wos* in (60) are clausal functional heads, although they are omophonous to *wh*-pronouns.

49. Moreover, in (59) and (62) an apparently declarative complementizer occurs below the interrogative particle; this issue will be discussed in § VII.2–3.

50. This sequence of elements is attested in Hungarian (Puskás 1992), Spanish (Plann 1982) and Catalan (Vallduví 1992; Hirschbühler—Rivero 1981):

(1) *Nem tudom hogy mit làtott Jànos.* (Hungarian)
(I) dont't know that what_{ACC} saw Jànos_{NOM}
'I don't know what Janos saw.'

(2) *Repetiò que qué libros querian comprar.* (Spanish)
(he) repeated that which books (they) wanted to buy
'He asked repeatedly what books they wanted to buy.'

(3) *Preguntan que el ganivet on el ficaràs.* (Catalan)
(they) ask that the knife where will you put it
'They ask where you will put the knife.'

The difference between these structures and the English/Italian indirect questions cannot be addressed here.

51. A possible exception is constituted by the multiple complementizers in Dutch dialects, e.g. the interrogative *of* in (59), which seems to be semantically expletive.
52. This proposal presupposes a certain degree of flexibility in the selectional relations between the heads of the complementizer system, so that any head may be freely select either IP or any lower projection of the system that may happen to be structurally licensed.
53. This hypothesis has been independently proposed by Nakajima (1996), who, however, does not take into account relative clauses, but only sentential complementation. His proposal differs from ours in one important respect: he argues that the null head Topic° must adjoin in LF to the selecting head for "selectional feature checking", whereas we assume (crucially for our analysis) that Topic° need only be head governed but it does not undergo abstract incorporation (cf. § 8 below).
Since Nakajima's proposal only holds for s-selected clauses, its predictions for zero relatives are unclear. Zero relatives seem to be subject to the ECP-like requirement which Nakajima reduces to selection checking (cf. the impossibility of extraposition); however, the assumption that zero relatives are subject to selection checking is incompatible with the adjunct analysis.
Nakajima also examines the contrast between the interrogative particles *whether* and *if* which will be discussed in § VII.6, reaching conclusions partially similar to ours.
54. In some verb-second languages, declarative complements to bridge verbs may be introduced either by the declarative complementizer or by verb second (cf. the discussion in Vikner 1995):

(1) *Er sagt, daß die Kinder diesen Film gesehen haben.*
he says that the children this movie seen have
(2) *Er sagt, diesen Film haben die Kinder gesehen.*
he says this movie have the children seen
'He says that the children have seen that movie.'

This suggests that the functional head licensing verb-second topicalization may substitute for the declarative complementizer.
55. The selection of an infinitival IP by a lexical head has been proposed for raising and Exceptional Case Marking constructions. Kayne (1984: 110–114) and Pesetsky (1995: 8–9) instead propose that ECM clauses are introduced by a phonetically null complementizer.
56. Though a TopicP can be superposed to a subordinator Focus/Wh°, as mentioned in fn. 43 above.

In contemporary Italian, complementizer deletion is possible only in complement clauses in the subjunctive or conditional mood, or in the future tense. For some speakers complementizer deletion is further constrained to subjunctive complements that have a null subject (with the sole exception of the second person singular pronoun *tu*, which cannot be replaced by *pro* in subjunctive clauses):

(1) *Credevo* pro *fosse partito.*
I thought (he) had left

(2) *Non sapevo tu fossi malato.*
I didn't know you were sick

(3) *Credo sarà interessante.*
(I) think (it) will be interesting

Contrary to English *that* deletion, complementizer deletion licensed by the subjunctive mood does not take place in relative clauses:

(4) * *Cerco un libro tu possa leggere in vacanza.*
I'm looking for a book you can read on holidays

Giorgi—Pianesi (1996) analyse the null head introducing the subjunctive complement clause as a syncretic category Mood/Agr°. However, they do not extend this approach to complementizer deletion in English (cf. their fn. 68 to chapter 5).
Alternatively, Poletto (1994) proposes an account of this phenomenon based on the assumption of a head AgrC° occurring between C° and IP (cf. Shlonsky 1994). She argues that in (1)–(2) the subjunctive verbal form raises to a head Mod° and then to AgrC°: this accounts for the fact that sentential adverbs (which she takes to be adjoined to IP) cannot occur between the matrix verb and the embedded verbal form:

(5) * *Credo sicuramente arrivi.*
(I) think (he) surely come$_{SUBJ}$

Since in her idiolect complementizer deletion is compatible with a preverbal lexical subject, Poletto also argues that the subject DP can be licensed by raising to Spec,AgrC° to the left of the verb:

(6) Credo [$_{AgrCP}$ Gianni$_i$ [$_{AgrCP}$ arrivi [$_{IP}$ t$_i$ t$_V$ domani]]].
(I) think Gianni arrive$_{SUBJ}$ tomorrow

A problem with this hypothesis is that the deletion of the complementizer *che* cannot be directly derived from the movement of the verb to AgrC°, unless *che* is generated in AgrC° rather than in C°. Under the hypotheses proposed in the preceding paragraph, it may be assumed that the

incorporation of the subjunctive verb to a head of the Comp area requires the omission of the higher CP projection. Then (3) is excluded because the Spec of this head is not available as a landing site for the relative phrase, and there is no higher available Spec. Alternatively, (3) may be excluded because the external D° is specified as [+N] and it disallows the raising of a lexical head [+V] into the head of its complement.
This hypothesis implies that the subjunctive mood licenses a low head in the Comp area which can act as a subordinator. The exact properties of this head remain to be determined. Poletto (1994) suggests a parallelism with the phenomenon of "double complementizers" of Occitan and Piedmontese, which will be briefly discussed in § VII.3.

57. This hypothesis will play a crucial role in the analysis of appositive relatives (§ 9).
58. Rochemont (1989: 155) argues the contrary, but the grammaticality judgements he reports for his Italian examples are inaccurate.
59. Vikner gives convincing arguments to the effect that this particle is a complementizer and not an expletive subject pronoun.
60. In order to rule out (78), Vikner proposes that *der* requires agreement between its Spec and the Spec of its complement IP. The0refore, if a constituent other than the subject moves to the Spec of *der*, an inconsistent indexing arises:

(1) * *Vi kender en* $[_{CP}$ *bog*$_k$ $[_{CP}$ *der* $[_{IP}$ *denne lingvist*$_j$ [I [... t_k ...]]]]]

61. Recall that by Kayne's (1994) definition of c-command, Spec,DP c-commands outside DP. The hypothesis that Spec,DP is an operator position accounts for the fact that this position is available for A' binding of a pronoun outside DP in (1), but not for A-binding of an anaphor outside DP in (2):

(1) *Every girl*$_j$*'s father thinks he knows what's best for her*$_j$.
(2) * *Every girl*$_j$*'s father admires herself*$_j$.
(Kayne 1994: 25–26)

62. Recall that the NP "head" cannot stay in the complement position of the relative D°, because it must check the features of the external D°. On the alternative checking discussed in § III.2.3, cf. § 8 below.
63. Note that in the *wh*-relative (49) the relative determiner cooccurs with a negative constituent in Spec,FocusP. Since, contrary to TopicP, FocusP is not recursive (Rizzi 1995), the relative determiner cannot be in the Focus position. This too justifies the hypothesis that the landing site of the relative DP in (86) is a topic position.
64. The same possibility is attested by the examples of "postposition of the relative pronoun" in Latin, discussed in § III.4.5. The relevant examples are repeated here:

(1) *Spectatores optumos, fidem qui facitis maxumi...*
(Plautus, *Casina* 1–2)
spectators excellent, faith$_{ACC}$ who$_{NOM}$ make of the maximum value
'excellent spectators, who hold good faith in the highest honour'

(2) *Meus vicinus, meo viro qui liberum praehibet locum...*
(Plautus, *Casina* 536)
my neighbour$_{NOM}$, my husband$_{DAT}$ who$_{NOM}$ free offers place
'my neighbour, who offers to my husband a free place...'

The lack of a Relativized Minimality effect in the text examples (87)–(88) and in the Latin postposition examples suggests that in the relevant languages the landing site of relativization is not of the same type as the Topic position.

65. The order <NP, complementizer, relative determiner> is also found in Indonesian according to Downing (1978: 383).

66. Cf. Ross (1967: 130–131), Corver (1992), and Schäufele (1988) on Vedic. The proper analysis of this scrambling phenomenon is unclear. One possibility is that the relative and interrogative determiners are "adjectival" maximal projections sitting in a Spec, rather than functional heads selecting NP; then they can be scrambled from the Spec position, as proposed by Corver (1992: 70–74) for a similar scrambling of *wh*-determiners and other modifiers in Czech and Polish. The alternative possibility is that the preposed determiners are actually "remnant DPs" moved leftwards after the NP complement has scrambled. A similar phenomenon is French leftwards quantifier movement exemplified in (1), in which a quantifier moves to the left of its cliticized complement (cf. Kayne 1984: 90–93):

(1) *Marie a tous voulu les revoir*
Mary has all wanted them-see again
'Mary wanted to see them all again.'

(2) * *Marie a tous voulu revoir ses amis.*
Mary has all wanted to see again her friends.

Assuming the Q°-DP structure proposed in § II.3.5 for Italian *tutti*, the derivation of (1) would be the following: the DP complement of the Q° *tous* cliticizes, and the "remnant QP" moves leftward:

(3) ... [$_{QP}$ tous t$_i$] ..... les$_i$...

The contrast between (2) in French and (92) in Latin follows if Latin allows the scrambling of the complement of *quas*, whereas French has no rule scrambling a nominal subconstituent, apart from cliticization.

67. The status of the (90f) is less clear: it can perhaps be analysed as a correlative structure with the NP "head" deleted in the dependent clause and realised in the main clause.

68. The structure would then be:

(1) [$_{DP1}$ [$_{XP}$ [$_{DP2}$ ἧς t_{NP}] [$_{XP}$ κατεστρέφετο t_{DP2}]] [$_{DP1}$ D° [$_{CP}$ [$_{NP}$ χώρας] [$_{CP}$ C° t_{XP}]]]]

with the relative clause preposed to Spec of a phonetically null external D°. Another possibility is that in this example the N° of the relative "head" has incorporated to the external D°, and the remnant CP has moved to Spec,DP. However, it is unclear whether the incorporation of N° to the external D° would be compatible with Kayne's (1994: n. 24 to ch. 8) proposal that the movement of the relative clause to Spec,DP requires the incorporation of C° to the external D°. Assuming this proposal, the derivation of (90) would require the incorporation of two heads to D°, which is ruled out by the Liner Correspondence Axiom (see § 8 for further discussion of this point).

69. Let us assume that Force° is specified as [- V]. This assumption will be exploited in the analysis of subject relativization (§ VII.7.)

70. Cf. in part Pesetsky's (1994) principle Telegraph, discussed in § 1.2.

71. As for interrogative clauses, if the *wh*-phrase moves to Spec,FocusP it should be followed by the interrogative particle realizing Focus° (*if, se*):

(1) * *Non so* [$_{FocP}$ *chi se* [$_{IP}$ *ha invitato*].
I don't know who if he has invited

But in indirect constituent questions Focus° is obligatorily null. This can be derived from the assumption that the interrogative particle realizes a Focus° which functions as a clausal operator conveying the interpretation of a yes/no question. In (1), instead, the operator is the *wh*-phrase in Spec: the realization of the interrogative particle is then ruled out by the Full Interpretation principle.

72. Even if in English the pied piped phrase were to target Spec,TopP rather than Spec, Force Phrase, the relative D° would still be prevented from incorporating by relativized minimality.

73. This is inconsistent with hypothesis (85), whereby Spec,DP is an operator position.

74. The analysis also extends to pied piping relatives, with the further assumption that the pied piped material is reconstructed within IP before IP is scrambled in LF (cf. § V.4.1).

Notes to Chapter VII

1. The representations in (1)–(3) abstract from the process of PF deletion through incorporation of the relative D° and Force° (§§ VI.2.3, VI.8).

2. In other terms, topicalization and relativization are similar types of A' movement. In a recent paper, Grosu—Landman (1996b) propose that relativization induces "...an informational partition such that the 'relativized' element [i.e. the relative pronoun] is construed as a focus-

related topic, and the remainder of the relative, as an INFORMATIVE comment on it, which must therefore include a focus". The topical status of the relative pronoun is shown by the impossibility of focussing it (cf. also Rizzi 1995):

(1) *This is the fellow whose FATHER I like, and that is the fellow whose MOTHER / *WHOM I like.*

This proposal is consistent with the present hypothesis about the non-distinctness of the landing sites of topicalization and relativization (in English; Italian Clitic Left Dislocation probably involves a different type of topicalization). It must be noted that in Grosu—Landman's approach the informational partition corresponds to a semantic partition whereby the relative clause is interpreted as predicated of the "head"; this semantic partition is not straightforwardly compatible with the conclusion, reached in chapter IV, that in restrictives the "head" is reconstructed in the trace position, eliminating the potential configuration of predication in LF (cf. §§ III.2.1, IV.2). However, the semantic interpretation sketched in § IV.2 is only tentative, and the raising structure may ultimately turn out to be compatible with a different semantic analysis. For the moment, the problem remains open.

3. Note that if *whether* is a *wh*-phrase, it is unclear why it does not occur in direct questions triggering auxiliary-subject inversion. Furthermore, Ross (1970: 146) points out that contrary to *wh*-phrases, *whether* cannot license a *wh-in-situ.*
4. Kiparsky (1995: 142) independently argues that Old English *hwæðer* was a complementizer.
5. Actually, Giorgi—Pianesi propose two types of scattering, one that takes place after a bundle of features has been inserted in the numeration (scattering A) and another, a pre-syntactic one, that applies before the numeration is built up (scattering B). Here the term "scattering" is meant to indicate the first type of process.
6. The universal ordering constraint (13) is derivational; for the purposes of the present discussion, it is possible to assume a representational equivalent: $F_1 > F_2$ implies that (the head bearing) F_1 cannot asymmetrically c-command (the head bearing) F_2. The derivational-representational statement of the principle is neutral with respect to the deeper question of why features must be ordered, and why a specific order obtains (e.g. why (14) below, instead of any other possible ordering of the same features).
7. This binary feature is obviously a simplification. Among [-declarative] clauses various different subtypes must be distinguished; this refinement will not be attempted here.
8. Perhaps the features [presupp] and [wh] too establishes a privative opposition, singling out a specific type of [+declarative] and [-declarative] clause respectively.

9. The feature [±presupp.] is not specified on Top/Fin°. Note in fact that Force° (= *that*) cannot be omitted in factive (+presupposed) complement clauses.
10. Note that contrary to (17), in the *wh*-relative (3) two heads are realised, Top/Fin° and Force° (the latter bears an extra feature [±presupposed]) This seems to constitute a violation of the economy condition on scattering; which could explain the highly marked status of the *wh*-relative in contemporary English. Alternatively, in the *wh*-relative Top/Fin° may bear a feature [topic], as mentioned in § VI. 8.1.
    On the impossibility for Top/Fin° to host topicalization in a structure like (17) see below § 6.
11. On the realization of [+finite] in Italian cf. § 1.3.
12. It is also possible to maintain the assumption of chapter VI that Top° and Force° optionally bear a feature [relative]. However, at this point the status of the specifiers positions with respect to the PUB suffices to determine the availability of Spec,TopP for relativization in English, but not in Italian.
13. The feature [+finite] is then realised on Focus/Wh°, attracting the finite verb. There may also be a higher Topic Phrase hosting a [topic] constituent, as in the following German example:

    (1) *Den Karl, wer hat den gesehen?*
    the Karl, who has him seen?
    'As for Karl, who saw him?' (Cardinaletti 1983: 113)

    By the ordering of features in (14), [+finite] must be syncretically realized on the lowest head of the complementizer system, and the verb stops there.
14. There is a considerable debate on the status of subject-verb root clauses in Germanic. These can be taken to involve (1) raising of the subject to a topic-like position in the complementizer system, or (2) raising of the verb to a clause initial Infl°, with the subject in Spec,IP: cf. among others Branigan (1995); Cardinaletti (1990: 154–159); Vikner (1995: chapter 4); Wilder—Cavar (1994: 71–72); Zwart (1993) for relevant discussion. We remain agnostic on this problem; the formulation of the V/2 parameter in (19) below is compatible with either analysis: in fact, with respect to the ordering of features in (14), [+finite] can in principle be syncretically realized either on Infl° or on a structurally licensed head in the complementizer system.
15. Note that the highest head of a root clause cannot be licensed by clause (1) of (11), since it is not selected. Thus, principle (11) may derive Kayne's (1994: 31) hypothesis that the highest element of a chain of heads must have a specifier. We disregard here the abstract node adjoined to the root proposed by Kayne (1994: 36–37).
16. But see the qualification in fn. 13.
17. In Chomsky's (1995: ch. 4) theory of features the property of strenght is an unanalysed primitive, and it can only be stipulated (but see Pollock 1995). Holmberg—Platzack (1988) and Platzack (1995) assume that a [+finite] C° must be lexicalized in order to license Nominative case under government. The interaction of hypothesis (20) below with the licensing of Nominative case is unclear.

On V/2 as triggered by "morphosyntactic attraction" of the verb to C° see also Tomaselli (1990: ch. 5).

18. The residual verb second attested in contemporary English is not triggered by a strong feature [+finite], but by other features subject to the F-criterion: the relevant feature is specified on the verb, and the verb must raise to a head position in the complementizer system in order to set up a Spec/head agreement relation with a constituent matching the same feature in Spec (Rizzi 1991). A different view is proposed by Wilder—Cavar (1994).
19. The particle *ðe* apparently is not realised in embedded questions. It may be assumed that in these the presence of a fronted wh-phrase (presumably hosted by an "active" Focus/WhP) suffices to mark the clause as subordinate.
20. As argued in detail by Mitchell (1985: §§ 2134–2139).
21. Various languages allow declarative complement clauses to be introduced by a determiner preceding Force°: this is what Zaring (1992) calls a "clausal determiner" (cf. § II.2.2). The hybrid D/F° is perhaps a syncretic category merging a clausal determiner and Force°.
22. Also evidence for a topic position to the left of *that* is provided by (1) (from Shakespeare), where a topicalized adverb intervenes between the relative "head" and *that*:

    (1) *...for he to-day that sheds his blood with me shall be my brother...* (*Henry V* 4.3.61–62)

23. In fact, the cases of reduplication of *that* are sporadic and belong to the earliest stages of Middle English, probably corresponding to the stage in which the reanalysis was taking place (Allen 1977: 245).
24. For the notion of tense "anchoring" see Enç (1987), Giorgi—Pianesi (1996) and Stowell (1993).
25. Actually, the dependent clause seems to be completely external to the main clause in that, contrary to the subordinate clauses of modern V/2 languages, it does not count as the first position of the main clause with respect to verb second (Kiparsky 1995: 157–158).
26. The particle *ðe/that* seems to be optional, but its distribution is not entirely clear. The absence of the particle may perhaps correspond to the omission of Top/Fin° in the complementizer system of a dependent clause (with the feature [+finite] syncretically realised on another head).
27. It is possible to speculate that the low head *que* is intrinsically specified as [+V], and it cannot incorporate to the external D°, which is [-V]; thus, its PF deletion cannot be licensed in relative clauses (cf. § VI.8).
    R. Kayne (personal communication) points out that in these varieties of French the infinitival prepositional complementizers pattern with the standard French declarative complementizer, disallowing the violation of the Doubly Filled Comp Filter:

    (1) * *Elle m'a demandé où de aller.* (Burchert 1993: 184)
    she asked me where C° to go

(2) * *Je ne sais pas lequel à choisir.* (Kayne 1976: 290)
I don't know which one C° to choose

(3) *Je cherche un livre à lire.*
I'm looking for a book C° to read

(4) * *Je cherche quelqu'un avec qui à parler.*
I'm looking for somebody with whom C° to speak

In this respect, the prepositional complementizer of infinitival relatives is more similar to Standard French *que* than to Quebec French *que*. This may be due to the fact that it is specified as [-V], and hence it can incorporate to the external D°, so that its PF deletion is licensed. Alternatively, *de* can be analysed as a spellout of a negatively specified Fin° (cf. Rizzi 1995 on Italian *di*), which receives the specification for [+declarative] by a default rule. Its absence in infinitival questions and relatives may depend on the possibility of syncretically realizing the feature [-finite] on a [+wh] Focus° and a [+relative] Force° respectively.
For a radically different conception of *di*, see Kayne (1995b).
Burchert (1992) argues that whereas *que* can be generated in the lower layer of a CP-recursion structure in these varieties of French, the prepositional complementizer *de* is selected by the governing verb in the higher level of CP-recursion. Therefore, whereas in (34) the *wh*-phrase and the particle *que* are in two distinct layers of a CP recursion structure, in (1) they are necessarily in a Spec/head configuration, violating the Wh Criterion (since *de* is [-wh]).

28. Perhaps related to this is the Italian emphatic structure:

(1) *Lui/ Gianni/ *egli/* nessuno sì che è intelligente!*
he/Gianni/he/nobody yes that is intelligent

The ungrammaticality of the weak pronoun *egli* (Cardinaletti—Starke 1994; Cardinaletti 1995) and of the negative quantifier suggests that the subject DP is in a topic position; this conclusion is supported by the observation that in the same structure an object DP can be dislocated with an obligatory resumptive clitic:

(2) *Lui sì che lo conosco!*
him yes that I know

The affirmative particle would be in Spec,FocusP (or perhaps a form of Focus° itself: cf. Laka 1990 for the suggestion that the affirmative particle is in the S projection, corresponding to FocusP), and *che* in a lower head.

(3) $[_{TopP}$ *lui* $[_{FocP}$ *sì* $[_{XP}$ *che* $[_{IP}$ *lo conosco*$]]]]$

The hypothesis can perhaps be extended to the *che* appearing in the cleft structure, assuming that the clefted constituent is in Spec,FocusP:

(4) *E'* $[_{FocP}$ *Gianni* $[_{XP}$ *che* $[_{IP}$ *ho incontrato*$]]]$.
it is Gianni that I met

The generalization would then be that in Standard Italian (and French) the lower *che* is only licensed by certain instances of Focus°.
A possibly related phenomenon is the *che* appearing in exclamative clauses in some northern Italian varieties:

(5) *Che bello che è!*
How beautiful that (he) is!

29. Cf. also Wheeler (1988: 273–274).
30. Poletto (1994) instead analyses the lower *que* as Shlonsky's (1994) AgrC°, thus explaining its ability to host the subject in its Spec.
31. The reduplication of the declarative complementizer with an intervening topic is also attested in early stages of Italian (Palermo 1994: 189–190):

(1) *... voria che io e 'l mio fratello che stesera in pacie...*
(I) wish that I and my brother that would stay in peace
'I would like for my brother and I to be on good terms with each other'

Castellani's texts also contain a formula where a predicate nominals appears to be topicalized between the NP "head" and the complementizer *che*:

(2) *...madona Barocingha moglie ke fue di meser Salvagio...*
lady B., wife that was of Sir S.
'lady B;, who was wife to sir S.' (*Nuovi Testi*, p. 175)

The topic deserves further investigation.
32. Cf. also Law (1991: 267).
33. Other Dutch dialects allow the cooccurrence of a *wh*-phrase and *of*:

(1) *Ik vraag me af* [$_{FP}$ *wie* [$_{FP}$ *of* [$_{IP}$ *hij* t *uitgenodigd heeft*]]].
I wonder who if he invited has
'I wonder who he invited.'
(2) *Hij weet* [$_{FP}$ *welke jongen* [$_{FP}$ *of* [$_{XP}$ *dat* [$_{IP}$ *je* t_i *gezien hebt*]]]].
He knows which boy if that you seen have
'He knows which boy you saw.' (Hoekstra 1992: 191)

These dialects must allow an expletive use of the interrogative particle.
Hoekstra (1992:197) reports another interesting violation of the Doubly Filled Comp Filter in relative clauses. Dutch has two types of relative pronoun: a *wh* one and a "demonstrative" one. In the dialects that allow (1)–(2) the *wh* relative pronoun, but not the demonstrative one, can cooccur with the "interrogative" particle *of*:

(3) ? *het hujs wat of ik gezien heb*
the house which if I seen have
'the house I have seen'

(4) * *het hujsie dat of ik gezien heb*
the house this if I seen have

This suggests that the two types of relative determiner move to two distinct positions in the Comp system; *wh*-determiners land in the same position as interrogative phrases, namely the Spec of *of.*

34. M. T. Guasti (personal communication) has observed the use of the *wh*-locative adverb *où* as a complementizer in early stages of the acquisition of relative clauses in French. Since *où* appears in correlation with the deletion of a preposition selecting the relative DP (cf. § 9.3), she suggests that it is a form of C° agreeing with a DP marked with an oblique Case. Sometimes *où* is also extended to the relativization of DPs marked with a structural Case. A similar development could have given rise to Bavarian *wo*.
Similarly, French *dont* evolved from * *de unde,* an interrogative locative adverb reinforced by a preposition (Hadermann 1987). Burchert (1993: 179–184) argues that *dont* is a complementizer, that is, the form of *que* agreeing with an operator endowed with an inherent Case: this is very close to Guasti's suggestion. Hadermann (1987: 335) notes that in Old French *dont* could be used as a simple relative complementizer, like Bavarian *wo,* e.g. in the relativization of a direct object:

(1) *...pour une courtoisie dont il me fist jadis...*
for a favour that he to-me made at one time
'for a favour that he made me once'

35. The complementary distribution is attested in the Zurich dialect (Bayer 1984: n. 11). A further complicating factor is the presence of second person agreement morphology on Bavarian complementizers.

36. Similar to Bavarian is the Tyrol dialect, according to Alber (1994: 12):

(3) *Des Haus (des) (wos) mr gebaut hom ...*
this house which that we built have
'this house that we built'

37. Consider also the following contrast:

(1) * *John is looking for a rich woman who to marry.*
(2) *John is looking for a rich man whose daughter to marry.*

By our hypotheses, in (2) the NP "head" must be inside the pied piped phrase *whose daughter* in Spec,ForceP. The substructure of the pied piped DP is (3):

(3) $[_{D/PP}$ $[_{NP}$ *man*] [D/P° $[_{AgrP}$ $[_{DP}$ *whose* t] [Agr° $[_{NP}$ *daughter*]]]]

The highest head is D/P°, a head which licenses a possessor phrase in the lower Spec,AgrP. The NP "head" is extracted from the possessor phrase and moves to Spec,D/PP. By hypothesis, this position is accessible to the non-quantificational categories NumP/NP: cf. Kayne's (1994: 85–86) analysis of the complex possessive structure *two pictures of John's*, repeated here.

(4) $[_{D/PP}$ $[_{NumP}$ *two pictures*] $[_{D/PP}$ *of* $[_{AgrP}$ *John's* [Agr° t_{NumP}]]]]

38. The parallelism is also noted and discussed by Nakajima (1996).

39. This condition is satisfied in the *wh*-relative, where Top° is head-governed by Force° (deleted in PF).
It is unclear whether the condition can be satisfied in root clauses. One possibility is to assume that the head government requirement only holds in those structures where it can in principle be satisfied, e.g. in non-root clauses. Otherwise, it may be assumed that in root clauses Top° is head-governed by an abstract element, perhaps along the lines of Kayne (1994: 36–37) or of Chomsky (1995: 292).

40. Authier (1992) points out that the same constraint holds for negative preposing, as discussed in § I.2.1.

41. However, Lasnik—Saito (1992: 77) report the following example as grammatical:

(1) [*That this solution*$_j$, *I proposed* t_j *last year*] *is widely known.*

42. Authier's constraint is formulated in terms of CP recursion:
A head of type C may optionally select a CP that bears the index of a topic iff that head is lexically governed.
Müller—Sternefeld's is instead formulated in terms of 'matching projections' CP and TopP:
A designated embedded T node must be head-governed by a lexical category

43. In the following examples, which Kiss (1996: 140) attributes to R. Kayne, topicalization is possible in an island clause:

(1) *I resent it* [that [*to Tom*] *she gave only an hour of her time*].
(2) [*That* [*to Tom*] *she would dare to say such a thing*] *is beyond belief.*

44. It must be assumed that this abstract incorporation of AgrS° differs from the overt movement of an auxiliary to a complementizer head (Aux-to-Comp); in this case, the incorporated auxiliary does not act as a proper governor. Rizzi (1990: 40–41) argues that an incorporated head cannot properly govern because it is not immediately dominated by its own projection, but by the projection of the host head. This problem does not arise in case of abstract incorporation, assuming that the two abstract functional heads can undergo an operation of "fusion" (in the sense of § VI.2.3) which unifies them in a single head.

45. Cf. Müller—Sternefeld (1993: 486), and the diachronic origin of *that* discussed in § 2.
46. This explanation for the contrast between argument and adverbial topics is independently proposed by Browning (1996: 243).
47. Cf. § II.6.
48. Actually, some dialectal varieties allow the omission of the complementizer in case of local subject relativization, as discussed by Erdmann (1980):

    (1) % *We had a chap used to come into the shop...* (Erdmann 1980: 142)

    However, this possibility is limited to some presentational contexts, where the DP containing the relative clause is indefinite and it is selected by the copula or the verb *have.*
    The grammaticality of (1) in some dialects may be determined by a different status of AgrS° in these dialects with respect to standard English. Recall that according to Hoekstra's (1993) generalization, number agreement on C° is only possible if the number morphology is "neutral" with respect to the categorial features [±N], [±V]. As shown in (77), the Standard English number morphology is distinguished in the verbal and in the nominal paradigms. A possible conjecture is that in the dialects that allow (1), the agreement morphology of the present tense does not express the number feature, but some other nominal feature; this could make AgrS° compatible with an external determiner, allowing the structure in (1). However, this approach does not explain the restriction to presentational contexts.
    An alternative possibility is to assume that in presentational contexts the copula may select a TopicP rather than a DP. The indefinite determiner following the copular verb is generated within the DP raised to Spec,TopP:

    (2) BE $[_{TopP}$ $[_{DP}$ *a man*$]_i$ $[_{TopP}$ Top° $[_{AgrSP}$ t_i ...]]]

    In this case, TopicP would be c-selected by a [+V] head and the incorporation of AgrS° would be unproblematic.
49. Cf. Cardinaletti—Starke (1994) for the relevance of this feature in the syntax of pronouns.
50. The *wh*-relative corresponding to (83) is grammatical, although the animate NP "head" moves to Spec,ForceP:

    (1) *I'm looking for* $[_{DP}$ *some* $[_{FP}$ *body* $[_{TopP}$ *who* t_{NP} [Agrs°$_i$+Top° [t_i t_{Agr} ...]]]]

    In this structure the Agr features on Force° are not activated: by relativized minimality, the subject trace must be properly governed through incorporation of AgrS° to Top°.
51. In contemporary Mainland Scandinavian the *wh*-relative is disappearing, and in contemporary English it is restricted to the formal style, whereas the

colloquial style is developing the *inwhich* construction, discussed by Rudin (1986), with the insertion of a dummy preposition. The markedness of the *wh*-relative may be related to the high cost of its two-steps derivation (cf. fn. 10), in a way that remains to be investigated.

52. The data are taken from Dino Compagni, *Cronica*, edited by G. Luzzatto, Turin, Einaudi 1968.
53. Cf. also Wanner (1981) and Scorretti (1981). As shown in (87b), the omission of *che* does not interfere with subject relativization; this is due to the null subject character of this dialect, which allows subject movement from the postverbal position (cf. Rizzi 1990: 70–71).
54. Benincà (1995) provides interesting evidence to the effect that the two topicalization patterns involved two distinct positions. In fact, nonresumptive topicalization cooccurs with proclisis on the finite verb (1), whereas resumptive topicalization consistently cooccurs with enclisis:

(1) [*Quello che tu vorrai*] *mi renderai* t, *e* [*gli altri*] *ti terrai* t. (Novellino IX; Benincà 1995: 335)
that which you will want (you) me_{CL}, will give back and the others (you) will keep
'You'll give me what you want to and keep the rest'

(2) [*A voi*] [*le mie poche parole ch'avete intese*] *olle dette con grande fede.* (Benincà 1995: 337)
to you my few words that (you) have heard, (I) have $them_{CL}$ said with great faith
'The few words that you heard, I told them to you in good faith.'

Benincà analyses enclisis as the movement of the verb to C° to the left of the clitic. She suggests that movement to C° only takes place when Spec,CP is empty; this is not the case in (1), because the topic is in Spec,CP; the condition is instead satisfied in (2), because resumptive topicalization does not involve Spec,CP, but a higher recursive Topic projection. From the perspective of the present analysis, the head hosting nonresumptive topicalization can be tentatively assimilated to the English Top/Fin°. However, it is unclear why this head attracts the finite verb only when its Spec is empty.

55. The *wh*-relative is also attested in the nonliterary texts edited by A. Castellani (*Nuovi testi fiorentini del Dugento*, Florence, Sansoni 1952; *La prosa italiana delle origini. Testi toscani di carattere pratico*, Bologna, Patron 1982), and the *wh*-resumptive in the early XV century prose of Filippo degli Agazzari from Siena (in G. Maranini—G. Baldassarri, eds., *Racconti esemplari del Due e Trecento*, Roma, Salerno Editrice, 1993); *wh*- and zero relatives are also attested as late as in the XVII century texts edited by A. Petrucci (*Scrittura e popolo nella Roma barocca, 1585–1721*, Roma, Quasar 1982). The date of the change of parameter (4) remains to be determined through a systematic study of a coherent corpus.
56. On the V/2 character of ancient Tuscan cf. the discussion of the Tobler-Mussafia Law by Benincà (1995).

57. Cf. also Acquaviva (1990) on Old Irish. Duffield (1991) instead analyses the particle *a* inducing nasalization as a C°, and the particle inducing lenition as a lower functional head (Tense). In his account, the asymmetry between the two types of relative clause concerns the presence versus absence of the CP level (cf. the analysis proposed in § 5 for *that*- and zero relatives). The licensing of a resumptive pronoun apparently requires the presence of the CP level.

58. In this case, the resumptive pronoun appears incorporated to the P°.

59. It appears from McCloskey's discussion that a resumptive pronoun not embedded in a PP cannot be fronted in the same way. In the proposed approach, this cannot be explained by the hypothesis that rules out *wh*-relatives in Italian: if the Spec of the resumptive D° is an operator position, it should be able to host the *wh*-QP in its Spec. Note also that from this position the QP would have the whole CP in its scope.

60. The hypothesis that a *wh*-pronoun is a nominal projection QP, lower than DP, is also proposed by Kayne (1994: 108) for French *qui* in his analysis of the following structure:

(1) *Qui de sérieux as-tu rencontré?*
who of serious have you met

Kayne's analysis is as follows:

(2) D° [$_{D/PP}$ [$_{QP}$ *qui*] [$_{D/PP}$ *de* [$_{IP}$ [$_{AP}$ *sérieux*] [$_{IP}$ I° t_{QP} ...]]]]

The preposition *de* is a "nominal complementizer" selected by a null D°. In turn, *de* selects an Inflectional Phrase whose Spec is filled by the adjectival phrase, and whose complement includes the QP headed by the interrogative pronoun *qui*. The structure is derived by moving QP to the Spec of *de*. Crucially, the interrogative pronoun must be generated within the complex DP; hence it is not a D°, but a lower functional head.
The parallelism between Irish *cé* and French *qui* is confirmed by McCloskey's observation that the structure in (94) is impossible with a more complex interrogative phrases, like *cén fear* 'which man', which mirrors the impossibility of the structure in (2) with the interrogative phrase *quel homme* (Kayne 1994:108–109):

(3) * *Quel homme d'intelligent connais-tu?*
which man of intelligent know you

Both cases are excluded if the larger *wh*-phrases *cén fear* and *quel homme* are constituents larger than QP. (McCloskey 1992: 227–228 instead proposes for Irish a phonological constraint that excludes polisyllabic interrogative phrases form the structure in (94).)

61. Alternatively, it is possible to analyse *leis* in (94) as an inflected preposition and the *wh*-phrase as its complement raised to Spec,PP. However, this

analysis does not account for (95), nor for the contrast between *cé* 'who' and *cén fear* 'which man' mentioned in the preceding note.

62. In Hebrew too, resumptive pronouns fall under the Highest Subject constraint; see Shlonsky (1992) for discussion.
63. The filter against empty Comps accounts for the absence of zero relatives in Hebrew.
64. The fronting of a resumptive pronoun to a position immediately following the complementizer is also observed in Swiss German resumptive relatives according to van Riemsdijk (1989b).
65. Note that the pronoun can also undergo "partial A' movement" to an intermediate position:

    (1) *ha-ʔish she xana ʔamra* [*she-*ʔoto *dalya maʔamina* [*she-kobi pagash*]]
    the man that Xana said that him Dalya believes that Kobi met

    This shows that the fronting of the pronoun is not necessarily associated to the Comp of the relative clause, but it resembles a process of topicalization.
66. The following violation of the Coordinate Structure Constraint is reported by Doron (1982: 17):

    (1) *ha-ish she raʔiti ʔet dani ve ʔoto*
    the man that I saw Dani and him

    It is perhaps possible to assume that in Hebrew the "head" is generated in a Topic position and is connected to the resumptive pronoun through a relation of predication that does not involve movement (cf. Chomsky's (1977) analysis of Left Dislocation in English). The problem remains open.
67. This ambiguity is confirmed by the observation that the violations of strong islands and of the Coordinate Structure constraint are possible only in headed relatives but not in free relatives.
68. These Authors discuss many restrictions on the distribution of resumptive clitics. In some dialects, e.g. Paduan, they only appear in appositive relatives; in many dialect, they appear only in some argument position, typically the lower ones in Keenan—Comrie's (1977) NP Accessibility Hierarchy. These problems exceed the limits of the present discussion.
69. The example has a subordinate clause, in order to avoid confusion with another topic construction of root clauses, the "hanging topic", where the preposition can be omitted (cf. Cinque 1977: fn. 12).
70. As mentioned in § V.4, Kayne (1994: 78–83) assimilates to clitic doubling Romance right dislocation, arguing that the doubled direct object has not moved in overt syntax, but it moves to a Topic position in LF:

    (1) *L'ho dato a Gianni, il libro.*
    (I) it_{CL} have given to Gianni, the book
    'The book, I gave it to Gianni.'

71. Cf. Kayne (1972: 90) for the hypothesis that in French complex inversion the clitic and the lexical subject form a single constituent.
72. Note however that in the doubling relative clauses and in Clitic Left Dislocation a doubled direct object is not introduced by a dummy preposition.
73. This relativization strategy is also attested in Brazilian Portuguese (M. C. Figueiredo Silva, personal communication). Haberland—van der Auwera (1990: 148–149) point out that this strategy is also used in Modern Greek in the relativization of prepositional objects: the language lacks resumptive clitics for this position, and the "unrecoverable deletion" strategy is forced by the impossibility of preposition stranding.
74. Note that the *d*-pronoun need not move to Spec,CP in order to be licensed. In the following example, for instance, it appears in an IP-internal position:

    (1) *Gestern las er das ohne Zweifel nicht.*
    yesterday read he it without doubt not
    'yesterday he surely didn't read it' (from Vikner 1991b)

75. Van Riemsdijk (1985) proposes the generalization that internal *wh* movement is a property of OV languages. In his fn. 19 he reports that Ancient Tuscan also had internal *wh*-movement (an observation due to G. Cinque), and he suggests that in that stage Italian was at least partially OV. In fact, Ancient Tuscan also had object scrambling.
76. It remains to be explained why internal *wh*-movement is obligatory in German. This aspect of the problem is discussed in Bianchi (1995: chapter VI), within a more general attempt to reduce all pied piping phenomena to the general locality condition on A' movement.
77. Cf. also the following contrast, reported by von Stechow (1979: 231):

    (1) *Die Frau, mit der Klaus-Jürgen zu schlafen hofft.*
    the woman with whom K.J. to sleep hopes
    (2) * *Die Frau, mit der K.J. hofft daß er schläft.*
    the woman with whom K.J. hopes that he sleep

78. This is another phenomenon with respect to which *that*-relatives and *wh*-relatives are opposed to zero relatives in English.

Notes to Chapter VIII

1. Cf. (abstracting from the restrictive/appositive status of the relative clause), his representation (143) on p. 114, reproduced as (11) below.
2. This extension of the minimal domain is inconsistent with Chomsky's (1995: 179) restriction of the minimal domain to two-membered chains.

Alternatively, it may be assumed that the incorporated D° itself checks the features of the host D°, in the way suggested in § VI.8.1.

3. Note that the head F° of FP_1 has been deleted by incorporating to the D° head of DP_j, but FP_1 then moves to the left of that determiner; this is another violation of the proper binding condition, rescued in LF by the reconstruction of FP_1.
4. In (8), the relative FP1 is actually pied piped by the NP "head" in its highest Spec.
5. Specifically, Kayne (1994: 92; 98) reduces the impossibility of omitting the second *that* in (3) to a surface recursion restriction whereby the constituent in the Spec of a null relative C° cannot have an overt complement.
The hypothesis that what excludes Top° in the rightmost relative of (3) is not the head government requirement is supported by Kayne's observation that the omission of the complementizer in (3) is less deviant if the immediately lower subject is pronominal:

(1) ? *I just read the book that's about your ancestors Ø you gave me last year.*

This rescuing effect of the pronominal subject is not found in subject clauses, where Top° is ungoverned:

(2) [* *(That) it's out now*] *is wonderful.*

Bever—Langendoen (1972) also note that the optional omission of the complementizer (i.e. the selection of Top° in the place of Force°) mostly cooccurs with a pronominal subject. This suggests that Top/Fin° may be licensed by pronominal subjects, in a way that remains to be investigated. (On the different distribution of weak pronominal subjects and "strong" subjects, cf. Cardinaletti (1996)).

6. Cf. the discussion in § V.1.4.
7. According to Kayne, in order for IP to skip Spec,FP Force° must abstractly incorporate to the external D°, making Spec,FP equidistant to Spec,DP. As discussed in § VI.9, this abstract incorporation takes place only in *wh-* and pied piping relatives.
8. In § VI.9, it was instead proposed that the appositive IP moves to a topic-like position within the DP, where it receives a backgrounded interpretation. The analysis carries over only if this topic position is higher than D° in the nominal "complementizer system". This problem cannot be addressed here.
9. In fn. 1 to chapter IV, an alternative analysis of stacking was suggested: the rightmost relative clause may be taken to be extraposed (i.e. stranded) within the leftmost relative clause. For instance, the structure of (1) would be (2):

(1) [*The girl that* $Bill_j$ *had been dating that* he_j *finally married*] *was Italian.*

(2) [$_{DP}$ *the* [$_{FP}$ *girl*$_j$ [$_{FP}$ *that Bill*$_i$ *had been dating* [$_{DP}$ D° [$_{FP}$ t$_j$ [$_{FP}$ *that he*$_i$ *finally married* t$_j$]]]]] ...

The problem with this approach is that it cannot subsume the stacking of an appositive relative to the right of a restrictive one, as in (9a), because appositive relatives in general disallow extraposition, for reasons that remain unclear. On extraposition see Kayne (1994: chapter 9) and § 3 below.

10. The problem does not raise in the approach to checking sketched in § VI.2.3, whereby the NP "head" is always extracted from the relative clause.
11. In *that*-relatives the relative D° is deleted by abstract incorporation to the external D° (§ VI.2.3). In (17), incorporation only takes place from the first conjunct, violating the Coordinate Structure Constraint. The status of this constraint in Munn's analysis is unclear.
12. The reverse order is possible:

    (1) *the picture that Bill liked and Mary hated*

    but it can be taken to involve the coordination of two IPs, rather than of a Force Phrase and a Topic Phrase.
13. The discussion in this section was inspired by Borsley (1996).
14. This analysis presupposes that Spec,FP is not included in the relative clause island: cf. §§ II.5.2. With respect to the representation (103) of chapter VI, repeated here as (1), the stranding hypothesis incorrectly predicts the grammaticality of an extraposed *wh*-relative in Italian; in fact, the QP "head" checks the features of the external Agr_D on its way to a higher position:

    (1) QP ... [$_{DP}$ D° [$_{AgrP}$ t$_{QP}$ [Agr_D [$_{FP}$ [D_{REL} t$_{QP}$] [Force° IP]]]]]

    Then, the deletion through incorporation of either Force° or D_{REL} yields equally optimal derivations. The second option gives rise to the *wh*-relative.
15. However, the structure is overtly attested with numerals:

    (1) [$_{DP}$ *which* [$_{QP}$ *two* [$_{NP}$ *books*]]]

    This suggests that the structure in (21) is licit, but the underlying indefinite determiner cannot be spelled out. The PF deletion of Q° does not apply in cases like (1), presumably because the substantive features of a numeral cannot be subsumed under those of D°. For a similar idea cf. Zamparelli (1995: 118–119).
16. One problem with (20) is that the indefinite determiner is internal to the raised "head", and hence it is expected to allow reconstruction in the relative clause. This wrongly predicts that it may fall in the scope of the universal quantifier in (1) (cf. § II.3.4):

    (1) *I was looking for* [*a book*] *yesterday* [*that everybody could read* t].
17. Kayne (1994: fn. 4 to chapter 9) proposes that (22a) is excluded by the constraint against complex specifiers mentioned in fn. 5.

18. The relevant interpretation of the structure is the one in which the auxiliary and the past participle form a present perfect verbal form.
19. In (22b) and (27b), instead, the empty category is the complement of the Q° *tous*, and it can be properly governed by it.
20. We shall assume that the phi-features of a determiner count as L-features.
21. "A position is L-related if it is in a local relation to an L-feature".
22. The following extraposition example, discussed by Kayne (1994: fn. 15 to chapter 9), is correctly excluded in the present analysis of the *wh*-relativ

    (1) * *A man who just walked in* [[e] *we knew in high school*].

23. Nakajima reduces the nesting requirement to Relativized Minimality and remains neutral between a movement and a base-generation approach to extraposition. Rochemont—Culicover and Wiltschko instead explicitly assume base generation; Wiltschko reduces the nesting requirement to a locality principle restricting the relation between the extraposed clause and its antecedent. For related discussion see also Haider (1993).
24. Koster also assumes that the phrase headed by the colon bears the category label of the two conjuncts.
25. The coordination analysis can also account for extraposition of a relative clause from a PP, which is problematic for the stranding analysis (see Kayne 1994: 126):

    (1) [*A man*]$_i$ *came* [$_{PP}$ *into* [*the room*]$_j$] *last night* [*that I had just finished painting*]$_j$ [*who had blond hair*]
    (Nakajima 1992: 314)

References

Abney, S. P.
1987 The English Noun Phrase in Its Sentential Aspects. PhD Dissertation, MIT.

Acquaviva, P.
1989 Aspetti della complementazione frasale. Master Dissertation, University of Pisa.

Acquaviva, P.
1990 "La funzione sintattica delle marche morfologiche di lenizione e nasalizzazione in antico irlandese", *Annali della Scuola Normale Superiore di Pisa* 20: 695-728.

Adams, D. Q.
1972 "Relative Clauses in Ancient Greek", in: P. M. Peranteau—J.N. Levi—G.C. Phares (eds.), *The Chicago Which Hunt. Papers from the Relative Clause Festival*. Chicago: Chicago Linguistic Society, 9-22.

Åfarli, T. A.
1994 "A Promotion Analysis of Restrictive Relative Clauses", *The Linguistic Review* 11: 81-100.

Ageno, F.
1956 "Particolarità nell'uso antico del relativo", *Lingua Nostra* XVII: 4-7.

Alber, B.
1994 "Indizi per l'esistenza di uno split-CP nelle lingue germaniche". [Unpublished manuscript, University of Venice and Padua].

Allen, C. L.
1977 Topics in Diachronic English Syntax. PhD Dissertation, MIT.

Allen, C. L.
1980 "Movement and Deletion in Old English", *Linguistic Inquiry* 11: 261-323.

Baker, M.
1988 *Incorporation. A Theory of Grammatical Function Changing*. Chicago: The University of Chicago Press.

Baltin, M. R.
1982 "A Landing Site Theory of Movement", *Linguistic Inquiry* 13: 1-38.

Barss, A.—H. Lasnik
1986 "A Note on Anaphora and Double Objects", *Linguistic Inquiry* 17: 347-354.

Barwise, J.—R. Cooper
1981 "Generalized Quantifiers and Natural Language", *Linguistics and Philosophy* 4: 159-219.

Belletti, A.
1988 "The Case of Unaccusatives", *Linguistic Inquiry* 19: 1-34.

Belletti, A.
1990 *Generalized Verb Movement: Aspects of Verb Syntax.* Turin: Rosenberg—Sellier.

Belletti, A.
1994 "Case Checking and Clitic Placement: Three Issues on (Italian/ Romance) Clitics", *Geneva Generative Papers* 1.2: 101-118.

Belletti, A.
1995 "Italian/Romance Clitics: Structure and Derivation". Conference held at the Department of Cognitive Sciences (DIPSCO), Istituto Scientifico S. Raffaele, Milan, May 1995.

Belletti, A.—L. Rizzi
1988 "Psych Verbs and Theta Theory", *Natural Language and Linguistic Theory* 6: 291-352.

Benincà, P.
1994 "Contesti del complementatore *che*". Conference held at the University of Pisa, February 1994.

Benincà, P.
1995 "Complement Clitics in medieval Romance: The Tobler-Mussafia Law", in: A. Battye—I. Roberts (eds.), *Clause Structure and Language Change*. New York: Oxford University Press, 325-344.

Benincà, P.—L. Vanelli
1982 "Appunti di sintassi veneta", in: M. Cortelazzo (ed.), *Guida ai dialetti veneti* 4. Padua: CLEUP, 9-38.

Benveniste, E.
1966 *Problèmes de linguistique générale*. Paris: Gallimard.

Bever, T.—D. T. Langendoen
1971 "A Dynamic Model of the Evolution of Language", *Linguistic Inquiry* 3: 433-464.

Bianchi, V.
1992 "Sulla struttura funzionale del sintagma nominale italiano", *Rivista di Grammatica Generativa* 17: 39-64

Bianchi, V.
1993 "An Empirical Contribution to the Study of Idiomatic Expressions", *Rivista di Linguistica* 5: 349-385.

Bianchi, V.
1995 Consequences of Antisymmetry for The Syntax of Headed Relative Clauses. Doctoral Dissertation, Scuola Normale Superiore di Pisa.

Bokamba, E. G.—M. Dramé
1976 "Where do Relative Clauses Come from in Mandingo?" *Chicago Linguistic Society* 14: 28-43.

Borer, H.
1984 "Restrictive Relatives in Modern Hebrew", *Natural Language and Linguistic Theory* 2: 219-260.

Borsley, R. D.
1996 Relative Clauses and the Theory of Phrase Structure. [Manuscript, University of Bangor. Revised version published in *Linguistic Inquiry* 28: 629-647.]

Branigan, P.
1996 "Verb Second and the A-bar Syntax of Subjects", *Studia Linguistica* 50: 50-79.

Brody, M.
1995 *Lexico-Logical Form*. Cambridge, Massachusetts: The MIT Press.

Browning, M.
1987 Null Operator Constructions. PhD Dissertation, MIT.

Brucart, J. M.
1992 "Some Asymmetries in the Functioning of Relative Pronouns in Spanish", *Catalan Working Papers in Linguistics* 2: 113-143.

Brugé, L.—G. Brugger
1996 "On the Accusative *a* in Spanish", *Probus* 8: 1-51.

Brunel. J.
1977 "La relative greque à la lumière de ses origines", *Bulletin de la Société Linguistique de Paris* LXXII: 211-240.

Bundrick, C.
1989 "An Inference-Based Account of Restrictive Relative *Which* and *That*", *Studies in The Linguistic Sciences* 19: 19-31.

Burchert, F.
1993 "On the French Complementizers *que, dont* and *de*", *Probus* 5: 179-191.

Campbell, A.
1987[8] *Old English Grammar*. Oxford: Clarendon Press.

Campos, H.
1992 "Enunciative Elements in Gascon", *Linguistics* 30: 911-940.

Campos, H.—M. Zampini
1990 "Focalization Strategies in Spanish", *Probus* 2: 47-64.

Cardinaletti, A.
1983 "Lo status dei pronomi *d*- e la ricostruzione nella 'dislocazione a sinistra' in tedesco", *Rivista di Grammatica Generativa* 8: 111-125.

Cardinaletti, A.
1990 Pronomi nulli e pleonastici nelle lingue germaniche e romanze. Saggio di sintassi comparata. Doctoral Dissertation, University of Venice and Padua.

Cardinaletti, A.
1994 "On the Internal Structure of Pronominal DPs", *The Linguistic Review* 11: 195-219.

Cardinaletti, A.
1995 "Subjects and Clause Structure", *University of Venice Working Papers in Linguistics* 6: 55-96.

Cardinaletti, A.—G. Giusti
1991 "Partitive *ne* and the QP-Hypothesis", in: E. Fava (ed.), *Proceedings of the XVII Meeting of Generative Grammar*. Turin: Rosenberg—Sellier, 121-141.

Cardinaletti, A.—M. Starke
1994 "The Typology of Structural Deficiency. On the Three Grammatical Classes". [Unpublished manuscript, University of Venice and University of Geneva.]

Carkeet, D.
1976 "Old English Correlatives: An Exercise in Internal Syntactic Reconstruction", *Glossa* 10: 44-63.

Carlson, G.
1977 "Amount Relatives", *Language* 53: 520-542.

Castellani, A.
1952 *Nuovi testi fiorentini del Dugento.* Firenze: Sansoni.

Castellani, A.
1982 *La prosa italiana delle origini. Vol I: testi toscani di carattere pratico*. Bologna: Patron.

Cecchetto, C.
1995 Grammatica e sintassi della Forma Logica. Doctoral Dissertation, State University of Milan.

Cennamo, M.
forthcoming "Relative clauses", in: M. Maiden—M. Parry (eds.), *Dialects ofItaly*. London: Routledge.

Chomsky, N.
1965 *Aspects of the Theory of Syntax*. Cambridge, Massachusetts: The MIT Press.

Chomsky, N.
1977 "On Wh Movement", in: P. W. Culicover—T. Wasow—A. Akmajian (eds.), *Formal Syntax*. New York: Academic Press, 71-132.

Chomsky, N.
1981 *Lectures on Government and Binding*. Dordrecht: Foris.

Chomsky, N.
1982 *Some Concepts and Consequences of the Theory of Government and Binding,* Cambridge, Massachusetts: The MIT Press.

Chomsky, N.
1986a *Knowledge of Language. Its Nature, Origin, and Use.* New York: Praeger.

Chomsky, N.
1986b *Barriers*. Cambridge, Massachusetts: The MIT Press.

Chomsky, N.
1993 "A Minimalist Program for Linguistic Theory", in: K. Hale—S.J. Keyser (eds.), *The View from Building 20.* Cambridge Massachusetts: The MIT Press, 1-52.

Chomsky, N.
1995 *The Minimalist Program*. Cambridge, Massachusetts: The MIT Press.

Chomsky, N.—H. Lasnik
1977 "Filters and Control", *Linguistic Inquiry* 8: 425-504.

Chomsky, N.—H. Lasnik
1991 Principles and Parameters Theory. [Manuscript.] Printed in Chomsky (1995).

Cinque, G.
1978 "La sintassi dei pronomi relativi *cui* e *quale* nell'italiano moderno", *Rivista di Grammatica Generativa* 3: 31-126.

Cinque, G.
1980 "On Extraction from NPs in Italian", *Journal of Italian Linguistics* 5: 47-99.

Cinque, G.
1981 "On Keenan and Comrie's Primary Relativization Constraint", *Linguistic Inquiry* 12: 293-308.

Cinque, G.
1982 "On the Theory of Relative Clauses and Markedness", *The Linguistic Review* 1: 247-296.

Cinque, G.
1988 "La frase relativa", in: L. Renzi (ed.), *Grande grammatica italiana di consultazione*, volume I. Bologna: Il Mulino.

Cinque, G.
1990 *Types of A' Dependencies*. Cambridge, Massachusetts: The MIT Press.

Cinque, G.
1990a Accordo e movimento da testa a testa nel sintagma nominale romanzo. Paper presented at the XVI Incontro di Grammatica Generativa. Pisa, February 1990.

Cinque, G.
1993 On the Evidence for Partial N Movement in the Romance DP. In: *Italian Syntax and Universal Grammar*. Cambridge: Cambridge University Press, 287-309.

Cinque, G.
1995 "Sintassi del participio e struttura della frase". Conference held at the Scuola Normale Superiore, Pisa, March 1995.

Cole, P.
1987 "The Structure of Internally Headed Relative Clauses", *Natural Language and Linguistic Theory* 5: 277-302.

Cole, P.—W. Harbert—G. Hermon
1978 "Headless Relative Clauses in Quechua", *Studies in The Linguistic Sciences* 8: 26-41.

Cole, P.—G. Hermon
1994 "Is There LF Wh-Movement?", *Linguistic Inquiry* 25: 239-262.

Cornilescu, A.
1993 "Remarks on the Determiner System of Rumanian: The Demonstratives *al* and *cel*", *Probus* 4: 189-260.

Corver, N.
1992 "On Deriving Certain Left Branch Extraction Asymmetries: A Case Study in Parametric Syntax", *NELS* 22: 67-84.

Corver, N.—D. Delfitto
1994 Feature Asymmetry and the Nature of Pronoun Movement. [Unpublished manuscript, University of Tilburg and University of Utrecht.]

Crisma, P.
1993 "On Adjective Placement in Romance and Germanic Event Nominals", *Rivista di Grammatica Generativa* 18: 61-100.

Culicover, P. W.
1993 "Evidence against ECP Accounts of the That-Trace Effect", *Linguistic Inquiry* 24: 557-561.

Culicover, P. W.
1996 "On Distinguishing A' Movements", *Linguistic Inquiry* 27: 445-463.

Culy, C. D.
1990 The Syntax and Semantics of Internally Headed Relative Clauses. PhD Dissertation, Stanford University.

Dayal Srivastav, V.
1995 "Quantification in Correlatives", in: E. Bach—E. Jelinek—A. Kratzer—B. Partee (eds.), *Quantification in Natural Languages*. Dordrecht: Kluwer, 179-205.

Dekeyser, X.
1986 "Relative Markers in the Peterborough Chronicle: 1070-1154", *Folia Linguistica Historica* 7: 93-105.

Déprez, V.
1990 "Two Ways of Moving the Verb in French", *MIT Working Papers in Linguistics* 13: 47-85.

de Rijk, R. P. G.
1972 "Relative Clauses in Basque: A Guided Tour", in: P. M. Peranteau—J. N. Levi— G. C. Phares (eds.), *The Chicago Which Hunt. Papers from the Relative Clause Festival.* Chicago: Chicago Linguistic Society, 115-135.

Dobrovie Sorin, C.
1994 *The Syntax of Romanian. Comparative Studies in Romance.* Berlin: Mouton de Gruyter.

Doetjes, J.
1992 "Rightward Floating Quantifiers Float to the Left", *The Linguistic Review* 9: 313-332.

Donati, C.
1994 "Il *que* relativo spagnolo", *Lingua e Stile* XXX: 565-595.

Doron, E.
1982 "On the Syntax and Semantics of Resumptive Pronouns", *Texas Linguistic Forum* 19: 1-48.

Downing, B. T.
1978 "Some Universals of Relative Clause Structure", in: J. H. Greenberg (ed.), *Universals of Human Language. Volume IV: Syntax*, Stanford, Stanford University Press: 375-418.

Duffield, N.
1991 "On Negation, Minimality and Irish Relative Clauses", *Chicago Linguistic Society 27: The Parasession on Negation*. Chicago: Chicago Linguistic Society, 30-47.

Emonds, J.
1979 "Appositive Relatives Have No Properties", *Linguistic Inquiry* 10: 211-43.

Erdmann, P.
1980 "On the History of Subject Contact Clauses in English", *Folia Linguistica Historica* 1: 139-170.

Enç, M.
1991 "The Semantics of Specificity", *Linguistic Inquiry* 22: 1-25.

Engdahl, E.
1985 "Parasitic Gaps, Resumptive Pronouns and Subject Extraction", *Linguistics* 23: 3-44.

Fabb. N.
1990 "The Difference between English Restrictive and Appositive Clauses", *Journal of Linguistics* 26: 57-77.

Fanselow, G.
1990 "Scrambling as NP Movement", in: G. Grewendorf—W. Sternefeld (eds.), *Scrambling and Barriers*. Amsterdam: John Benjamins, 115-140.

Fassi Fehri, A.
1976 "Relatives et adjectifs en Arabe: le problème de la détermination", *Lingua* 38: 125-152.

Frampton, J.
1991 "Relativized Minimality: A Review", *The Linguistic Review* 8: 1-46.

Franks, R.—Y. S. Lee—O. Rambow
1992 "Scrambling as Non-Operator Movement and The Special Status of Subjects", in: S. Barbiers—M. den Dikken—C. Levelt (eds.), *Proceedings of the Third Leiden Conference of Junior Linguists*: 135-154.

Fukui, N.—M. Speas
1986 "Specifiers and Projection", *MIT Working Papers in Linguistics* 8: 128-72.

Gapany, J.—D. Apothéloz
1993 "Autour des relatives non standard", *Travaux Neuchâtelois de Linguistique* 20: 125-137.

Gazdar, G..—E. Klein.— G..Pullum.— I. Sag
1985 *Generalized Phrase Structure Grammar*. Oxford:Basil Blackwell.

Giorgi, A.
1984 "Toward a Theory of Long Distance Anaphors. A GB Approach", *The Linguistic Review* 3: 307-361.

Giorgi, A.—G. Longobardi
1990 *Noun Phrases: Configuration, Parameters and Empty Categories*. Cambridge: Cambridge University Press.

Giorgi, A.—F. Pianesi
1996 *Tense and Aspect: From Semantics to Morphosyntax.* New York: Oxford University Press.

Giusti, G.
1991 "The Categorial Status of Quantified Nominals", *Linguistische Berichte* 136: 438-454.

Giusti, G.
1993 *La sintassi dei determinanti.* Padova: Unipress.

Giusti, G.—D. Vulchanova
1994 "Quantified Noun Phrase Structure in Bulgarian", *University of Venice Working Papers in Linguistics* 5.1: 43-65.

Go£ab, Z.—V. A. Friedman
1972 "The Relative Clause in Slavic", in: P. M. Peranteau—J. N. Levi—G. C. Phares (eds.), *The Chicago Which Hunt. Papers from the Relative Clause Festival.* Chicago: Chicago Linguistic Society, 30-46.

Goodall, G.
1983 "A Three-Dimensional Analysis of Coordination", *Chicago Linguistic Society* 19: 146-154.

Gonda, J.
1954 "The Original Character of The Indo-European Relative Pronoun **io-*", *Lingua* 4: 1-41.

Gonda, J.
1955 "Notes on the Indo-European *k^{u}i- and *k^{u}o- Pronouns", *Lingua* 4: 241-285.

Greenberg, J.H.
1963 "Some Universals of Grammar with Particular Reference to the Order of Meaningful Elements", in: J. H. Greenberg (ed.), *Universals of Language.* Cambridge, Massachusetts: MIT Press.

Greenberg, J. H.
1978 "How Does a Language Acquire Gender Markers?", in: J.H. Greenberg—C. A. Ferguson—E. Moravcik (eds.), *Universals of Human Language III.* Stanford: Stanford University Press, 47-82.

Grewendorf, G.—C. Poletto
1989 "La costruzione scissa: un'analisi contrastiva", *Rivista di Grammatica Generativa* 14: 105-142.

Grimshaw, J. B.
1975 "Evidence for Relativization by Deletion in Chaucerian Middle English", *NELS* 5: 216-224.

Grimshaw, J. B.
1990 *Argument Structure.* Cambridge, Massachusetts: The MIT Press.

Grimshaw, J. B.
1991 Extended Projection. [Unpublished manuscript, Brandeis University.]

Grosu, A.
1974 "On the Nature of the Left Branch Condition", *Linguistic Inquiry* 5: 308-319.

Grosu, A.
1996 "The Proper Analysis of 'Missing-P" Free Relative Constructions", *Linguistic Inquiry* 27: 257-293.

Grosu, A.—F. Landman
1996a Strange Relatives of the Third Kind. [Unpublished manuscript, Tel Aviv University.]

Grosu, A.—F. Landman
1996b "Carlson's Last Puzzle: Will It Go the Way of Fermat's Last Theorem?", to appear in: *Proceedings of the Annual Meeting of the Israeli Association for Theoretical Linguistics.*

Guasti, M. T.—U. Shlonsky
1993 "The Acquisition of French Relative Clauses Reconsidered", *Geneva Generative Papers*: 54-62.

Guéron, J.
1984 "Topicalization Structures and Constraints on Coreference", *Lingua* 63: 139-174.

Guiraud, P.
1966 "Le système du relatif en français populaire", *Langages* 3: 40-48.

Haberland. H.—J. van der Auwera
1990 "Topics and Clitics in Greek Relatives", *Acta Linguistica Hafniensia* 22: 127-157.

Hadermann, P.
1987 "Histoire d'un pronom relatif: *dont*", *Revue Roumaine de Linguistique* 32: 333-353.

Haegeman, L.
1992 *Theory and Description in Generative Syntax*. Cambridge: Cambridge University Press.

Haider, H.
1993 Detached Clauses - The Later the Deeper. [Unpublished manuscript, University of Stuttgart.]

Halle, M.—A. Marantz
1993 "Distributed Morphology and the Pieces of Inflection", in: K. Hale—S. J. Keyser (eds.), *The View from Building 20.* Cambridge, Massachusetts: The MIT Press, 111-176.

Harbert, W.
1982/3 "On the Nature of the Matching Parameter", *The Linguistic Review* 2: 237-284.

Harbert, W.
1983 "A Note on Old English Free Relatives", *Linguistic Inquiry* 14: 549-553.

Haudry, J.
1973 "Parataxe, hypotaxe et corrélation dans la phrase latine", *Bulletin de la Société Linguistique de Paris* 68:147-186.

Haudry, J.
1981 "Les deux flexions de l'adjectif germanique", *Bulletin de la Société Linguistique de Paris* 76: 191-200.

Heim, I.
1982 The Semantics of Definite and Indefinite Noun Phrases. PhD Dissertation, University of Massachusetts at Amherst.

Heim, I.
1987 "Where Does the Definiteness Restriction Apply? Evidence from the Definiteness of Variables", in: E. Reuland—A. ter Meulen (eds.), *The Representation of (In)definiteness*. Cambridge, Massachusetts: The MIT Press, 21-42.

Higginbotham, J.
1983 "Logical Form, Binding, and Nominals", *Linguistic Inquiry* 14: 395-420.

Higginbotham, J.
1985 "On Semantics", *Linguistic Inquiry* 16: 547-593.

Hirschbühler, P.—M. L. Rivero
1981 "Catalan Restrictive Relatives: Core and Periphery", *Language* 57: 591-625.

Hirschbühler, P.—M. L. Rivero
1983 "Remarks on Free Relatives and Matching Phenomena", *Linguistic Inquiry* 14: 505-520.

Hock, H. H.
1988 "Relative Clauses and Rebracketing in Old English", in: E.H. Antonsen—H.H. Hock (eds.), *Germanic Linguistics II: Papers from the IInd Symposium on Germanic Linguistics*. University of Illinois at Urbana-Champaign: Indiana University Linguistic Club, 35-55.

Hock, H. H.
1989 "Conjoined We Stand: Theoretical Implications of Sanskrit Relative Structures", *Studies in The Linguistic Sciences* 19: 93-126.

Hoekstra, E.
1992 "X-bar Theory and Licensing Mechanisms", *The Linguistic Review* 8: 47-73.

Hoekstra, E.
1992b "On the Parametrization of Functional Projections in CP", *NELS* 23: 191-204.

Hoekstra, E.
1993 "Some Implications of Number Agreement in Comp", in: F. Drijkoningen—K. Hengeveld (eds.), *Linguistics in the Netherlands 1993*. Amsterdam: John Benjamins, 61-68.

Holmberg, A.—C. Platzack
1991 "On the Role of Inflection in Scandinavian Syntax", in: W. Abraham,—W. Kosmejier—E. Reuland (eds.), *Issues in Germanic Syntax*. Berlin: Mouton de Gruyter, 93-118.

Hooper, J.—S. Thompson
1973 "On the Applicability of Root Transformations", *Linguistic Inquiry* 4: 465-497.

Hornstein, N.—A. Weinberg
1981 "Case Theory and Preposition Stranding", *Linguistic Inquiry* 12: 55-91.

Horvath, J.
1986 *Focus in the Theory of Grammar and the Syntax of Hungarian*. Dordrecht: Foris.

Huang, C.T.J.
1993 "Reconstruction and the Structure of VP: Some Theoretical Consequences", *Linguistic Inquiry* 24: 103-138.

Huot, H.
1978 "Appositions et relatives appositives", *Recherches Linguistiques* 5-6: 103-143.

Ingria, R.
1990 "The Limits of Unification", *Proceedings of the 28th Annual Meeting of the ACL*. Pittsburgh, PA, 194-204.

Ishihara, R.
1984 "Clausal Pied Piping: A Problem for GB", *Natural Language and Linguistic Theory* 2: 397-418.

Iwakura, K.
1980 "On Wh-Movement and Constraints on Rules", *Linguistic Analysis* 6: 53-95.

Jackendoff, R.
1977 *X-Bar Syntax: A Study of Phrase Structure.* Cambridge, Massachusetts: The MIT Press.

Jackendoff, R.
1990 "On Larson's Treatment of the Double Object Construction", *Linguistic Inquiry* 21: 427-56.

Kayne, R. S.
1972 "Subject Inversion in French Interrogatives", in: J. Casagrande—B. Saciuk (eds.), *Generative Studies in Romance Languages* Rowley, Massachusetts: Newbury House Publishers, 70-126.

Kayne, R. S.
1975 *French Syntax. The Transformational Cycle.* Cambridge, Massachusetts: The MIT Press.

Kayne, R. S.
1976 "French Relative 'que'", in: M. Luján—F. Hensey (eds.), *Current Studies in Romance Linguistics.* Washington, D.C.: Georgetown University Press, 255-299.

Kayne, R. S.
1984 *Connectedness and Binary Branching*. Dordrecht: Foris.

Kayne, R. S.
1989 "Facets of Romance Past Participle Agreement", in: P. Benincà (ed.), *Dialect Variation and the Theory of Grammar*. Dordrecht: Foris, 55-84.

Kayne, R. S.
1989b Notes on English Agreement. [Unpublished manucript, City University of New York.]

Kayne, R. S.
1991 "Romance Clitics, Verb Movement, and PRO", *Linguistic Inquiry* 22: 647-686.

Kayne, R. S.
1993 "Toward a Modular Theory of Auxiliary Selection", *Studia Linguistica* 47: 3-31.

Kayne, R. S.
1994 *The Antisymmetry of Syntax*. Cambridge, Massachusetts: The MIT Press.

Kayne, R. S.
1995 Lo statuto del clitico riflessivo. Conference held at the Scuola Normale Superiore, January 1995.

Keenan, E.
1985 "Relative Clauses", in: T. Schopen (ed.), *Language Typology and Syntactic Description*. Cambridge: Cambridge University Press, vol. II, 141-170.

Keenan, E.—B. Comrie
1977 "Noun Phrase Accessibility and Universal Grammar", *Linguistic Inquiry* 8: 63-99.

Kemenade, A. van
1993 "Syntactic Changes in Late Middle English", in: H. Aertsen—R. J. Jeffers (eds.), *Historical Linguistics 1989*. Amsterdam: John Benjamins, 235-248.

Kenesei, I.
1994 "Subordinate Clauses", in: F. Kiefer—K. É. Kiss (eds.), *Syntax and Semantics 27: The Syntactic Structure of Hungarian*. New York: Academic Press.

Kester, E.
1993 "The Inflectional Properties of Scandinavian Adjectives", *Studia Linguistica* 47: 139-153.

Kimball, J.—J. Aissen
1971 "I Think, You Think, He Think", *Linguistic Inquiry* 2:242-246.

Kiparsky, P.
1995 "Indo-European Origins of Germanic Syntax", in: A. Battye—I. Roberts (eds.), *Clause Structure and Language Change*. New York: Oxford University Press, 140-169.

Kiss, K. E.
1996 "Two Subject Positions in English", *The Linguistic Review* 13: 119-142.

Koopman, H.
1984 *The Syntax of Verbs*. Dordrecht: Foris.

Kornai, A.—G. Pullum
1990 "The X-bar Theory of Phrase Structure", *Language* 66: 24-50.

Koster, J.
1995 Extraposition as coordination. Conference held at the Scuola Normale Superiore di Pisa, November 1995.

Kunstmann, P.
1991 "Création et diffusion du relatif/interrogatif lequel en ancien français. Comparaison avec d'autres langues romanes", in: *Actes du XVIII Congrès International de Linguistique et Philologie Romane*. Tübingen: Max Niemeyer, vol. II, 660-670.

Ladusaw, W. A.
1980 "On the Notion Affective and the Analysis of Negative Polarity Items", *Journal of Linguistic Research* 1.2: 1-16.

Law, P.
1991 "Verb Movement, Expletive Replacement, and Head Government", *The Linguistic Review* 8: 253-285.

Laenzlinger, C.
1993 "A Syntactic View of Romance Pronominal Sequences", *Probus* 5: 241-270.

Larson, R.
1987 "'Missing Prepositions' and the Analysis of English Free Relative Clauses", *Linguistic Inquiry* 18: 239-266.

Larson, R. K.
1988 "On the Double Object Construction", *Linguistic Inquiry* 19: 335-91.

Lasnik, H.—M. Saito
1992 *Move Alpha*. Cambridge, Massachusetts: The MIT Press.

Lasnik, H.—T. Stowell
1991 "Weakest Crossover", *Linguistic Inquiry* 22: 687-720.

Lebeaux, D.
1991 "Relative Clauses, Licensing, and the Nature of the Derivation", in: S. Rothstein (ed.), *Syntax and Semantics 25: Perspectives on Phrase Structure*. New York: Academic Press, 209-239.

Lenerz, J.
1985 "Diachronic Syntax: Verb Position and Comp in German", in: J. Toman (ed.), *Studies in German Grammar*. Dordrecht: Foris, 104-132.

Link, G.
1983 "The Logical Analysis of Plural and Mass Terms: a Lattice-Theoretical Approach", in: R. Bäuerle—C. Schwarze—A. von Stechow (eds.), *Meaning, Use and Interpretation of Language*. Berlin: Walter de Gruyter, 302-323.

Loetscher, A.
1972 "Some Problems Concerning Standard German Relative Clauses", in: P. M. Peranteau—J. N. Levi—G. C. Phares (eds.), *The Chicago Which Hunt. Papers from the Relative Clause Festival*. Chicago: Chicago Linguistic Society, 47-58.

Longobardi, G.
1980 "Les relatives nominales indo-européennes", in: P. Ramat (ed.), *Linguistic Reconstruction and Indo-European Syntax*. Amsterdam: John Benjamins, 171-182.

Longobardi, G.
1985 "Connectedness, Scope and C-Command", *Linguistic Inquiry* 16: 163-192.

Longobardi, G.
1991 "In Defense of the Correspondence Hypothesis", in: J. Huang—R. May (eds.), *Logical Structure and Linguistic Structure*. Dordrecht: Kluwer, 149-196.

Longobardi, G.
1994 "Reference and Proper Names: A Theory of N-Movement in Syntax and Logical Form", *Linguistic Inquiry* 25: 609-665.

Maling, J. M.
1978 "The Complementizer in Middle English Appositives", *Linguistic Inquiry* 9: 719-725.

Maling, J. M.
1978a "An Asymmetry with Respect to Wh-Islands", *Linguistic Inquiry* 9: 75-89.

Manzini, M. R.
1992 *Locality. A Theory and Some of Its Empirical Consequences*. Cambridge, Massachusetts: The MIT Press.

Manzini, M. R.
1993 "Triggers for V-2: Germanic and Romance", *Quaderni del Dipartimento di Linguistica dell'Università di Firenze* 4: 1-17.

Manzini, M. R.
1994 "Syntactic Dependencies and Their Properties: A Note on Strong Islands", *University College of London Working Papers in Linguistics* 6: 205-218.

Manzini, M. R.
1994a "Locality, Minimalism and Parasitic Gaps", *Linguistic Inquiry* 25: 481-508.

Manzini, M. R.
1995 From Move to Form Dependency. Conference held at the University of Siena, April 1995.

Manzini, R.—K. Wexler
1987 "Parameters, Binding Theory, and Learnability", *Linguistic Inquiry* 18: 413-444.

Marcantonio, A.
1985 "The Definite and Indefinite Conjugation in Hungarian: A Typological and Diachronic Analysis", *Acta Linguistica Academiae Scientiarum Hungaricae* 35: 267-298.

May, R.
1985 *Logical Form: Its Structure and Derivation*. Cambridge, Massachusetts: The MIT Press.

McCawley, J. D.
1981 "The Syntax and Semantics of English Relative Clauses", *Lingua* 53: 99-149.

McCawley, J. D.
1982 "Parentheticals and Discontinuous Constituent Structure", *Linguistic Inquiry* 13: 91-106.

McCloskey, J.
1990 "Resumptive Pronouns, A'-Binding, and Levels of Representation in Irish", in: R. Hendrick (ed.), *Syntax and Semantics 23: The Syntax of Modern Celtic Languages*. New York: Academic Press, 199-248.

McCloskey, J.
1991 "Clause Structure, Ellipsis and Proper Government in Irish", *Lingua* 85: 259-302.

Milsark, G.
1977 "Toward an Explanation of Certain Peculiarities of the Existential Construction in English", *Linguistic Analysis* 3: 1-30.

Mitchell, B.
1985 *Old English Syntax*. Oxford: Clarendon Press.

Morgan, J. L.
1972 "Some Aspects of Relative Clauses in English and Albanian", in: P. M. Peranteau—N. Levi—G. C. Phares (eds.), *The Chicago Which Hunt. Papers from the Relative Clause Festival.* Chicago: Chicago Linguistic Society: 63-72.

Moritz, L.—D. Valois
1994 "Pied-Piping and Specifier-Head Agreement", *Linguistic Inquiry* 25: 667-707.

Moro, A.
1994 *I predicati nominali e la struttura della frase*. Padova: Unipress.

Motapanyane, V.
1994 "An A Position for Romanian Subjects", *Linguistic Inquiry* 25: 729-734.

Müller, G.
1992) "In Support of Dative Movement", in: S. Barbiers—M. den Dikken—C. Levelt (eds.), *Proceedings of LCJL 3*, 201-18.

Müller, G.—W. Sternefeld
1993 "Improper Movement and Unambiguous Binding", *Linguistic Inquiry* 24: 461-507.

Müller, G.—W. Sternefeld
1994) "Scrambling as A' Movement", in N. Corver—H. van Riemsdijk (eds.), *Studies on Scrambling*. Berlin: Mouton de Gruyter.

Munn, A.
1992 "A Null Operator Analysis of ATB Gaps", *The Linguistic Review* 9: 1-26.

Nakajima, H.
1992 "Another Type of Antecedent Government", *Linguistic Inquiry* 23: 313-328.

Nakajima, H.
1996 "Complementizer selection", *The Linguistic Review* 13: 143-164.

Nanni, D.—J. Stillings
1978 "Three Remarks on Pied Piping", *Linguistic Inquiry* 9: 310-318.

Ojeda, A. E.
1982 "Degree Relatives and the Neuter Article in Spanish", *Chicago Linguistic Society* 18: 407-418.

Ojeda, A. E.
1990) "Definite Descriptions and Definite Generics", *Linguistics and Philosophy* 14: 367-397.

Ortiz de Urbina, J.
1990 "Operator Feature Percolation and Clausal Pied Piping", *MIT Working Papers in Linguistics* 13: 193-208.

Palermo, M.
1994 *Il carteggio vaianese (1537-1539)*. Firenze: Accademia della Crusca.

Perlmutter, D. M.
1972 "Evidence for Shadow Pronouns in French Relativization", in: P.M. Peranteau—J.N. Levi—G.C. Phares (eds.), *The Chicago Which Hunt. Papers from the Relative Clause Festival*, Chicago: Chicago Linguistic Society, 73-195.

Pesetsky, D.
1994 Telegraphic Effects and Optimality. Conference held at the Department of Cognitive Sciences (DIPSCO), Istituto Scientifico S. Raffaele, Milan, May 1994.

Pesetsky, D.
1995 *Zero Syntax: Experiencers and Cascades*. Cambridge, Massachusetts: The MIT Press.

Petrucci, A.
1982 *Scrittura e popolo nella Roma barocca 1585-1721*. Roma: Edizioni Quasar.

Picallo, M. C.
1994 "Catalan Possessive Pronouns: The Avoid Pronoun Principle Revisited", *Natural Language and Linguistic Theory* 12: 259-299.

Pittner, K.
1995 "The Case of German Relatives", *The Linguistic Review* 12: 197-232.

Plann, S.
1980 "On the Non-Evidence for Cross relativization from Spanish", *Linguistic Inquiry* 11: 607-613.

Plann, S.
1992 "Indirect Questions in Spanish", *Linguistic Inquiry* 13: 297-312.

Platzack, C.
1994 Syntactic Differences in the Comp Domain: Evidence for Split Comp. Paper presented at the XVII Glow Colloquium, Wien.

Platzack, C.
1995 "The Loss of V/2 in English and French", in A. Battye—I. Roberts (eds.), *Clause Structure and Language* Change. New York: Oxford University Press, 200-226.

Poletto, C.
1994 "Complementizer Deletion and Verb Movement in Italian." *University of Venice Working Papers in Linguistics* 5.2: 49-79.

Pollock, J, Y.
1989 "Verb Movement, Universal Grammar, and the Structure of IP", *Linguistic Inquiry* 20: 365-424.

Pollock, J. Y.
1995 On the Historical Syntax of V Movement: PRO-drop and the Morphological Basis of the [± Strong] Infl Parameter. Conference held at the Scuola Normale Superiore di Pisa, November 1995.

Postal, P. M.
1972 "On Some Rules That Are Not Successive Cyclic", *Linguistic Inquiry* 3: 211-222.

Postal, P. M.
1993 "Remarks on Weak Crossover Effects", *Linguistic Inquiry* 24: 539-556.

Postal, P. M.
1993b "Parasitic Gaps and the ATB Phenomenon", *Linguistic Inquiry* 24: 735-754.

Postal, P. M.
1994 "Contrasting Extraction Types", *Journal of Linguistics* 30: 159-186.

Puskàs, G.
1992 "The Wh Criterion in Hungarian", *Rivista di Grammatica Generativa* 17: 141-186.

Reinhart, T.
1981 "A Second COMP Position", in: A. Belletti—L. Brandi—L. Rizzi (eds.), *Theory of Markedness in Generative Syntax*. Pisa: Scuola Normale Superiore, 517-557.

Reinhart, T.
1984 *Anaphora and Semantic Interpretation*. London: Croom Helm.

Reinhart, T.
1987 "Specifier and Operator Binding", in: E. Reuland—A. ter Meulen (eds.), *The Representation of (In)definiteness*. Cambridge, Massachusetts: The MIT Press, 130-167.

Reinhart, T.—E. Reuland
1991 "Anaphors and Logophors: An Argument Structure Perspective", in: J. Koster—E. Reuland (eds.), *Long-Distance Anaphora*, Cambridge: Cambridge University Press: 283-321.

Reinhart, T.—E. Reuland
1993 "Reflexivity", *Linguistic Inquiry* 24: 657-720.

van Riemsdijk, H.
1985 "On Pied Piping Infinitives in German Relative Clauses", in: J. Toman (ed.), *Studies in German Grammar*. Dordrecht: Foris, 165-192.

van Riemsdijk, H.
1989 "Movement and Regeneration", in: P. Benincà (ed.), *Dialect Variation and the Theory of Grammar*. Dordrecht: Foris, 105-136.

van Riemsdijk, H.
1989b "Swiss relatives", in D. Jaspers—W. Klooster—Y. Putseys—P. Seuren (eds.), *Sentential Complementation and the Lexicon*. Dordrecht: Foris, 343-354.

Rigau, G.
1988 "Strong Pronouns", *Linguistic Inquiry* 19: 503-511.

Ritter, E.
1992 "Crosslinguistic Evidence for Number Phrase", *Canadian Journal of Linguistics* 37: 197-218.

Rivero, M.L.
1980 "Theoretical Implications of the Syntax of Left Branch Modifiers in Spanish", *Linguistic Analysis* 6: 407-461.

Rivero, M. L.
1986 "La sintaxis diacrónica del español y la adquisición del lenguaje", in: *Linguistique générale et linguistique romane. Actes du XVIIme CiLPhR*, Université deProvence, vol. I:227-240.

Rizzi, L.
1986 "Null Objects in Italian and the Theory of *pro*", *Linguistic Inquiry* 17: 501-557.

Rizzi, L.
1987 "Sull'uniformità categoriale delle strutture sintattiche", *Rivista di Grammatica Generativa* 12: 61-81.

Rizzi, L.
1990 *Relativized Minimality*. Cambridge, Massachusetts: MIT Press.

Rizzi, L.
1991 "Residual Verb Second and the Wh Criterion", *Technical Reports in Formal and Computational Linguistics* 2, University of Geneva.

Rizzi, L.
1994 The Fine Structure of Comp. Conference held at the University of Florence, April 1994.

Rizzi, L.
1995 The Fine Structure of the Left Periphery. [Manuscript, University of Geneva. Revised version published in: L. Haegeman (ed.), *Elements of Grammar*. Dordrecht: Kluwer, 281-337.]

Rizzi, L.—I. G. Roberts
1989 "Complex Inversion in French", *Probus* 1: 1-30.

Roberts, I. G.
1991 "Excorporation and Minimality", *Linguistic Inquiry* 22: 209-18.

Roberts, I.G.
1993 *Verbs and Diachronic Syntax*. Dordrecht: Kluwer.

Roberts, I.G.
1995 Some Implications of an Analysis of VSO Order. Paper presented at the XXI Incontro di Grammatica Generativa, Milan, 1995.

Rochemont, M.
1989 "Topic Islands and the Subjacency Parameter", *Canadian Journal of Linguistics* 34: 145-170.

Rochemont, M.—P. W. Culicover
1990 *English Focus Constructions and the Theory of Grammar*. Cambridge: Cambridge University Press.

Roussou, A.
1992 "Factive Complements and Wh Movement in Modern Greek", *University College of London Working Papers in Linguistics* 4: 123-147.

Roussou, A.
1993 "I to C Movement and the That-t Filter", *University College London Working Papers in Linguistics* 5: 103-127.

Roussou, A.
1995 Subject Extraction and Head Dependencies. Paper presented at the XXI Incontro di Grammatica Generativa, Milan, 1995.

Ross, J. R.
1967 Constraints on Variables in Syntax. PhD Dissertation, MIT.

Ross, J. R.
1970 "Whether deletion", *Linguistic Inquiry* 2: 146.

Rudin, C.
1990-91 "Topic and Focus in Bulgarian", *Acta Linguistica Hungarica* 40: 429-447.

Rudin, C.
1986 "Changing the Rules: 'Extra' Prepositions in Relative Clauses", *Chicago Linguistic Society* 22: 277-285.

Safir, K.
1984 "Multiple Variable Binding", *Linguistic Inquiry* 15: 603-638.

Safir, K.
1986 "Relative Clauses in a Theory of Binding and Levels", *Linguistic Inquiry* 17: 663-689.

Saito, M.
1989 "Scrambling as Semantically Vacuous A' Movement", in: M. R. Baltin—A. Kroch (eds.), *Alternative Conceptions of Phrase Structure*. Chicago: The University of Chicago Press, 182-200.

Sandfeld, K. R.
1965[2] *Syntaxe du Français contemporain. Les propositions subordonnées*. Geneva: Librairie Droz.

Santelmann, L.
1993 "The Distribution of Double Determiners in Swedish: Den Support in D°", *Studia Linguistica* 47: 154-176.

Sauzet, P.
1989 "Topicalisation et Prolepse en Occitan", *Revue de Langues Romanes* 93: 235-273.

Schachter, P.
1973 "Focus and Relativization", *Language* 49: 19-46.

Schachter, P.
1977 "Constraints on Coordination", *Language* 53: 86-103.

Schäufele, S.
1988 "Where's my NP? Nontransformational Analyses of Vedic Pronominal Fronting", *Studies in The Linguistic Sciences* 18: 129-162.

Schroten, J.
1992 "On Spanish Definite Determiners: Personal Pronouns and Definite Articles", *Recherches de linguistique romane et française d'Utrecht* XI: 9-24.

Scorretti, M.
1981 "Complementizer Ellipsis in 15th Century Italian", *Journal of Italian Linguistics* 6: 35-46.

Shlonsky, U.
1991 "Quantifiers as Functional Heads: A Study of Quantifier Float in Hebrew", *Lingua* 84: 159-180.

Shlonsky, U.
1992 "Resumptive Pronouns as a Last Resort", *Linguistic Inquiry* 23: 443-468.

Siloni, T.
1991 On Participial Relatives and Complementizer D°: A Case Study in Hebrew and French. [Revised version published in *Natural Language and Linguistic Theory* 13: 445-487.]

Sportiche, D.
1988 "A Theory of Floating Quantifiers and its Corollaries for Constituent Structure", *Linguistic Inquiry* 19: 425-449.

Sportiche, D.
1992 Clitic Constructions. [Manuscript, University of California at Los Angeles. Revised version published in: J. Rooryck—L. Zaring (eds.), *Phrase Structure and The Lexicon.* Dordrecht: Kluwer, 213-276.]

Srivastav, V.
1988 "Relative Clauses in Hindi and Learnability", *Cornell Working Papers in Linguistics* 8: 133-161.

Srivastav, V.
1991 "The Syntax and Semantics of Correlatives", *Natural Language and Linguistic Theory* 9: 637-686.

von Stechow, A.
1979 "Visiting German Relatives", in: R. Bäuerle—U. Egli—A. von Stechow (eds.), *Semantics from Different Points of View.* Springer Verlag, 226-265.

Stockwell, R. P.—D. Minkova
1991 "Subordination and Word Order Change in the History of English", in: D. Kastovsky (ed.), *Historical English Syntax.* Berlin: Mouton de Gruyter, 367-408.

Stockwell, R.P.—P. Schachter—B. Hall Partee
1973 *The Major Syntactic Structures of English.* New York: Holt, Rinehart and Winston.

Stowell, T.
1982 "The Tense of Infinitives", *Linguistic Inquiry* 13: 561-570.

Stowell, T.
1983 "Subjects across Categories", *The Linguistic Review* 2: 285-310.

Stowell, T.
1989 "Subjects, Specifiers, and X-bar Theory", in: Baltin, M. R.—A. Kroch (eds.), *Alternative Conceptions of Phrase Structure.* Chicago: The University of Chicago Press, 232-262.

Stowell, T.
1993 The Syntax of Tense. [Unpublished manuscript, UCLA.]

Stroik, T. S.
1994 "Saturation, Predication, and the DP Hypothesis", *Linguistic Analysis* 24: 39-61.

Suñer, M.
1988 "The Role of Agreement in Clitic-Doubled Constructions", *Natural Language and Linguistic Theory* 6: 391-434.

Szabolsci, A.
1994 "The Noun Phrase", in F. Kiefer—K. É. Kiss (eds.), *Syntax and Semantics 27: The Syntactic Structure of Hungarian.* New York: Academic Press.

Szabolsci, A.
1994b All Quantifiers Are Not Equal: The Case of Focus. [Unpublished manuscript, Hungarian Academy of Sciences and University of California at Los Angeles.]

Szamosi, M.
1976 "On a Surface Structure Constraint in Hungarian", in: J. McCawley (ed.), *Syntax and Semantics 7: Notes from the Linguistic Underground*. New York: Academic Press: 409-425.

Tomaselli, A.
1990 *La sintassi del verbo finito nelle lingue germaniche*. Padova: Unipress.

Toribio, A. J.
1992 "Proper Government in Spanish Subject Relativization", *Probus* 4: 291-304.

Tallerman, M.
1993 "Fronting Constructions in Welsh", *Newcastle and Durham Working Papers in Linguistics* 1: 251-275.

Traugott, E. Closs
1972 *A History of English Syntax*. New York: Holt, Rinehart and Winston.

Uriagereka, J.
1995 "Aspects of the Syntax of Clitic Placement in Western Romance", *Linguistic Inquiry* 26: 79-123.

Vergnaud, J. R.
1974 French Relative Clauses. PhD Dissertation, MIT.

Vergnaud, J. R.—M. L. Zubizarreta
1992 "The Definite Determiner and the Inalienable Constructions in French and English", *Linguistic Inquiry* 23: 595-

Vallduví, E.
1992 "A Preverbal Landing Site for Quantificational Operators", *Catalan Working Papers in Linguistics*: 319-343.

Vikner, S.
1991 "Relative *der* and Other C° Elements in Danish", *Lingua* 84: 109-136.

Vikner, S.
1991b Scandinavian Object Shift and West Germanic Scrambling. Conference held at the University of Geneva, February 1991.

Vikner, S.
1995 *Verb Movement and Expletive Subjects in the Germanic Languages*. New York: Oxford University Press.

Viotto, A.
1994 Le frasi relative nel dialetto veneziano: alcune implicazioni teoriche. Master Dissertation, University of Venice.

Wanner, D.
1981 "Surface Complementizer Deletion: Italian che ~ Ø", *Journal of Italian Linguistics* 6: 47-82.

Watanabe, A.
1993 "Larsonian CP Recursion, Factive Complements, and selection", *NELS* 23: 523-537.

Watanabe, A.
1995 "Conceptual Basis of Ciclicity", *MIT Working Papers in Linguistics* 27: 269-291.
Webelhuth, G.
1985 "German is Configurational", *The Linguistic Review* 4: 203-246.
Weisler, S.
1980 "The Syntax of That-less Relatives", *Linguistic Inquiry* 11: 624-631.
Wheeler, M. W.
1988 "Occitan", in: M. Harris—N. Vincent (eds.), *The Romance Languages*. London: Croom Helm, 246-278.
Wilder, C.
1991 "Tough Movement Constructions", *Linguistische Berichte* 132: 115-132.
Wilder, C.— D. Cavar
1994 "Word Order Variation, Verb Movement, and Economy Principles", *Studia Linguistica* 48: 46-86.
Wiltschko, M.
1994 "Extraposition in German", *Wiener Linguistische Gazette* 48: 1-30.
Williamson, J. S.
1987 "An Indefiniteness Restriction for Relative Clauses in Lakhota", in: E. Reuland— A. ter Meulen (eds.), *The Representation of (In)definiteness*. Cambridge, Massachusetts:The MIT Press, 168-190.
Zamparelli, R.
1994 Correspondences in Functional Structures Across Major Categories. [Unpublished manuscript, University of Rochester.]
Zamparelli, R.
1994b Partitions in DP. Paper presented at the XVII Glow Colloquium, Wien.
Zamparelli, R.
1995 Layers in the Determiner Phrase. PhD Dissertation, University of Rochester.
Zaring, L.
1992 "French *ce* as a Clausal Determiner", *Probus* 4: 53-80.
Zubizarreta, M. L.
1994 Word Order, Prosody, and Focus. [Unpublished manuscript, University of Southern California.]
Zwart, C. J. W.
1994 "Dutch is Head-Initial", *The Linguistic Review* 11: 377-406.
Zwart, C. J. W.
1994a "Verb Movement and Complementizer Agreement", *MIT Working Papers in Linguistics* 18: 297-341.

Subject Index

Bei Fragen zur Produktsicherheit wenden Sie sich bitte an:
If you have any questions regarding product safety,
please contact:

Walter de Gruyter GmbH
Genthiner Straße 13
10785 Berlin
productsafety@degruyterbrill.com